# Adjoint Equations and Perturbation Algorithms in Nonlinear Problems

Guri I. Marchuk
Valeri I. Agoshkov
Victor P. Shutyaev

Institute of Numerical Mathematics
Moscow, Russia

CRC Press
Taylor & Francis Group
Boca Raton London New York

CRC Press is an imprint of the
Taylor & Francis Group, an **informa** business

CRC Press
Taylor & Francis Group
6000 Broken Sound Parkway NW, Suite 300
Boca Raton, FL 33487-2742

© 1996 by Taylor & Francis Group, LLC
CRC Press is an imprint of Taylor & Francis Group, an Informa business

First issued in paperback 2019

No claim to original U.S. Government works

ISBN-13: 978-0-367-44858-5 (pbk)
ISBN-13: 978-0-8493-2871-8 (hbk)

Visit the Taylor & Francis Web site at
http://www.taylorandfrancis.com

and the CRC Press Web site at
http://www.crcpress.com

**Library of Congress Cataloging-in-Publication Data**

Catalog record is available from the Library of Congress.

# Preface

The last years have witnessed the appearance of new statements of problems which require thorough analysis of complex systems on the basis of the theory of adjoint equations. This range of problems first of all includes problems of global climate changes on our planet, state of environment and protection of environment against pollution, preservation of the biosphere in conditions of vigorous growth of population, intensive development of industry, and many others. Investigations of these complex systems based on the sensitivity theory and perturbation method give a new impulse to the development of the theory of adjoint equations. One can add here the problem of analysis of observational data, application of adjoint equations to retrospective study of processes governed by imitation models, and to the study of the models themselves, realized with the help of contemporary computing technology. One should pay special attention to inverse problems, whose statements and solutions are possible on the basis of adjoint equations.

Originally defined by Lagrange, adjoint operators have since been thoroughly substantiated theoretically and broadly applied in solving many problems in mathematical physics. They became part of a golden fund of science through quantum mechanics, theory of nuclear reactors, optimal control, and finally helped in solving many problems on the basis of perturbation method and sensitivity theory.

For some time, nonlinear problems became an object of much research work; they were extremely difficult for analysis and interpretation. Naturally, these or other generalizations appear in addition to the theory of adjoint equations which are of high importance for the classes of problems of mathematical physics. One such generalization was considered in the work by G. I. Marchuk. Approaches formulated by V. P. Maslov, V. S. Vladimirov and I. V. Volovich arouse great interest. A theory of nonlinear adjoint equations proposed by M. M. Veinberg is likely to be as productive. These authors are also responsible for a number of recent results in these areas. G. I. Marchuk and V. I. Agoshkov in their works have suggested a principle of constructing the adjoint operators which has a sense for a broad class of nonlinear operators.

We would like to point out one more feature of the adjoint problems theory. Adjoint problems provide the closest approach to problems of optimal control, as developed by R. Bellman, L. S. Pontryagin, N. N. Krasovsky, J.-L. Lions,

R. Glowinski, A. Balakrishnan, A. Bensoussan, and many other researchers, as well as to problems of sensitivity of main problems' solutions, or functionals of those, to problem inputs.

In this connection, we must mention the works by J.-L. Lions and his pupils, which became fundamental, dedicated to investigation of problems on insensitive optimal control, nonlinear sentinels for distributed systems. The general approach (Hilbert Uniqueness Method) developed by J.-L. Lions makes it possible to prove the existence of insensitive control in nonlinear systems. This approach was developed with the method of a small parameter for some classes of systems with nonlinear equation of state in the works by V. I. Agoshkov and V. M. Ipatova.

The problem of obtaining and processing measurement data in various fields of knowledge grows more important nowadays. This problem may be modelled mathematically as a multidimensional (space and time) data assimilation and data processing problem which is one of the optimal control problems. Statements of these problems proved to be formulated on the bases of adjoint equations selected in a proper manner. The problems of this type attracted the attention of many specialists applying optimal control methods to practical solving various problems, and were studied by J.-L. Lions, R. Glowinski, G. I. Marchuk, V. V. Penenko and N. N. Obraztsov, F. X. Le Dimet, O. Talagrand, I. Navon, C. Wunsh, G. I. Marchuk and V. I. Agoshkov, G. I. Marchuk and V. B. Zalesny, G. I. Marchuk and V. P. Shutyaev.

So, there is a real basis for generalization of results connected with the theory of adjoint equations in nonlinear problems. This book presents an attempt to systemize and theoretically generalize the results obtained by various authors while solving nonlinear problems of mathematical physics with the methods of adjoint equations.

The monograph contains the presentation of the theory of adjoint equations in nonlinear problems and their applications to perturbation algorithms for solution of nonlinear problems in mathematical physics. To the best of the authors' knowledge, there is no systematic description of these questions in scientific literature. More than that, even the concept of an adjoint operator itself is not acknowledged while considering nonlinear problems. Therefore, the authors give a detailed survey of the development of the theory of adjoint equations in nonlinear problems of mathematical physics, and formulate a series of principles of construction of adjoint operators in nonlinear problems.

We consider in this book properties of adjoint operators corresponding to nonlinear operators, conditions of solvability of adjoint equations constructed according to various principles. A separate chapter is dedicated to the questions of usage of the calculus of variations, transformation groups, and conservation laws for construction of adjoint operators in nonlinear problems. The book deals also with the application of adjoint equations to the construction and justification of perturbation algorithms for nonlinear problems. A number of chapters are dedicated to the applications of the theory of adjoint equations and perturbation theory to the solution of concrete applied problems such as nonlinear elliptic problems, problems of the transport of particles, boundary value problems for a quasilinear equation of motion, nonlinear mathematical models of movement of a substance in a medium, and nonlinear data assimilation problems.

The material is illustrated both with simple examples and with more complex results of numerical experiments. So, the authors of the monograph try to attract the attention of a more or less wide circle of researchers to new approaches which can result in some impulse to the creation of new technologies of planning the experiments, and the investigation of complex systems while solving applied problems. It is difficult to overappreciate the role of adjoint equations here. The results described in this book open new possibilities in using the adjoint equations in nonlinear problems of mathematical physics.

Guri Marchuk, Valeri Agoshkov, and Victor Shutyaev

# The authors

**Guri Marchuk, academician,** is a director of the Institute of Numerical Mathematics, Russian Academy of Sciences.

In 1962–1980 he worked in the Siberian Branch of the USSR Academy of Sciences, first as a director of the Computer Center and then as Chairman of this Branch and Vice-President of the USSR Academy of Sciences. In 1980–1986 he was a Deputy Prime Minister of USSR and Chairman of the State Committee of Science and Technology. In 1986–1991 he was the President of the USSR Academy of Sciences.

Guri Marchuk is a prominent scientist in numerical and applied mathematics; Fridman, Keldysh and Carpinski prizes winner; member of the Academies of Sciences of Bulgaria, Czechoslovakia, Europe, Finland, France, Germany, India, Poland and Rumania; Honorary Professor of Calcutta, Houston, Karlov, Tel-Aviv, Toulouse, and Oregon Universities, Budapest and Dresden Polytechnic Universities; member of the Editorial board of many foreign (France, Germany, Italy, Sweden, three in USA) and several Russian journals, and Editor-in-Chief of Russian Journal of Numerical Analysis and Mathematical Modelling published by the Institute of Numerical Mathematics RAS in the Netherlands.

Guri Marchuk is the author of the series of monographs on numerical mathematics, numerical simulation of nuclear reactors, numerical technique for problems of the atmosphere and ocean dynamics, immunology, medicine and environment protection. For the notable progress in scientific and organizational activities Guri Marchuk was awarded with prestigious state rewards.

**Valeri Agoshkov, Doctor of Sciences,** is a Professor of Mathematics and a Leading Researcher of the Institute of Numerical Mathematics, Russian Academy of Sciences.

From 1970 until 1980 he was a Researcher at the Computing Center of the Siberian Division of the USSR Academy of Sciences in Novosibirsk. He defended a Kandidat Thesis (Ph.D.) on the theme "Variational Methods for Neutron Transport Problems" in 1975 in the field of Numerical Mathematics. From 1981 until now he's been working at the Institute of Numerical Mathematics, Russian Academy of Sciences (Moscow). He's the head of Adjoint Equations and Perturbation Theory Group of the Institute of Numerical Mathematics. He defended a Doctoral Thesis on the theme "Functional Spaces, Generalized Solution of Transport Equations and Their Regularity Properties" in 1987 in the field of Differential Equations.

His research interests are: principles of construction of adjoint operators in non-linear problems, solvability of equations with adjoint operators; domain decomposition methods and the Póincare–Steklov operator theory; numerical methods for partial differential equations; functional spaces, boundary-value problems for transport equations and regularity of solutions; optimal control theory and its applications in the data assimilation processes.

V. Agoshkov's scientific results in the above fields have been published in 130 papers, 7 books and presented at various international conferences (in the former USSR, USA, France, Italy, Germany, Poland and elsewhere). V.I. Agoshkov prepared 6 courses of lectures for the students of the Novosibirsk University and the Moscow Institute of Physics and Technology.

**Victor Shutyaev, Ph.D.,** is a Senior Researcher of the Institute of Numerical Mathematics, Russian Academy of Sciences. He defended a Kandidat Thesis (Ph.D.) on the theme "Some questions of perturbation theory for neutron transport problems" in 1983 in the field of Numerical Mathematics. From 1982 until now he's been working as a Researcher at the Institute of Numerical Mathematics (Russian Academy of Sciences, Moscow).

His research interests are: adjoint equations and perturbation theory for linear and nonlinear problems of mathematical physics, numerical methods for partial differential equations, mathematical transport theory, optimal control problems. He is the author of more than 50 papers and co-author of 3 monographs.

V.P. Shutyaev prepared and delivered 3 courses of lectures for students of the Moscow Institute for Physics and Technology.

*To Olga, Evelina, and Larisa*

# Contents

# Chapter 1

# Principles of construction of adjoint operators in non-linear problems

## 1. DUAL SPACES AND ADJOINT OPERATORS

The concept of 'adjoint operator' is widely used in the theory of differential equations for a formally adjoint operator which often governs just a sequence of differentiation operators. In the context of functional analysis, the definition of the adjoint operator is more sophisticated and essentially depends on boundary conditions. But to introduce an adjoint operator and the corresponding adjoint equation one needs first to formulate the concept of 'dual space'. It turns out that, in doing so, there exist several possibilities at our disposal. To make clear how the adjoint operators and equations are treated in the subsequent text we give the definition of the dual space, beginning with the case of Banach spaces.

**1.1.** Let $X$ be a Banach space with a norm $\| \cdot \|_X$ and elements $g, f, l, \ldots$ Consider a set of continuous functionals $g^*, f^*, l^*, \ldots$ defined on the elements of $X$ with the values $g^*(f), \ldots$ Often by $\langle f, g^* \rangle_X$ one denotes the value $g^*(f)$ and the expression $\langle f, g^* \rangle_X$ is said to be a duality relation of $f$ and $g^*$. Hereinafter we assume that $g^*(f) \equiv \langle f, g^* \rangle_X$.

Now we can require that the functionals be either linear, i.e. the following condition is satisfied

$$g^*(\alpha f + \beta l) = \alpha g^*(f) + \beta g^*(l), \tag{1.1}$$

where $\alpha$ and $\beta$ are, in general, complex numbers, or antilinear (semi-linear, conjugate linear), i.e.

$$g^*(\alpha f + \beta l) = \bar{\alpha} g^*(f) + \bar{\beta} g^*(l). \tag{1.2}$$

The definition of the dual space $X^*$ depends on deciding between these two requirements. From here on we assume the functionals to be linear[316], but the antilinear functionals could be also considered.[98]

Introduce now the operations of addition and multiplication by a number for the elements $g^*, h^*, \ldots$ These operations can be defined either by the formula[316]

$$(\alpha g^* + \beta h^*)(f) = \alpha g^*(f) + \beta h^*(f) \tag{1.3}$$

or by the formula

$$(\alpha g^* + \beta h^*)(f) = \bar{\alpha} g^*(f) + \bar{\beta} h^*(f). \tag{1.4}$$

From this point on we assume (1.3) to hold. As a result the set of linear functionals is formed into a linear space which is denoted by $X^*$ and it is said to be a space dual to $X$. The functional identically equal to zero on the entire space $X$ is bound to be a zero point of $X^*$. We can introduce a norm in $X^*$ of the form

$$\|g^*\|_{X^*} = \sup_{0 \neq f \in X} \frac{|g^*(f)|}{\|f\|_X} = \sup_{f \in X, \|f\|_X = 1} |g^*(f)|, \tag{1.5}$$

rendering $X^*$ to be a normed space (and, in addition, a Banach one). This definition of the norm in $X^*$ yields a generalized Cauchy–Bunyakovsky inequality

$$|\langle f, g^* \rangle_X| \leq \|f\|_X \|g^*\|_{X^*}. \tag{1.6}$$

Owing to (1.6) and the above-mentioned properties, the expression $\langle f, g^* \rangle_X$ is also said to be a scalar product.

*Remark* 1.1. It should be realized that the concept of 'scalar product' for $\langle f, g^* \rangle_X$ is a matter of convention, since we find

$$\langle f, \alpha g^* + \beta h^* \rangle_X = \alpha \langle f, g^* \rangle_X + \beta \langle f, h^* \rangle_X$$

whereas for the true scalar product

$$\langle f, \alpha g^* + \beta h^* \rangle_X = \bar{\alpha} \langle f, g^* \rangle_X + \bar{\beta} \langle f, h^* \rangle_X.$$

*Remark* 1.2. Along with the definition of the norm $\|g^*\|_{X^*}$ by (1.5), note that

$$\|f\|_X = \sup_{0 \neq g^* \in X^*} \frac{|g^*(f)|}{\|g^*\|_{X^*}} = \sup_{\|g^*\|_{X^*} = 1} |g^*(f)| \tag{1.7}$$

if $f \in X$.

**1.2.** After having introduced the dual space, we give the definition of the adjoint operator in Banach spaces which is commonly accepted. Let $X$ and $Y$ be two Banach spaces and $A$ be a linear operator with a domain $D(A) \subset X$ dense in $X$. The operator $A$ is assumed to map $X$ into $Y$ (i.e., its range $R(A)$ belongs to $Y$). Let $X^*$ and $Y^*$ be spaces dual to $X$ and $Y$, respectively. Consider an arbitrary linear continuous functional $g^* \in Y^*$. It satisfies the inequality $|g^*(Af)| \leq c\|Af\|_Y$ for any $f \in D(A)$. This functional can be treated as a functional on $X$ (for a fixed $g^*$ and varying $f \in D(A) \subset X$). But in this case the functional may appear to be unbounded, i.e. the inequality $|g^*(Af)| \le c\|f\|_X$ may fail. We thus consider just the functionals $g^* \in Y^*$ which satisfy the inequality $|g^*(Af)| \le c\|f\|_X$ for any $f \in D(A)$. By $D(A^*)$ we denote the set of these functionals. Clearly $D(A^*)$ is not empty, since $g^* \equiv 0$ belongs to $D(A^*)$. For a fixed $g^* \in D(A^*)$ we find that $g^*(Af)$ is a linear continuous functional on $X$. Denote it by $h^*(f)$, i.e.

$$g^*(Af) = h^*(f), \quad h^* \in X^*. \tag{1.8}$$

Therefore, every element $g^* \in D(A^*)$ has an element $h^* \in X^*$ corresponding to it. Since $D(A)$ is dense in $X$, the functional $h^*$ is uniquely determined. The one-to-one correspondence $g^* \to h^*$ defines an operator $A^*$ which is said to be adjoint to $A$. It has the domain $D(A^*) \subset Y^*$ and the range $R(A^*) \subset X^*$. Thus,

$$A^* g^* = h^*, \quad g^* \in D(A^*), \quad h^* \in R(A^*), \tag{1.9}$$

$$A^*: Y^* \to X^*.$$

If $f \in D(A)$ and $g^* \in D(A^*)$, then, in accord with the definition of $A^*$, the following relation holds:

$$g^*(Af) = h^*(f),$$

i.e.

$$g^*(Af) = A^* g^*(f), \tag{1.10}$$

or, in terms of duality relations,

$$\langle Af, g^* \rangle_Y = \langle f, A^* g^* \rangle_X. \tag{1.11}$$

Note that if $A^*$ and $B^*$ are two adjoint operators introduced by the above-given definition and corresponding to some operators $A$ and $B$, then

$$(\alpha A + \beta B)^* = \alpha A^* + \beta B^*$$

on $D(A^*) \cap D(B^*)$. The set of adjoint operators appears thus to be linear.

**1.3.** Let now $X$ and $Y$ be Hilbert spaces with the scalar products $(\cdot, \cdot)_X$, $(\cdot, \cdot)_Y$ and the norms $\| \cdot \|_X = (\cdot, \cdot)_X^{1/2}$, $\| \cdot \|_Y = (\cdot, \cdot)_Y^{1/2}$, respectively. Then all the above holds true. Denote by $X^*$ and $Y^*$ the adjoint spaces corresponding to $X$ and $Y$. Note that, when considering Hilbert spaces, the spaces $X$ and $X^*$ (as well as $Y$ and $Y^*$) can be identified, i.e. one can assume that $X = X^*$ ($Y = Y^*$). The basis of this important property is the Riesz theorem on the representation of a linear bounded functional.

**Theorem 1.1** (F. Riesz). *For every element $g^* \in X^*$ there is a unique element $g \in X$ such that $g^*(f) = (f, g)_X$ for any $f \in X$ and $\|g\|_X = \|g^*\|_{X^*}$.*

According to this theorem, there exists an isometry $J_X g^* = g$ between $X^*$ and $X$ which is antilinear

$$J_X(\alpha g^* + \beta h^*) = \bar{\alpha} J_X(g^*) + \bar{\beta} J_X(h^*). \tag{1.12}$$

This results from the properties of a scalar product which suggest that the element $(\alpha g^* + \beta h^*)$ has the element $\bar{\alpha} g + \bar{\beta} h$ corresponding to it, given the correspondence $g^* \to g$, $h^* \to h$. Then the value $g^*(f)$ can be represented in the following manner:

$$g^*(f) = (f, g)_X = (f, J_X g^*)_X = J_X^{-1} g(f), \tag{1.13}$$

and the equality $\|g^*\|_{X^*} = \|g\|_X$ holds. This enables the spaces $X^*$ and $X$ to be identified as abstract sets (but not as linear spaces). This is precisely the identification we keep in mind in the subsequent text when putting $X \equiv X^*$. If $X \equiv X^*$, then the duality relation $\langle f, g \rangle_X$ is assumed to coincide with the scalar product $(f, g)_X$ for $f, g \in X$. The Hilbert space $X$ in this case is said to be basic.

**1.4.** Let us consider the definition of the adjoint operator in Hilbert spaces $X$ and $Y$ which are yet to be identified with their dual spaces. Let $A$ be a linear operator mapping $X$ into $Y$ with the domain $D(A)$ dense in $X$. By the definition of the adjoint operator in Banach spaces, we arrive at the operator $A^*$ mapping $Y^*$ into $X^*$ with the domain $D(A^*)$, satisfying the relationship

$$\langle Af, g^* \rangle_Y = \langle f, A^* g^* \rangle_X, \tag{1.14}$$

where $f \in D(A)$, $g^* \in D(A^*)$.

If $X$ and $Y$ are identified with $X^*$ and $Y^*$, respectively, the operator adjoint to $A$ differs from $A^*$. To define this operator we assume that $X \equiv X^*$, $Y \equiv Y^*$ and consider relation (1.14) where $g^* \in Y \equiv Y^*$, $A^* g^* \equiv h^* \in X \equiv X^*$. Let $g \in Y$ and $h \in X$ be the elements corresponding to $g^*$ and $h^*$ by Theorem 1.1, i.e. $J_Y g^* = g$, $J_X h^* = h$. Then (1.14) can be represented in the form

$$(Af, g)_Y = (f, h)_X. \tag{1.15}$$

According to this equality, each element $g$ of the set $J_Y D(A^*) \equiv D(\tilde{A}^*) \subset Y$ is matched one to one with an element $h$ of the set $J_X R(A^*) \equiv R(\tilde{A}^*) \subset X$. This correspondence defines an operator $\tilde{A}^*$:

$$\tilde{A}^* g = h, \quad g \in D(\tilde{A}^*), \tag{1.16}$$

$$\tilde{A}^*: Y \to X.$$

The operator $\tilde{A}^*$ will be also referred to as the adjoint operator. (Sometimes $\tilde{A}^*$ is said to be Hilbert adjoint to $A$.) We find from (1.15) and (1.16)

$$(Af, g)_Y = (f, \tilde{A}^* g)_X, \quad f \in D(A), \quad g \in D(\tilde{A}^*). \tag{1.17}$$

This gives the relation between $A^*$ and $\widetilde{A}^*$:

$$A^* = J_X^{-1} \widetilde{A}^* J_Y. \tag{1.18}$$

*Remark* 1.3. We obtain from (1.17) the equality

$$\widetilde{C}^* = \bar{\alpha}\widetilde{A}^* + \bar{\beta}\widetilde{B}^*, \tag{1.19}$$

where $C = \alpha A + \beta B$. The right-hand side of (1.19) contains the conjugate numbers $\bar{\alpha}$ and $\bar{\beta}$ instead of $\alpha$ and $\beta$ (as is the case in the equality $(\alpha A + \beta B)^* = \alpha A^* + \beta B^*$).

It is readily seen that the operators $A^*$ and $\widetilde{A}^*$ are distinct. For example, $A^*$ maps $Y^*$ into $X^*$, whereas $\widetilde{A}^*$ maps $Y$ into $X$, i.e. they operate in different spaces. Henceforward $\widetilde{A}^*$ is considered to mean the operator adjoint to $A$ in self-adjoint Hilbert spaces (i.e. when identifying Hilbert spaces with their duals). It is precisely this operator that is of frequent use in the theory of functional equations in Hilbert spaces. Let us formulate its definition in the following form.

**Definition 1.1.** *Let $A$ be a linear operator mapping $X \equiv X^*$ into $Y \equiv Y^*$ with a domain $D(A)$ dense in $X$. Denote by $D(\widetilde{A}^*) \subset Y$ a set of elements of $Y$ such that for each $g \in D(\widetilde{A}^*)$ there exists a unique element $h \in X$ satisfying the equality $(Af, g)_Y = (f, h)_X$ for any $f \in D(A) \subset X$. Let $\widetilde{A}^*$ be an operator with the domain $D(\widetilde{A}^*)$ such that $\widetilde{A}^*g = h$ or, in equivalent form,*

$$(Af, g)_Y = (f, \widetilde{A}^*g)_X \tag{1.20}$$

*for any $f \in D(A)$, $g \in D(\widetilde{A}^*)$. The operator $\widetilde{A}^*$ is said to be adjoint to $A$.*

Note that Definition 1.1 of the adjoint operator in Hilbert spaces is best suited to the case when $X = Y$, i.e. when the operator $A$ maps $X$ into the same $X$. In this case, for example, one can compare $A$ with $\widetilde{A}^*$ and introduce the definitions of symmetrical and self-adjoint operators. If $A$ and $\widetilde{A}^*$ operate in the same space $Y$ and $A = \widetilde{A}^*$ (which implies $D(A) = D(\widetilde{A}^*)$), then the operator $A$ is said to be self-adjoint. If $D(A) \subseteq D(\widetilde{A}^*)$ and for any $f, g \in D(A)$ the equality $(Af, g)_Y = (f, Ag)_Y$ holds, then the operator $A$ is said to be symmetrical and $A \subseteq \widetilde{A}^*$.

**1.5.** From the above discussion it stems that one of the basic assumptions, when introducing the operator $\widetilde{A}^*$, is the identification of Hilbert spaces with their dual spaces. However, broadly speaking, this identification is not necessary. The spaces are often identified when only appropriate. Furthermore, if there are several Hilbert spaces, only a few of them can be identified with their dual spaces to make them basic. This is not necessary to do for other spaces. In this case one more adjoint operator (distinct from $A^*$ and $\widetilde{A}^*$) can be introduced. Let $X^{(1)}$ and $Y^{(1)}$ be two Hilbert spaces, $X^{(1)}$ being densely imbedded into $Y^{(1)}$. Then there exists[135] a self-adjoint positive definite operator $\Lambda$ mapping $Y^{(1)}$ into $Y^{(1)}$ with the domain $D(\Lambda) = X^{(1)}$ and the range $R(\Lambda) = Y^{(1)}$ such that

$$\|f\|_{X^{(1)}} = \|\Lambda f\|_{Y^{(1)}}, \quad f \in X^{(1)}. \tag{1.21}$$

Starting from $\Lambda$, one can introduce the power operator $\Lambda^\alpha$, $-1 \leq \alpha \leq 1$, the domain of $\Lambda^\alpha$ being a Hilbert $H^\alpha$. Hilbert spaces $H^\alpha$, $-1 \leq \alpha \leq 1$, satisfy the relations

$$H^0 = Y^{(1)}, \quad H^{(1)} = X^{(1)},$$

$$H^{\alpha_2} \subset H^{\alpha_1}, \quad -1 \leq \alpha_1 < \alpha_2 \leq 1, \tag{1.22}$$

$$\|f\|_{H^\alpha} = \|\Lambda^\alpha f\|_{Y^{(1)}}, \quad f \in H^\alpha, \quad -1 \leq \alpha \leq 1.$$

Consider the space $H^\alpha$ for a given $\alpha \in [0, 1]$. The scalar product and the norm in $H^\alpha$ are denoted by $(\cdot, \cdot)_{H^\alpha}$ and $\|\cdot\|_{H^\alpha} = (\cdot, \cdot)_{H^\alpha}^{1/2}$, respectively. Denote by $(H^\alpha)^*$ the space dual to $H^\alpha$. The space $(H^\alpha)^*$ can be shown[135] to be isometric to $H^{-\alpha}$. This isometry is defined by the equality

$$g^*(f) = (f, g)_{Y^{(1)}}, \quad g^* \in (H^\alpha)^*, \quad g \in H^{-\alpha}, \quad f \in H^\alpha. \tag{1.23}$$

The spaces $(H^\alpha)^*$ and $H^{-\alpha}$ can be thus identified. In this case the value $g^*(f)$ of a functional $g^*$ ($g^* \in (H^\alpha)^*$, $f \in H^\alpha$) may be written as $(f, g)_{Y^{(1)}}$ ($f \in H^\alpha$, $g \in H^{-\alpha}$). If now we identify the space $H^0 = Y^{(1)}$ with its dual space, then the only space $Y^{(1)}$ will be self-adjoint (basic). The other spaces $H^\alpha$, $0 < \alpha \leq 1$, have the dual spaces $H^{-\alpha}$ (which, obviously, do not coincide with $H^\alpha$) and the value of a functional $g \in H^{-\alpha} \equiv (H^\alpha)^*$ is given by $(f, g)_{Y^{(1)}}$, $f \in H^\alpha$. The norm $\|g\|_{H^{-\alpha}}$ can be defined also (along with (1.22)) by the formula

$$\|g\|_{H^{-\alpha}} = \sup_{0 \neq f \in H^\alpha} \frac{|(f, g)_{Y^{(1)}}|}{\|f\|_{H^\alpha}} = \sup_{\|f\|_{H^\alpha} = 1} |(f, g)_{Y^{(1)}}|. \tag{1.24}$$

Hereafter, when identifying the spaces $(H^\alpha)^*$ and $H^{-\alpha}$, we identify the elements $g^*$ and $g$ from (1.23) and keep the previous notation $g^*$ for $g$:

$$g^*(f) \equiv (f, g^*)_{H^0}.$$

A set of Hilbert spaces $\{H^\alpha\}$, $-1 \leq \alpha \leq 1$, introduced above is said to be a scale of Hilbert spaces. Note that the case $H^0 = Y^{(1)} = Y^{(1)*}$, $(H^\alpha)^* \equiv H^{-\alpha}$ (but $H^\alpha \not\equiv (H^\alpha)^*$ ) is widely met in the boundary value problem theory. In the subsequent text $H^{-\alpha} \equiv (X)^{-1}$ is considered to mean the space dual to a

Hilbert space $X$ in the scale $\{H^\alpha\}$ (i.e. $X = H^\alpha$ for some $\alpha$) implying that there is a basic space $H_0 \equiv H^0$.

The consideration of the spaces $H^{-\alpha}$, $0 \le \alpha \le 1$, adjoint to $H^\alpha$ makes it possible to introduce another definition of the adjoint operator. Let $X$ and $Y$ be two spaces in the scale of Hilbert spaces $\{H^\alpha\}$ introduced above, among which only $H_0 \equiv H^0$ is identified with its dual space, that is, constitutes the basic space and $X^* \equiv X^{-1}$, $Y^* \equiv Y^{-1}$.

**Definition 1.2.** *Let $A$ be a linear operator mapping $X$ into $Y$ with the domain $D(A)$ dense in $X$. Denote by $D(\widetilde{\widetilde{A}}^*) \subset Y^*$ a set of elements of $Y$ such that for each $g \in D(\widetilde{\widetilde{A}}^*)$ there exists an element $h \in X^*$ satisfying the equality*

$$(Af, g)_{H_0} = (f, h)_{H_0} \tag{1.25}$$

*for any $f \in D(A)$. Let $\widetilde{\widetilde{A}}^*$ be an operator with the domain $D(\widetilde{\widetilde{A}}^*)$ such that $\widetilde{\widetilde{A}}^* g = h$ or, in equivalent form,*

$$(Af, g)_{H_0} = (f, \widetilde{\widetilde{A}}^* g)_{H_0} \tag{1.26}$$

*for any $f \in D(A)$, $g \in D(\widetilde{\widetilde{A}}^*)$. The operator $\widetilde{\widetilde{A}}^*$ is said to be adjoint to $A$.*

*Remark* 1.4. Strictly speaking, instead of $X^*$ and $Y^*$, it should be written $X^{-1}$ and $Y^{-1}$ in Definition 1.2. But in view of the identification we keep the previous notations.

For the sake of simplicity we often denote the operators $\widetilde{A}^*$ and $\widetilde{\widetilde{A}}^*$ also by $A^*$. Here, to avoid a misunderstanding, we will always underline the spaces the operators $A$ and $A^*$ operate in. If, for example, $A$ operates in $X = Y = X^* = Y^*$, then the adjoint operator $A^*$ is given by Definition 1.1. If the operator $A$ maps $X$ into $Y$, where $X, Y \in \{H^\alpha\}$, then the adjoint operator $A^*$ is defined by Definition 1.2 and it maps $Y^* \equiv Y^{-1}$ into $X^* \equiv X^{-1}$.

*Remark* 1.5. Of particular interest is the case that $X = H^\alpha$ and $Y = X^* \equiv H^{-\alpha}$. Having defined $\widetilde{\widetilde{A}}^*$ by (1.26), we can thus introduce again a concept of the self-adjoint operator, since $A$ and $\widetilde{\widetilde{A}}^*$ map $H^\alpha$ into $H^{-\alpha} \equiv (H^\alpha)^*$. If $Af = \widetilde{\widetilde{A}}^* g$ for any $f, g \in D(A)$ and $D(A) = D(\widetilde{\widetilde{A}}^*)$, then the operator $A$ is said to be self-adjoint.

Now, following the above-given concepts of the theory of dual spaces and adjoint operators, a general remark needs to be made. The presentation in this book, as a rule, is given in Hilbert spaces. However, many of the statements on the properties of adjoint operators obtained in the original literature for

Banach spaces (and hence, possibly, for other adjoint operators) are widely used in the book. Such situations are not specified in the subsequent text, since the statements mentioned hold also true in Hilbert spaces for $\widetilde{A}^*$, $\widetilde{\widetilde{A}}^*$ and for other operators, and the relationships of the form (1.18) between these operators and $A^*$ in Banach spaces with the isometries $J_Y, J_X, \ldots$ are valid.

## 2. CONSTRUCTION OF ADJOINT OPERATORS BASED ON USING THE LAGRANGE IDENTITY

Let $X$ and $Y$ be two spaces in a sequence $\{H^\alpha\}$ of real Hilbert spaces, among which only $H_0 \equiv H^0$ is identified with its dual space, that is, constitutes the basic space. We also assume that the space $(H^\alpha)^*$ dual to $H^\alpha$, $0 \le \alpha \le 1$, is identified with $H^{-\alpha}$.

When considering the spaces $X$ and $Y$, we denote the scalar products, norms and duality relations by

$$(f,g)_X, \quad \|f\|_X = (f,f)_X^{1/2}, \quad \langle f, g^* \rangle_X \equiv (f, g^*)_{H_0}, \quad f, g \in X, \quad g^* \in X^*,$$

$$(f,g)_Y, \quad \|f\|_Y = (f,f)_Y^{1/2}, \quad \langle f, g^* \rangle_Y \equiv (f, g^*)_{H_0}, \quad f, g \in Y, \quad g^* \in Y^*,$$

where $X^* \equiv X^{-1}$, $Y^* \equiv Y^{-1}$. These are the assumptions under which the forthcoming analysis is carried out in the present book.

Consider now the first principle of constructing the adjoint operators in non-linear problems. Let $\Phi$ be a non-linear operator mapping $X$ into $Y$. The domain $D(\Phi) \subset X$ of this operator is assumed to be a convex set (but not necessarily linear), and zero may happen not to be an element of $D(\Phi)$. Let us take an element $U_0 \in D(\Phi)$ and introduce a new operator $F(u)$:

$$F(u) = \Phi(U_0 + u) - \Phi(U_0), \tag{2.1}$$

where $u = U - U_0$, $U, U_0 \in D(\Phi)$, with the domain

$$D(F) = \{v \in X: \ v + U_0 \in D(\Phi)\}.$$

The domain $D(F)$ is assumed to be independent of $u$ and dense in $X$. It is easy to see that $0 \in D(F)$. We assume also that $D(F)$ is a linear set.

A considerable part of the book concerns the operators of the form $F(u)$ defined according to (2.1). The authors have at least two reasons for it. The first reason is that the property of 'global solvability' is not typical for non-linear equations $\Phi(U) = 0$ in the general case. Sometimes there is only the 'local solvability' property. By this it is meant that the equation $\Phi(U) = 0$ has a solution in the neighbourhood of an element $U_0 \in D(\Phi)$ or, what is the same, the equation $F(u) = f$ for $f \equiv -\Phi(U_0)$ has a solution $u \in D(F)$ in the neighbourhood of zero. Hence, when considering the solution existence problem for the equation $\Phi(U) = 0$, it is appropriate to turn to the solution existence problem for the equation involving the operator $F(u) = \Phi(U_0 + u) - \Phi(U_0)$.

Another reason for the advisability of considering the operators of the form (2.1) is in the authors' opinion that perturbation algorithms are applicable routinely to the equation of the form $F(u) = f$, the element $U_0$ being a 'non-perturbed' solution and $U = U_0 + u$ a 'perturbed' one in the neighbourhood of $U_0$. Notice, however, that if $0 \in D(\Phi)$ and $\Phi(0) = 0$, then for $U_0 \equiv 0$ we have $u = U$ and $F(U) = \Phi(U)$. Hence, in this case all the considerations presented below are valid also for the operator $\Phi$.

So, let $F(u)$ be a non-linear operator mapping $X$ into $Y$. The domain $D(F) \subset X$ of this operator is assumed to be a linear set, dense in $X$. Let also $0 \in D(F)$ and assume that $F(0) = 0$. Define an adjoint operator corresponding to $F$ on the basis of the Lagrange identity.[77,192]

**Definition 2.1.** *An operator $A^*(u)$ with a domain $D(A^*) \subset Y^*$ and a range in $X^*$ is said to be an adjoint operator corresponding to $F$ if $A^*(u)$ satisfies the Lagrange identity*

$$(F(u), v)_{H_0} = (u, A^*(u)v)_{H_0} \qquad (2.2)$$

*for any $u \in D(F)$, $v \in D(A^*)$.*

*Remark* 2.1. The operator $A^*(u)$ defined by (2.2) is also referred to as associated.[310]

*Remark* 2.2. It should be emphasized that the operator $A^*(u)$ in Definition 2.1 is said to be 'an adjoint operator corresponding to $F$', and not 'an operator adjoint to $F$'.

Note that several operators $A^*$ may satisfy (2.2) for a non-linear operator $F$; that is, Definition 2.1 does not define $A^*$ in a unique manner. Assume, for example, that $F(u)$ is representable in the form

$$F(u) = A(u)\, u, \qquad (2.3)$$

where $A(u)$ is a linear operator dependent on $u$, with a domain $D(A) \supset D(F)$. Fix an element $u \in D(F)$ and in a manner standard in linear operator theory (see Definition 1.2) introduce the (Hilbert) adjoint operator $A^*(u)$ using the identity

$$(A(u)\,w, v)_{H_0} = (w, A^*(u)\,v)_{H_0}, \qquad (2.4)$$

where $w \in D(A)$, $v \in D(A^*)$. By putting $w = u$ in (2.4) we come to (2.2). Therefore, if the representation (2.3) admits several operators $A(u) = A_i(u)$, $i = 1, 2\ldots$, then, by introducing the adjoint operators $A_i^*(u)$ $i = 1, 2\ldots$, according to (2.4), we obtain several adjoint operators corresponding to $F$. Let us exemplify the foregoing.

*Example* 2.1. Let $(t, x) \in \Omega = (0, 1) \times (0, 1)$, $X = X^* = Y = Y^* = L_2(\Omega)$ be a space of real-valued functions periodic in $t$ and $x$, with a period equal to unity in each variable (of course, the functions are defined for any $(t, x) \in \mathbf{R}^2$). The norm in $L_2(\Omega)$ has an ordinary form:

$$\|u\|_{L_2} = \left( \int_\Omega |u(t, x)|^2 \ dt \ dx \right)^{1/2}$$

Consider a periodic problem of the form

$$F(u) \equiv \frac{\partial u}{\partial t} + u \frac{\partial u}{\partial x} + au = f, \quad a = \text{ constant} > 0, \tag{2.5}$$

where the domain of the operator $F$ is $D(F) = C^{(1)}(\Omega) \subset L_2(\Omega)$. We choose the following forms of $F(u)$:

$$F(u) = A_1(u) u, \quad \text{where } A_1(u) v = \frac{\partial v}{\partial t} + u \frac{\partial v}{\partial x} + av,$$

$$F(u) = A_2(u) u, \quad \text{where } A_2(u) v = \frac{\partial v}{\partial t} + \left( a + \frac{\partial u}{\partial x} \right) v, \tag{2.6}$$

$$F(u) = A_3(u) u, \quad \text{where } A_3(u) v = \frac{\partial v}{\partial t} + \frac{1}{2} \frac{\partial(uv)}{\partial x} + av.$$

Using operators $\{A_i(u)\}$ and equalities (2.4), we construct adjoint operators $\{A_i^*(u)\}$ (the usual method of integration by parts is suitable here):

$$A_1^*(u) w = -\frac{\partial w}{\partial t} - \frac{\partial(uw)}{\partial x} + aw = -\frac{\partial w}{\partial t} - u \frac{\partial w}{\partial x} + \left( a - \frac{\partial u}{\partial x} \right) w,$$

$$A_2^*(u) w = -\frac{\partial w}{\partial t} + \left( a + \frac{\partial u}{\partial x} \right) w, \tag{2.7}$$

$$A_3^*(u) w = -\frac{\partial w}{\partial t} - \frac{u}{2} \frac{\partial w}{\partial x} + aw.$$

Note that each of the domains $\{D(A_i^*)\}$ of the operators $\{A_i^*\}$ contains the Sobolev space $W_2^1(\Omega)$ of periodic functions. Therefore, the operators $\{A_i^*(u)\}$ have the form shown in (2.7), at least in $W_2^1(\Omega)$. We find that the ambiguity in the representation of $F(u)$ by $A(u)$ results in an ambiguity in the definition of the adjoint operator corresponding to $F$.

## 3. DEFINITION OF ADJOINT OPERATORS BASED ON USING TAYLOR'S FORMULA

**3.1.** Let $X$ and $Y$ be the spaces introduced in Section 2. Consider a nonlinear operator $\Phi$ mapping $X$ into $Y$. The domain $D(\Phi)$ of this operator is assumed to be a convex set (but not necessarily linear), and zero may happen not to be an element of $D(\Phi)$. Let $\Phi$ have the continuous Gâteaux derivative (that is, $\Phi$ is differentiable in the Frechet sense). Therefore, the following formula holds[290]:

$$\Phi(U) = \Phi(U_0) + \int_0^1 \Phi'(U_0 + tu) \; dtu, \quad u = U - U_0, \quad U, U_0 \in D(\Phi), \quad (3.1)$$

or

$$F(u) \equiv A(u) \, u = \Phi(U) - \Phi(U_0), \tag{3.2}$$

where the operator $A(u)$ is defined by the expression:

$$A(u) = \int_0^1 \Phi'(U_0 + tu) \; dt, \tag{3.3}$$

and the domain

$$D(A) \equiv D(F) = \{ v \in X : \; v + U_0 \in D(\Phi) \}. \tag{3.4}$$

The domain $D(A)$ is assumed to be dense in $X$. It is easily seen that $0 \in D(F)$, and $D(A)$ is independent of the element $u$ which defines the operator $A(u)$. We assume that $D(F)$ is also a linear set.

The following differentiability property of $F$ holds true.

**Lemma 3.1.** *Let the operator $\Phi$ have the continuous Gâteaux derivatives of the order $n = 1, \ldots, N$, $N < \infty$. Then the operator $F$ has the continuous Gâteaux derivatives up to the $N$-th order, and*

(1) *the domains of the derivatives $F^{(n)}$ and $\Phi^{(n)}$ coincide, i.e. $D(F^{(n)}) = D(\Phi^{(n)})$, $n = 1, \ldots, N$;*

(2) *the following equalities hold:*

$$F^{(n)}(u) \, h = \Phi^{(n)}(U_0 + u) \, h, \quad n = 1, \ldots, N,$$

$$\forall u \in D(F^{(n)}), \quad \forall h \in D(F^{(n)}); \tag{3.5}$$

(3) *Taylor's formula is true:*

$$F(u_0 + h) = \sum_{n=0}^{N-1} \frac{1}{n!} F^{(n)}(u_0) \, h^n$$

$$+ \frac{1}{(N-1)!} \int_0^1 (1-t)^{N-1} F^{(N)}(u_0 + th) \; dt \, h^N. \tag{3.6}$$

*Proof.* The differentiability of $F(u)$ is readily apparent from the relationships

$$\lim_{t \to 0} \frac{F(u + th) - F(u)}{t} = \lim_{t \to 0} \frac{\Phi(U_0 + u + th) - \Phi(U_0 + u)}{t} = \Phi'(U_0 + u) \, h.$$

Hence it follows also that $D(F') = D(\Phi')$. In a similar manner, the other equalities in (3.5) can be verified.

Formula (3.6) results from the relevant Taylor's formula for the operator $\Phi$ and equalities (3.5).

Let $N = 1$. Then (3.6) is one of the simplest Taylor formulae with integral remainder — the Lagrange formula:

$$F(u_0 + h) = F(u_0) + \int_0^1 F'(u_0 + th) \ dt \ h. \tag{3.7}$$

By putting $u_0 = 0$, $h = u$ and using $F(0) = 0$, we arrive at the following expression for $F(u)$:

$$F(u) = \int_0^1 F'(tu) \ dt \ u \equiv A(u) \ u, \tag{3.8}$$

where

$$A(u) = \int_0^1 F'(tu) \ dt = \int_0^1 \Phi'(U_0 + tu) \ dt$$

(see (3.2), (3.3)) is a linear operator, with the domain $D(A) = D(F)$, mapping $X$ into $Y$: $A(u)(t_1 v_1 + t_2 v_2) = t_1 A(u) v_1 + t_2 A(u) v_2$, $A(u)$: $X \to Y$, $v_1, v_2 \in D(A)$; $t_1$ and $t_2$ are arbitrary real numbers.

The definition of the adjoint operator will be based on formula (3.8). Thus, we fix an element $u \in D(F)$ and in an ordinary manner (see Definition 1.2) introduce an operator $A^*(u)$ adjoint to $A(u)$, mapping $Y^*$ into $X^*$, with a domain $D(A^*)$:

$$A^*(u) \ u^* \in X^*, \quad A^*(u): \ Y^* \to X^*, \quad u \in D(F). \tag{3.9}$$

**Definition 3.1.** *The operator* $A^*(u) = \left( \int_0^1 F'(tu) \ dt \right)^*$ *is said to be an adjoint operator corresponding to the non-linear operator* $F$.

Owing to the uniqueness of the Gâteaux derivative, of the integral $\int_0^1 F'(tu) \ dt$ of the operator function $F'(tu)$, and of the operator adjoint to a linear one whose domain is dense everywhere in the entire space, $A^*(u)$ is unique in the sense of Definition 3.1. Then it is obvious that any two elements $u \in D(A)$ and $u^* \in D(A^*)$ satisfy the identity

$$(F(u), u^*)_{H_0} = (u, A^*(u) \ u^*)_{H_0}, \tag{3.10}$$

or the Lagrange identity

$$(A(u) \ u, u^*)_{H_0} = (u, A^*(u) \ u^*)_{H_0} \tag{3.11}$$

(recall that $X^* \equiv X^{-1}$, $Y^* \equiv Y^{-1}$). If, in addition, the element $u$ is the solution of the original (main) equation

$$F(u) \equiv A(u)\, u = f, \quad f \in R(F) \subset Y \tag{3.12}$$

and $u^*$ is the solution of the adjoint equation

$$A^*(u)\, u^* = g, \quad g \in R(A^*) \subset X^* \tag{3.13}$$

(for a fixed solution $u$ of (3.12)), then (3.11) implies the adjointness relation

$$(f, u^*)_{H_0} = (u, g)_{H_0}.$$

*Remark* 3.1. If $F$ is a linear operator, $F(u) = Fu$, then Definition 3.1 yields an adjoint operator of the theory of linear operators in Hilbert spaces (see Definition 1.2), and we find $A^*(u) \equiv A^* = F^*$, while (3.11) transforms to $(Au, u^*)_{H_0} = (u, A^*u^*)_{H_0}$.

*Example* 3.1. Consider the operator $F$ and the spaces of Example 2.1. We find for this operator

$$F'(u)\, v = \frac{\partial v}{\partial t} + u\frac{\partial v}{\partial x} + \left(a + \frac{\partial u}{\partial x}\right) v. \tag{3.14}$$

Let us find the adjoint operator $(F'(u))^*$. Since we assume that $X = X^* = Y^* = Y = L_2(\Omega)$, integration by parts yields

$$(F'(u)\, v, u^*)_{L_2(\Omega)} = \left(v, -\frac{\partial u^*}{\partial t} - u\frac{\partial u^*}{\partial x} + au^*\right)_{L_2(\Omega)} \tag{3.15}$$

It is readily shown that if $u^* \in W_2^1(\Omega)$, the following inequality holds:

$$\left|(F'(u)\, v, u^*)_{L_2(\Omega)}\right| \le c\|u^*\|_{W_2^1(\Omega)}\|v\|_{L_2(\Omega)} \le c\|v\|_{L_2(\Omega)},$$

where $c = c(\|u\|_{C(\Omega)}) < \infty$. We thus find $W_2^1(\Omega) \subset D(A^*)$, and the restriction of the operator $(F'(u))^*$ to the set $W_2^1(\Omega)$ is

$$(F'(u))^* = -\frac{\partial u^*}{\partial t} - u\frac{\partial u^*}{\partial x} + au^*. \tag{3.16}$$

If we consider the operator $A^*(u)$ only on $W_2^1(\Omega)$, then $A^*(u)$ takes the form

$$A^*(u)\, u^* = \left(\int_0^1 F'(tu)\ dt\right)^* u^*$$

$$= \left(\frac{\partial}{\partial t} + \frac{u}{2}\frac{\partial}{\partial x} + \left(a + \frac{1}{2}\frac{\partial u}{\partial x}\right)\right) u^*$$

$$= -\frac{\partial u^*}{\partial t} - \frac{u}{2}\frac{\partial u^*}{\partial x} + au^*. \tag{3.17}$$

Therefore, here the adjoint operator in the sense of Definition 3.1 coincides on $W_2^1(\Omega)$ with one of the adjoint operators in the sense of Definition 2.1 (namely, with $A_3^*(u)$ in Example 2.1).

**3.2.** Consider the formula (3.6) for $N > 1$. Setting $u_0 = 0$, $h = u$, we come to

$$F(u) = \sum_{n=1}^{N} A_n(u)\, u \equiv A(u)\, u, \qquad (3.18)$$

where

$$A(u) = \sum_{n=1}^{N} A_n(u),$$

$$A_n(u) = \frac{1}{n!} F^{(n)}(0)\, u^{n-1}, \quad 1 \le n \le N-1, \qquad (3.19)$$

$$A_N(u) = \frac{1}{(N-1)!} \int_0^1 (1-t)^{N-1} F^{(N)}(tu) \; dt \; u^{N-1}.$$

By $A^*(u)$ we denote the operator adjoint to $A(u)$ and defined in a domain $D(A^*(u))$ with $A^*(u)\colon Y^* \to X^*$, $u \in D(F)$.

**Definition 3.2.** *An operator*

$$A^*(u) = \left( \sum_{n=1}^{N} A_n(u) \right)^*$$

*is said to be an adjoint operator corresponding to $F$.*

Let $A_n^*(u)$ be the operator adjoint to $A_n(u)$ defined in the domain $D(A_n^*(u))$. We introduce the set

$$D(\widetilde{A}^*(u)) \equiv \bigcap_{n=1}^{N} D(A_n^*(u)).$$

**Definition 3.3.** *An operator of the form*

$$\widetilde{A}^*(u) = \sum_{n=1}^{N} A_n^*(u)$$

*defined in the domain $D(\widetilde{A}^*(u))$ is said to be an adjoint operator corresponding to $F$.*

If we compare the operators $A^*(u)$ and $\widetilde{A}^*(u)$ in the sense of Definitions 3.2 and 3.3, we arrive at the conclusion that if the operators $\{A_n(u)\}_{n=1}^{N}$ (with a possible exception of one) are bounded, then $A^*(u) = \widetilde{A}^*(u)$. Otherwise the inclusion $\widetilde{A}^*(u) \subseteq A^*(u)$ takes place in a general case.

**3.2.** It is not difficult to rephrase the above definitions for the case of $F$ being an infinitely differentiable operator representable by the convergent Taylor series

$$F(u) = \sum_{n=0}^{\infty} \frac{F^{(n)}(0)}{n!} u^n, \quad u \in D(F). \tag{3.20}$$

Here the operators $A^*(u)$ and $\widetilde{A}^*(u)$ defined in their respective domains $D(A^*)$ and $D(\widetilde{A}^*)$ have the forms

$$A^*(u) = \left( \sum_{n=1}^{\infty} \frac{F^{(n)}(0)}{n!} u^{n-1} \right)^*,$$

$$\widetilde{A}^*(u) = \sum_{n=1}^{\infty} \left( \frac{F^{(n)}(0)}{n!} u^{n-1} \right)^*;$$

(it was taken into consideration that $F(0) = 0$).

**3.3.** Using the operators

$$A(u) = \int_0^1 F'(tu) \; dt$$

and $A^*(u)$, we can define the operator adjoint to $F$ and the concepts of the symmetrical and skew-symmetrical operators.

**Definition 3.4.** *If $D(F) \subset D(A^*)$, then an operator $F^*$ defined on $D(F)$ and given by the relation*

$$F^*(u) = A^*(u)\, u = \left( \int_0^1 F'(tu) \; dt \right)^* u \tag{3.21}$$

*is said to be adjoint to $F$.*

**Definition 3.5.** *An operator $F$ mapping $X$ into $Y \equiv X^*$ is said to be symmetrical if $A(u) = A^*(u)$ on $D(F)$ (in this case $F(u) = F^*(u)$), and it is said to be skew-symmetrical if $A(u) = -A^*(u)$ on $D(F)$ (then $F(u) = -F^*(u)$) for each element $u \in D(F)$.*

The conditions under which the operator $F(u)$ is symmetrical or skew-symmetrical will be formulated in the next chapter. Here we only point out the following simple fact. Namely, assume that $F(u)$ is a potential operator. Then the Gâteaux derivative of $F(u)$ is symmetrical[296]:

$$(F'(u)\,v, w)_{H_0} = (v, F'(u)\,w)_{H_0}. \tag{3.22}$$

Therefore, $A(u) = A^*(u)$ on $D(F)$ and $F$ is symmetrical.

## 4. OPERATORS OF THE CLASS $\mathcal{D}$ AND THEIR ADJOINT OPERATORS

The presentation in this section and the terminology are based on the book by M.Vainberg[296]. Let $X$ be a reflexive real Banach space, and $F$ be a non-linear continuously Gâteaux–differentiable operator mapping $X$ into $X^*$, $F(0) = 0$. Among the set of such operators, we single out the class $\mathcal{D}$ of such operators that each $F \in \mathcal{D}$ is put in correspondence with an operator $G$ belonging to this set, so that

$$\langle z, G'(x)\,y \rangle_X = \langle y, F'(x)\,z \rangle_X \tag{4.1}$$

for all $x, y, z \in X$ (where $G'(x)$ and $F'(x)$ are the Gâteaux derivatives of $G$ and $F$, respectively, and $\langle z, y \rangle_X \equiv y(z)$ is the value of the functional $y \in X^*$ on the element $z \in X$).

**Definition 4.1.** *An operator $G \in \mathcal{D}$ satisfying (4.1) for any $x, y, z \in X$ is said to be adjoint to $F$.*

If (4.1) holds, we will write

$$G'(x) = (F'(x))^*. \tag{4.2}$$

An operator adjoint in the sense of Definition 4.1 is unique. Actually, if an operator $\Phi \in \mathcal{D}$ satisfies (4.2) , then $G'(x) = \Phi'(x)$. While

$$\frac{d}{dt}\,G\,(tx) = G'(tx)\,x, \qquad \frac{d}{dt}\,\Phi\,(tx) = \Phi'(tx)\,x,$$

then using the Lagrange formula, we find

$$G(x) = \int_0^1 G'(tx)\,x\ dt = \int_0^1 \Phi'(tx)\,x\ dt = \Phi(x),$$

i.e. $G = \Phi$. Since the adjoint operator $G$ is unique, we may write

$$G \equiv F^* \tag{4.3}$$

or, as well,

$$(F')^* = (F^*)'. \tag{4.4}$$

**Definition 4.2.** *An operator $F \in \mathcal{D}$ is said to be symmetrical if $F = F^*$, and it is said to be skew-symmetrical if $F^* = -F$.*

Using Definition 4.1 and replacing the duality relation $\langle \cdot, \cdot \rangle_X$ by the scalar product $(\cdot, \cdot)_{H_0}$, we may introduce the set $\mathcal{D}$ and formulate the definitions of the adjoint operator and of the symmetrical and skew-symmetrical operators in the case that $X$ is in a sequence $\{H^\alpha\}$ of real Hilbert spaces (see Section 2). These operators are distinct from the corresponding operators introduced for the Banach spaces (nevertheless, respective isometries set up a correspondence between the operators). Thus to identify these operators we use the symbol $\mathcal{D}$ (for example, $F_{\mathcal{D}}^*$ is an operator adjoint to $F \in \mathcal{D}$ in the sense of Definition 4.1, but in Hilbert spaces). The operator $F_{\mathcal{D}}^*$ possesses all the properties of $F^* \equiv G \in \mathcal{D}$ (which are dealt with in the next chapter). In the subsequent consideration we study the relation between $F^*(u) = \left( \int_0^1 F'(tu) \ dt \right)^* u$ and, namely, $F_{\mathcal{D}}^*$.

*Example* 4.1. Let $X = \overset{\circ}{W}_2^1(0,1) = \{u \in W_2^1(0,1): u(0) = u(1) = 0\}$, $H_0 = L_2(0,1)$, $(\overset{\circ}{W}_2^1(0,1))^* = W_2^{-1}(0,1)$; that is, $H_0$ is the only basic space. Consider the problem

$$-\frac{d^2 u}{dx^2} + u^2(x) = f(x), \quad x \in (0,1),$$

$$u(0) = u(1) = 0. \tag{4.5}$$

Its generalized formulation is: find a function $u(x) \in \overset{\circ}{W}_2^1(0,1)$ satisfying the equality

$$a(u,v) \equiv \int_0^1 \left( \frac{du}{dx} \frac{dv}{dx} + u^2 v \right) dx = \int_0^1 fv \ dx \tag{4.6}$$

for an arbitrary function $v \in \overset{\circ}{W}_2^1(0,1)$. If we define by

$$(F(u), v)_{L_2} = a(u,v) \tag{4.7}$$

an operator $F$ mapping $\overset{\circ}{W}_2^1(0,1)$ into $W_2^{-1}(0,1)$ and defined in the domain $D(F) = \overset{\circ}{W}_2^1(0,1)$, the generalized formulation (4.6) can be written as the operator equation

$$F(u) = f. \tag{4.8}$$

Now it is easy to show for the operator $F$ that it is continuously Gâteaux-differentiable and $F'(u)$ is given by the equality

$$(F'(u) v, w)_{L_2} = \int_0^1 \left( \frac{dv}{dx} \frac{dw}{dx} + 2uvw \right) dx, \tag{4.9}$$

being an operator mapping $\overset{o}{W_2^1}(0,1)$ into $\overset{o}{W_2^{-1}}(0,1)$. Furthermore, the operator $F$ belongs here to set $\mathcal{D}$ and has an adjoint operator $F_{\mathcal{D}}^*$ identical to $F$; hence, $F$ is symmetrical.

# Chapter 2

# Properties of adjoint operators constructed on the basis of various principles

## 1. GENERAL PROPERTIES OF MAIN AND ADJOINT OPERATORS CORRESPONDING TO NON-LINEAR OPERATORS

**1.1.** Let $X$ and $Y$ be two spaces in a sequence $\{H^\alpha\}$ of real Hilbert spaces introduced in Chapter 1. Let $\Phi\colon X \to Y$ be a non-linear operator with a domain $D(\Phi) \subset X$ which is assumed to be a convex set. Assume that $\Phi$ is continuously Gâteaux-differentiable. Then the formula holds true:

$$\Phi(U) = \Phi(U_0) + \int_0^1 \Phi'(U_0 + tu)\ dtu,$$
$$u = U - U_0, \quad U, U_0 \in D(\Phi), \tag{1.1}$$

or

$$F(u) \equiv A(u)\,u = \Phi(U) - \Phi(U_0), \tag{1.2}$$

where $A(u)$ is an operator of the form

$$A(u) = \int_0^1 \Phi'(U_0 + tu)\ dt = \int_0^1 F'(tu)\ dt \tag{1.3}$$

with the domain

$$D(A) \equiv D(F) = \{v \in X\colon v + U_0 \in D(\Phi)\} \tag{1.4}$$

which is assumed to be a linear set. Fix an element $u \in D(A)$ and introduce the operator $A^*(u)$ using the identity

$$(A(u)\,v, g)_{H_0} = (v, A^*(u)\,g)_{H_0}, \tag{1.5}$$

where $u, v \in D(A)$, $g \in D(A^*)$. It is easy to see that the operator $A^*(u)$ satisfies the equality

$$(F(u), g)_{H_0} = (u, A^*(u)\,g)_{H_0} \quad \forall u \in D(F), \quad \forall g \in D(A^*) \tag{1.6}$$

resulting from (1.5) for $v = u$. Thus, having defined the operator $A(u)$ by formula (1.3), we may introduce a uniquely determined adjoint operator $A^*(u)$.

And yet this operator is one of the operators which satisfy the representation $F(u) = A(u) u$. Hereinafter, by $A(u)$ we always denote the operator defined according to (1.3). Any operator constructed on the basis of another principle will be denoted by $\widetilde{A}(u)$. Let thus $\widetilde{A}(u)$ be an operator (different from $A(u)$) mapping $X$ into $Y$, with a domain $D(\widetilde{A})$, and satisfying the equality:

$$F(u) = \widetilde{A}(u) u \quad \forall u \in D(\widetilde{A}). \tag{1.7}$$

For the operator $\widetilde{A}(u)$ we introduce also an adjoint operator $\widetilde{A}^*(u)$ defined by

$$(\widetilde{A}(u) v, g)_{H_0} = (v, \widetilde{A}^*(u) g)_{H_0}, \tag{1.8}$$

for any $u \in D(F)$, $v \in D(\widetilde{A})$, $g \in D(\widetilde{A}^*)$. This operator satisfies the equality of the form (1.6):

$$(F(u), g)_{H_0} = (u, \widetilde{A}^*(u) g)_{H_0} \quad \forall u \in D(F), \quad \forall g \in D(\widetilde{A}^*).$$

The representations $F(u) = A(u) u$ and $F(u) = \widetilde{A}(u) u$ imply that the intersection of $D(A)$ and $D(\widetilde{A})$ is non-zero. Thus, in the subsequent consideration, when comparing the operators $A(u)$ and $\widetilde{A}(u)$, we assume that these operators are defined on $D(A) \cap D(\widetilde{A})$ and denote this intersection again by $D(A)$. A similar assumption is taken to be satisfied when considering the adjoint operators $A^*$ and $\widetilde{A}^*$.

**1.2.** In the following, we consider the equations with the operator $A(u)$ and the non-linear operator $F$. Thus, of one the problems to be solved is the problem of differentiability of these operators. Lemma 3.1 of Chapter 1 gives the differentiability properties of $F$ if the operator $\Phi$ possesses the corresponding properties. Now consider the operator $A(u)$.

Let $u$ and $v$ be arbitrary fixed elements of $D(A)$. Consider the element $u + th$, $h \in D(A)$. If there exists an operator $B = B(u)$, bilinear in $h$ and $v$, such that

$$(B(u) h) v = \lim_{t \to 0} \frac{(A(u + th) - A(u)) v}{t},$$

then this operator is said to be a derivative of $A(u)$ and it is denoted by

$$A_u(u; \cdot) = \frac{\partial A}{\partial u}(u; \cdot).$$

Hence

$$\frac{\partial A}{\partial u}(u; h) v = \lim_{t \to 0} \frac{(A(u + th) - A(u)) v}{t}. \tag{1.9}$$

Using Lemma 3.1, we find for the operator $A_u(u; h)$

$$\frac{\partial A}{\partial u}(u; h) = F'(u) h - A(u) h,$$

$$\frac{\partial A}{\partial u}(u; h) = \Phi'(U_0 + u) h - A(u) h \tag{1.10}$$

for any $u \in D(A)$, $h \in D(F')$.

*Remark 1.1.* Taking (1.3) into account, we find, in addition to (1.10),

$$\frac{\partial A}{\partial u}(u; h) = \Phi'(U_0 + u)\, h - \int_0^1 \Phi'(U_0 + tu)\, dt h$$

for any $u \in D(A)$, $h \in D(F')$.

When studying properties of the operator $A(u)$, we may take advantage of the fact that some properties are given by the operator $F$ and its derivative $F'(u) = \Phi'(U_0 + u)$.

**Lemma 1.1.** *If the operator $F'(u) = \Phi'(U_0 + u)$ has an inverse operator for each*

$$u \in \bar{S}_R(0) = \{u \in D(F): \|u\| \le R\},$$

*the operator $A(u)$ also has an inverse operator for $u \in \bar{S}_R(0)$.*

*Proof.* Let $u \in \bar{S}_R(0)$. Then

$$A(u)\, v = F'(\tau u)\, v = F'(U)\, v, \quad 0 < \tau < 1,$$

where $U = \tau u$. But $\|U\|_X \le \|u\|_X$; therefore, $U \in \bar{S}_R(0)$, so that $(F'(U))^{-1}$ exists and, hence, $(A(u))^{-1}$ exists as well.

**Lemma 1.2.** *If the operator $F'(u)$ obeys the estimate*

$$\|F'(u)\, v\|_Y \ge m\, \|v\|_X \tag{1.11}$$

*for each $u \in \bar{S}_R(0)$ where $v$ is an arbitrary element in $D(F)$ and $m$ is a positive constant independent of $v$, then for any $u \in \bar{S}_R(0)$ we have*

$$\|A(u)\, v\|_Y \ge m\, \|v\|_X \tag{1.12}$$

*(with the same constant $m$).*

*Proof.* The estimate (1.12) is a corollary of (1.11) and the mean value theorem:

$$\|A(u)\, v\|_Y = \|F'(\tau u)\, v\|_Y \ge m\, \|v\|_X.$$

*Remark 1.2.* If we assume $u = 0$, then $A(0) = F'(0) = \Phi'(U_0)$ and, obviously, Lemmas 1.1 and 1.2 admit a reformulation for the case where their hypotheses are required to be satisfied only for $u = 0$.

Let us see whether the operator $A^*(u)$ is closed. It is well known in the linear adjoint operator theory that these operators are always closed. Naturally, this property also holds for $A^*(u)$.

**Lemma 1.3.** *The operator*

$$A^*(u) = \left( \int_0^1 F'(tu) \, dt \right)^* : Y^* \to X^*$$

*is closed for any* $u \in D(F)$.

*Proof.* Let $u_n^* \to u^*$ in $Y^*$, $u_n^* \in D(A^*)$, $A^*(u) u_n^* \equiv g_n \to g$ in $X^*$. Then for any $v \in D(F)$ we have

$$(v, A^*(u) u_n^*)_{H_0} \equiv (v, g_n)_{H_0} \to (v, g)_{H_0}.$$

By the construction of $A^*(u)$, we find

$$(v, A^*(u) u_n^*)_{H_0} \equiv (A(u) v, u_n^*)_{H_0} \to (A(u) v, u^*)_{H_0}.$$

Hence,
$$(A(u) v, u^*)_{H_0} = (v, g)_{H_0},$$

where $|(v, g)_{H_0}| \leq \|v\|_X \|g\|_{X^*} \leq c \|v\|_X < \infty$, i.e. $(v, g)_{H_0}$ is a bounded functional. By definition, this means that $u^* \in D(A)$ and $g = A^*(u) u^*$. Then, $A^*(u)$ is closed.

Consider Definition 2.1 of Chapter 1. As shown in the following example, the operator $A^*(u)$ introduced by Definition 2.1 is not necessarily bounded.

*Example* 1.1. Consider the operator, the sets and the spaces introduced in Example 2.1 of Chapter 1. Let

$$\widetilde{A}(u) v = \frac{\partial v}{\partial t} + \left( a + \frac{\partial u}{\partial x} \right) v,$$

$$D(A^*) = C^{(1)}(\Omega) \subset L_2(\Omega).$$

Then an operator of the form

$$\widetilde{A}^*(u) = -\frac{\partial}{\partial t} + \left( a + \frac{\partial u}{\partial x} \right)$$

satisfies the equality (2.2) of Chapter 1 for any $u \in C^{(1)}(\Omega)$. Since the coefficient $a + \partial u/\partial x$ in $\widetilde{A}^*(u)$ is bounded for $u \in C^{(1)}(\Omega)$, the operator $\widetilde{A}^*(u)$ is closed with the differentiation operator $\partial/\partial t$ being closed. But the latter operator mapping $L_2$ into $L_2$, with the domain $C^{(1)}(\Omega)$, is well-known to be not closed. (However, if we take $D(A^*) = W_2^1(\Omega)$ and treat differentiation in a generalized sense, then the operator $A^*(u)$ becomes bounded!) Thus, if we consider Definition 2.1 of an adjoint operator $A^*(u)$, then a choice of a domain

$D(A^*(u))$ is of great importance in questions whether the operator $A^*(u)$ is closed.

**Lemma 1.4.** *Let $Y = X^*$ and the operator $F$ be strictly monotone, i.e.*

$$\langle u - v, F(u) - F(v)\rangle_X \geq m\|u - v\|_X^2, \quad m = \text{constant} > 0.$$

*Then the operator $A(u)$ is positive definite on $D(F)$, i.e.*

$$\langle v, A(u)v\rangle_X \geq m\|v\|_X^2. \tag{1.13}$$

*Proof.* Using the mean value theorem and demanding that $t > 0$, we have

$$m\|tv\|_X^2 \leq \langle tv, F(u + tv) - F(u)\rangle_X$$

$$= \left\langle tv, \int_0^1 F'(u + tvt')\,dt'tv\right\rangle_X$$

$$= \left\langle tv, \int_0^t F'(u + t'v)\,dt'v\right\rangle_X$$

$$= t\int_0^t \langle v, F'(u + t'v)v\rangle_X\,dt'$$

$$= t^2\langle v, F'(u + \tau v)v\rangle_X, \quad \tau \in (0, t).$$

Dividing both parts of the inequality obtained by $t^2$ and then tending $t \to 0$, we find $\langle v, F^*(u)v\rangle_X \geq m\|v\|_X^2$; hence, $\langle v, A(u)v\rangle_X \geq m\|v\|_X^2$.

*Remark* 1.3 It is obvious that Lemma 1.4 holds if the operator $\Phi$ is assumed to be strictly monotone.

In the sequel, we will make use of the following representations of the operator $A(u)$:

$$A(u)\int_0^1 \Phi'(U_0 + tu)\,dt = \int_0^1 F'(tu)\,dt$$

$$= F'(0) + \int_0^1 (F'(tu) - F'(0))\,dt. \tag{1.14}$$

Hence, we get in particular the following conclusion. If $F'(v)$ is assumed to satisfy the Lipschitz condition

$$\|F'(u_1) - F'(u_2)\|_{X \to Y} \leq L\|u_1 - u_2\|_X \tag{1.15}$$

for any $u_1, u_2 \in \bar{S}_R(0) = \{u \in D(F): \|u\|_X \leq R\}$, then for $v \in D(F'(0))$ we also have $v \in D(A)$ for any $u \in \bar{S}_R(0)$, i.e. $D(F'(0)) \subset D(A(u))$.

The representation

$$F(u) \equiv A(u)\, u = F'(0)\, u + \int_0^1 [F'(tu) - F'(0)]\ dtu \qquad (1.16)$$

holds if the operator $F'(0)$ is not necessarily bounded. Actually, setting

$$F_1(u) \equiv F(u) - F'(0)\, u, \qquad (1.17)$$

we find $F_1'(u) = F'(u) - F'(0)$; hence, by (1.15),

$$\|F_1'(u)\|_{X \to Y} = \|F'(u) - F'(0)\|_{X \to Y} \leq L\|u\|_X.$$

For the non-linear operator $F_1$ the following representation takes place

$$F_1(u) = \int_0^1 F_1'(tu)\ dtu$$

and from (1.17) we get

$$F(u) = F'(0)\, u + F_1(u) = F'(0)\, u + \int_0^1 F_1'(tu)\ dtu$$

$$= F'(0)\, u + \int_0^1 [F'(tu) - F'(0)]\ dtu.$$

**1.3.** We have considered above the properties of the operator $A(u) = \int_0^1 F'(tu)\ dt$. Now we study the properties of the operator $\tilde{A}(u)$ in the representation

$$F(u) = \tilde{A}(u)\, u, \quad u \in D(F). \qquad (1.18)$$

In this case, a form of the operator $\tilde{A}(u)$ (or a principle of its construction) is not specified. Using this operator, we may also introduce the operator $\tilde{A}^*(u)$ and the corresponding adjoint equation $\tilde{A}^*(u)\, w = p$.

Let thus $F(u) = \tilde{A}(u)\, u$ and $\tilde{A}(u)$ be a linear operator where $D(\tilde{A}(u)) = D(F)$ for any $u \in D(F)$. When studying the properties of the operator $\tilde{A}(u)$, different types of its continuity are of great concern. Thus, the following lemma is valid.

**Lemma 1.5.** *If the operator $\tilde{A}(u)$ is continuous at the point $u = 0$ in the following sense:*

$$\lim_{t \to 0} \|\tilde{A}(tv)\, v - \tilde{A}(0)\, v\|_Y = 0 \qquad (1.19)$$

*for any $v \in D(F)$, then*

$$\tilde{A}(0) = F'(0) = \Phi'(U_0) = A(0).$$

*Proof.* Using the definition of the derivative $F'(u)$ and the condition $F(0) = 0$, we obtain

$$\|\widetilde{A}(0)\,v - F'(0)\,v\|_Y = \lim_{t \to 0} \left\| \widetilde{A}(0)\,v - \frac{F(tv) - F(0)}{t} \right\|_Y$$

$$= \lim_{t \to 0} \left\| \widetilde{A}(0)\,v - \frac{F(tv)}{t} \right\|_Y$$

$$= \lim_{t \to 0} \left\| \widetilde{A}(0)\,v - \frac{\widetilde{A}(tv)\,tv}{t} \right\|_Y$$

$$= \lim_{t \to 0} \|\widetilde{A}(0)\,v - \widetilde{A}(tv)\,v\|_Y = 0.$$

Since $\widetilde{A}(u)$ is an arbitrary operator in the representation $F(u) = \widetilde{A}(u)\,u$, the equalities $\widetilde{A}(0) = F'(0) = A(0)$ hold independently of the principle of constructing $\widetilde{A}(u)$, provided just the condition (1.19) is satisfied. Note that if the condition (1.19) is not true, there lies an ambiguity in choosing $\widetilde{A}(0)$. For example, one can consider the operator

$$\widetilde{A}(u)\,v = \begin{cases} A(u)\,v & \text{if } \|u\|_X > 0, \\ 0v & \text{if } u = 0. \end{cases}$$

Let now the operator $\widetilde{A}(u)$ be Gâteaux-differentiable in the sense of (1.9) and $\partial \widetilde{A}/\partial u \equiv \widetilde{A}'$ be its derivative. Then the following lemma holds.

**Lemma 1.6.** *If $\widetilde{A}(u)$ is Gâteaux-differentiable, then*

$$\widetilde{A}'(u; v)\,u = F'(u)\,v - \widetilde{A}(u)\,v \tag{1.20}$$

*for any $v \in D(F)$.*

*Proof.* Since $F(u + tv) = \widetilde{A}(u + tv)(u + tv)$, we find

$$\frac{\partial F}{\partial t}(u + tv) \bigg|_{t=0} = F'(u)\,v = \widetilde{A}'(u; v)\,u + \widetilde{A}(u)\,v.$$

*Remark* 1.4. By setting $u = 0$ in (1.20) we get immediately $F'(0)\,v = \widetilde{A}(0)\,v$.

**Lemma 1.7.** *If the operator $\widetilde{A}(u)$ is differentiable at the point $u = 0$ and, in addition, $F''(0, g)$ exists, then*

$$\widetilde{A}'(0; g)\,h + \widetilde{A}'(0; h)\,g = F''(0; g)\,h \tag{1.21}$$

*for any $g, h \in D(F)$.*

*Proof.* Since the operator $\widetilde{A}' \equiv \partial\widetilde{A}/\partial u$ is bilinear and $F(0) = 0$, we obtain

$$
\begin{aligned}
\widetilde{A}'(0;g)\,h + \widetilde{A}'(0;h)\,g &= \widetilde{A}'(0;g+h)(g+h) - \widetilde{A}'(0;g)\,g - \widetilde{A}'(0;h)\,h \\
&\equiv \lim_{t\to 0}\frac{1}{t}[\widetilde{A}(t(g+h))(g+h) - \widetilde{A}(0)(g+h) \\
&\quad -\widetilde{A}(tg)\,g + \widetilde{A}(0)\,g - \widetilde{A}(th)\,h + \widetilde{A}(0)\,h] \\
&= \lim_{t\to 0}\frac{\widetilde{A}(t(g+h))\,t(g+h) - \widetilde{A}(tg)\,tg - \widetilde{A}(th)\,th}{t^2} \\
&= \lim_{t\to 0}\frac{F(t(g+h)) - F(tg) - F(th) + F(0)}{t^2} \\
&\equiv F''(0;g)\,h.
\end{aligned}
$$

*Remark* 1.5. The equality (1.21) implies

$$
\widetilde{A}'(0;u)\,u = \frac{1}{2}F''(0;u)\,u \quad \forall u \in D(F). \tag{1.22}
$$

The operator $\widetilde{A}(u)$ will be said to be $n$ times Gâteaux-differentiable ($n \geq 2$), if there exists an $(n+1)$-linear operator $\widetilde{A}^{(n)}(u)$ (the $n$-th order derivative of $\widetilde{A}(u)$) such that

$$
\lim_{t\to 0}\left\|\frac{\widetilde{A}^{(n-1)}(u + th_{n+1}; h_n, \dots, h_2)\,h_1 - \widetilde{A}^{(n-1)}(u; h_n, \dots, h_2)\,h_1}{t}\right.
$$

$$
\left. -\widetilde{A}^{(n)}(u; h_{n+1}, h_n, \dots, h_2)\,h_1\right\|_Y = 0 \quad \forall\,\{h_i\} \subset D(F). \tag{1.23}
$$

The properties of the higher order Gâteaux derivatives suggest that the value $\widetilde{A}^{(n)}(u; h_{n+1}, h_n, \dots, h_2)\,h_1$ is invariant with respect to any depositioning of the elements $h_j$, $j = 2, \dots, n+1$.

**Lemma 1.8.** *If the operators $\widetilde{A}(u)$ and $F$ are $n$ times Gâteaux-differentiable in $u$, then for any $h_i \in D(F)$, $(i = 1, \dots, n; \; n \geq 2)$*

$$
\widetilde{A}^{(n)}(u; h_n, \dots, h_1)\,u = F^{(n)}(u; h_n, \dots, h_2)\,h_1
$$

$$
-\frac{1}{(n-1)!}\sum \widetilde{A}^{(n-1)}(u; h_{\nu_n}, \dots, h_{\nu_2})\,h_{\nu_1}, \tag{1.24}
$$

*where the summation is over all the permutations of the elements $h_i$, $i = 1, \dots, n$.*

*Proof.* We make use of the mathematical induction method. From (1.20) we find

$$F'(u + tg) v = \tilde{A}'(u + tg; v)(u + tg) + \tilde{A}(u + tg) v;$$

then

$$\frac{\partial}{\partial t} F'(u + tg) v \bigg|_{t=0} = F''(u; g) v = \tilde{A}''(u; g, v) u + \tilde{A}'(u; v) g + \tilde{A}'(u; g) v.$$

Let now (1.24) be satisfied for $n = k - 1$. In accordance with (1.23) we obtain

$$\tilde{A}^{(k)}(u; h_k, \ldots, h_1) u$$

$$= \lim_{t \to 0} \frac{\tilde{A}^{(k-1)}(u + th_k; h_{k-1}, \ldots, h_1) u - \tilde{A}^{(k-1)}(u; h_{k-1}, \ldots, h_1) u}{t}$$

$$= \lim_{t \to 0} \left[ \frac{\tilde{A}^{(k-1)}(u + th_k; h_{k-1}, \ldots, h_1)(u + th_k) - \tilde{A}^{(k-1)}(u; h_{k-1}, \ldots, h_1) u}{t} \right.$$

$$\left. - \tilde{A}^{(k-1)}(u + th_k; h_{k-1}, \ldots, h_1) h_k \right]$$

$$= \lim_{t \to 0} \left[ \frac{F^{(k-1)}(u + th_k; h_{k-1}, \ldots, h_2) h_1 - F^{(k-1)}(u; h_{k-1}, \ldots, h_2) h_1}{t} \right.$$

$$- \sum \frac{\tilde{A}^{(k-2)}(u + th_k; h_{\nu_{k-1}}, \ldots, h_{\nu_2}) h_{\nu_1} - \tilde{A}^{(k-2)}(u; h_{\nu_{k-1}}, \ldots, h_{\nu_2}) h_{\nu_1}}{t(k - 2)!}$$

$$\left. - \tilde{A}^{(k-1)}(u + th_k; h_{k-1}, \ldots, h_1) h_k \right],$$

where the summation is over all the permutations of the elements $h_j$, $j = 1, \ldots, k - 1$. Using the definition of the $k$-th order derivative of the operator $F$, the definition (1.23), and the continuity of the derivative $\tilde{A}^{(k-1)}$ in $t$ (which results from its differentiability), we arrive at

$$\tilde{A}^{(k)}(u; h_k, \ldots, h_1) u = F^{(k)}(u; h_k, \ldots, h_2) h_1$$

$$- \frac{1}{(k - 2)!} \sum \tilde{A}^{(k-1)}(u; h_k, h_{\nu_{k-1}}, \ldots, h_{\nu_2}) h_{\nu_1}$$

$$- \tilde{A}^{(k-1)}(u; h_{k-1}, \ldots, h_1) h_k$$

$$= F^{(k)}(u; h_k, \ldots, h_2) h_1$$

$$- \frac{1}{(k - 1)!} \sum \tilde{A}^{(k-1)}(u; h_{\nu_k}, \ldots, h_{\nu_2}) h_{\nu_1}.$$

**Lemma 1.9.** *Let the operator $\tilde{A}(u)$ be n times Gâteaux-differentiable for* $u \in \bar{S}_R(0)$ *and the derivative* $\tilde{A}^{(n)}(u)$ *be continuous at the point* $u = 0$ *in the following sense:*

$$\lim_{t \to 0} \|\tilde{A}^{(n)}(tv; g_n, \ldots, g_1)\, v - \tilde{A}^{(n)}(0; g_n, \ldots, g_1)\, v\|_Y = 0 \qquad (1.25)$$

*for any* $v \in D(F)$, $g_i \in D(F)$, $i = 1, \ldots, n$. *Then for any* $h_i \in D(F)$, $i = 1, \ldots, n+1$, *the equality*

$$\frac{1}{n!} \sum \tilde{A}^{(n)}\left(0; h_{\nu_{n+1}}, \ldots, h_{\nu_2}\right) h_{\nu_1} = F^{(n+1)}(0; h_{n+1}, \ldots, h_2)\, h_1 \qquad (1.26)$$

*holds, the summation being over all the permutations of the elements* $h_i$.

*Proof.* Using Lemma 1.8, we find

$$F^{(n+1)}(0; h_{n+1}, \ldots, h_2)\, h_1$$

$$= \lim_{t \to 0} \frac{F^{(n)}(th_{n+1}; h_n, \ldots, h_2)\, h_1 - F^{(n)}(0; h_n, \ldots, h_2)\, h_1}{t}$$

$$= \lim_{t \to 0} \left[ \frac{\tilde{A}^{(n)}(th_{n+1}; h_n, \ldots, h_1)\, th_{n+1}}{t} \right.$$

$$+ \frac{1}{(n-1)!}\, \frac{1}{t}\left\{ \sum \left( \tilde{A}^{(n-1)}(th_{n+1}; h_{\nu_n}, \ldots, h_{\nu_2})\, h_{\nu_1} \right. \right.$$

$$\left. \left. \left. - \tilde{A}^{(n-1)}(0; h_{\nu_n}, \ldots, h_{\nu_2})\, h_{\nu_1} \right) \right\} \right]$$

$$= \tilde{A}^{(n)}(0; h_n, \ldots, h_1)\, h_{n+1} + \frac{1}{(n-1)!} \sum \tilde{A}^{(n)}(0; h_{n+1}, h_{\nu_n}, \ldots, h_{\nu_2})\, h_{\nu_1}$$

$$= \frac{1}{n!} \sum \tilde{A}^{(n)}(0; h_{\nu_{n+1}}, \ldots, h_{\nu_2})\, h_{\nu_1}.$$

The last equality follows from the fact that $\tilde{A}^{(n)}$ is symmetrical in all the arguments, except the last one.

**Corollary.** $\tilde{A}^{(n)}(0, u, \ldots, u)\, u = \frac{1}{(n+1)} F^{(n+1)}(0; u, \ldots, u)\, u$.

*Remark* 1.6. Conceivably the hypothesis of existence of $\tilde{A}^{(n)}$ in a neighbourhood of zero and the fulfillment of (1.25) might not be necessary. (See, e.g., the proof of Lemma 1.7.)

Using Lemma 1.7 and Lemmas 1.1–1.4, it is easy to show that the following propositions are valid.

**Lemma 1.10.** *If the operator $F'(0) = \Phi'(U_0)$ has an inverse bounded operator and $\|\widetilde{A}(u) - \widetilde{A}(0)\|_{X \to Y} \to 0$ as $\|u\|_X \to 0$, then for a sufficiently small $R > 0$ the operator $\widetilde{A}(u)$ also has an inverse bounded operator for any $u \in \bar{S}_R(0)$.*

**Lemma 1.11.** *Let $X^* = Y$, $F(u)$ be a strictly monotone operator, and $\|\widetilde{A}(u) - \widetilde{A}(0)\|_{X \to Y} \to 0$ as $\|u\|_X \to 0$. Then there exists $R > 0$ such that the operator $\widetilde{A}(u)$ is positive definite on $D(F)$, i.e.*

$$\langle v, \widetilde{A}(u) v \rangle_X \geq m_1 \|v\|_X^2, \quad m_1 = \text{constant} > 0$$

*for any $u \in \bar{S}_R(0)$.*

**Theorem 1.1.** *Let the operator $F'(0)$ be closed, the operator $\widetilde{A}(u) - \widetilde{A}(0)$ be bounded, and (1.19) be satisfied. Then the operator $\widetilde{A}(u)$ is closed.*

*Proof.* Let $\lim_{n \to \infty} \|u_n - v\|_X = 0$ and $\lim_{n \to \infty} \|\widetilde{A}(u) u_n - y\|_Y = 0$. We find

$$F'(0) u_n = \widetilde{A}(0) u_n = (\widetilde{A}(0) - \widetilde{A}(u)) u_n + \widetilde{A}(u) u_n.$$

Hence,

$$\|F'(0) u_n - F'(0) u_m\|_Y = \|(\widetilde{A}(0) - \widetilde{A}(u))(u_n - u_m) + \widetilde{A}(u) u_n - \widetilde{A}(u) u_m\|_Y$$

$$\leq \|\widetilde{A}(0) - \widetilde{A}(u)\|_{X \to Y} \|u_n - u_m\|_X$$

$$+ \|\widetilde{A}(u) u_n - \widetilde{A}(u) u_m\|_Y.$$

For any $\varepsilon > 0$ and sufficiently large $n$ and $m$ the following inequalities hold:

$$\|u_n - u_m\|_X < \frac{\varepsilon}{2\|\widetilde{A}(u) - \widetilde{A}(0)\|_{X \to Y}}, \quad \|\widetilde{A}(u) u_n - \widetilde{A}(u) u_m\|_Y < \frac{\varepsilon}{2}.$$

Then

$$\|F'(0) u_n - F'(0) u_m\|_Y < \varepsilon.$$

Since $\varepsilon$ is arbitrary and $F'(0)$ is closed, we get $v \in D(F)$ and

$$\lim_{n \to \infty} F'(0) u_n = F'(0) v = \widetilde{A}(0) v.$$

Owing to the boundedness of $\widetilde{A}(u) - \widetilde{A}(0)$, we have

$$\lim_{n \to \infty} (\widetilde{A}(u) - \widetilde{A}(0)) u_n = (\widetilde{A}(u) - \widetilde{A}(0)) v.$$

Then

$$y = \lim_{n \to \infty} \widetilde{A}(u) u_n = \widetilde{A}(u) v - \widetilde{A}(0) v + \widetilde{A}(0) v = \widetilde{A}(u) v.$$

**1.4.** We have formulated above the properties of the operators $A(u)$ and $\widetilde{A}(u)$. Then some conclusions can be drawn, on the basis of these properties, on a number of characteristics of the operators $A^*(u)$ and $\widetilde{A}^*(u)$. Thus, by Lemma 1.3, the operator $A^*(u) = \left( \int_0^1 F'(tu)\, dt \right)^*$ is closed for any $u \in D(F)$. Lemma 1.4 implies the following proposition.

**Lemma 1.12.** *Let* $Y = X^*$, $D(F) \subset D(A^*)$, *and* $F(u)$ *be a strictly monotone operator. Then the operator* $A^*(u)$ *is positive definite on* $D(F)$, *i.e.*

$$\langle v, A^*(u)\, v \rangle_X \equiv (v, A^*(u)\, v)_{H_0} \geq m \|v\|_X^2, \quad m = \text{constant} > 0.$$

Let now $A^*(u)$ be an operator constructed of the basis of a linear operator $\widetilde{A}(u)$ in the representation $F(u) = \widetilde{A}(u)\, u$, the identity (1.8) being used in a manner standard in linear operator theory. Then $A^*(u)$ will also be closed. Lemma 1.5 gives the following

**Lemma 1.13.** *Let* $F(u) = \widetilde{A}(u)\, u$ *for any* $u \in D(F)$ *and*

$$\lim_{t \to 0} \|\widetilde{A}(tv) - \widetilde{A}(0)\|_Y = 0$$

*for any* $v \in D(F)$. *Then* $\widetilde{A}^*(0) = F'^*(0)$.

**Lemma 1.14.** *If the operator* $F'^*(0)$ *has a continuous inverse operator and* $\|\widetilde{A}(u) - \widetilde{A}(u)\|_{X \to Y} \to 0$ *as* $\|u\|_X \to 0$, *then for a sufficiently small* $R$ *the operator* $\widetilde{A}^*(u)$ *also has a continuous inverse operator for any* $u \in \bar{S}_R(0)$.

**Lemma 1.15.** *Let* $D(F) \subset D(\widetilde{A}^*(u))$, $F(u)$ *be a strictly monotone operator, and* $\|\widetilde{A}(u) - \widetilde{A}(u)\|_{X \to Y} \to 0$ *as* $\|u\|_X \to 0$. *Then the operator* $\widetilde{A}^*(u)$ *is positive definite on* $D(F)$, *i.e.*

$$\langle v, \widetilde{A}^*(u)\, v \rangle_X \geq m_1 \|v\|_X^2, \quad m_1 = \text{constant} > 0$$

*for any* $u \in \bar{S}_R(0)$, $R$ *being sufficiently small.*

## 2. PROPERTIES OF OPERATORS OF THE CLASS $\mathcal{D}$

First of all, note that the set $\mathcal{D}$ contains all the non-linear operators $F$ which are continuously Gâteaux-differentiable, vanish at zero point, and have the adjoint operators $F_{\mathcal{D}}^*$ with identical properties. Therefore, the investigation of the operator properties of the class $\mathcal{D}$ yields, at the same time, the properties of adjoint operators of $\mathcal{D}$.

We thus consider some properties of adjoint operators $F_{\mathcal{D}}^*$ (in the sense of Definition 4.1, Chapter 1) formulated by M.Vainberg.[296]

**Lemma 2.1.** *If $F$ has an adjoint operator, then*

(1) $F_{\mathcal{D}}^*$ *also has an adjoint operator and $F^{**} = (F_{\mathcal{D}}^*)^* = F$;*

(2) $F_{\mathcal{D}}^*(x) = \int_0^1 (F'(tx))^* x \, dt$, $x \in X$;

(3) $\langle x, F(x) \rangle_X = \langle x, F_{\mathcal{D}}^*(x) \rangle_X$;

(4) *if $G$ also has an adjoint operator, then $(aF + bG)^* = aF_{\mathcal{D}}^* + bG_{\mathcal{D}}^*$ for any real $a$ and $b$.*

**Lemma 2.2.** *If $F \in \mathcal{D}$, then the operator $F + F_{\mathcal{D}}^*$ is symmetrical and $F - F_{\mathcal{D}}^*$ is skew-symmetrical.*

**Theorem 2.1.** *If $F \in \mathcal{D}$, then it is: (1) symmetrical if and only if it is strictly potential (i.e. its potential is Fréchet-differentiable); and (2) skew-symmetrical if and only if it is linear and $\langle x, F(x) \rangle_X = 0$ for any $x \in X$.*

**Theorem 2.2.** *The necessary and sufficient condition for $F \in \mathcal{D}$ (i.e. for $F$ to allow the existence of an adjoint operator) is that a linear skew-symmetrical operator $A \in \mathcal{L}(X, X^*)$ must exist, such that the difference $F - A$ is a potential operator.*

*Remark* 2.1. The set $\mathcal{D}$ is thus a direct sum of all the symmetrical and skew-symmetrical operators, i.e. $F = F_S + A_F$, where $F_S = (F + F_{\mathcal{D}}^*)/2$ and $A_F = (F - F_{\mathcal{D}}^*)/2$; the operator $F_S$ being potential and $A_F$ being linear.

*Example* 2.1. Consider the operator $F$ and the spaces introduced in Example 2.1, Chapter 1. Let $T$ be an operator defined by the expression

$$T(u_0) v = \frac{\partial v}{\partial t} + u_0 \frac{\partial v}{\partial x} + \left( a + \frac{\partial u_0}{\partial x} \right) v$$

and the domain $D(T) = C^1(\Omega) \subset L_2(\Omega)$. Here $u_0$ is a function of $D(F) = C^1(\Omega)$. Then for $u_0$ and $u = u_0 + h$ we find

$$\|F(u) - F(u_0) - T(u - u_0)\|_{L_2(\Omega)} = t^2 \left\| h \frac{\partial h}{\partial x} \right\|_{L_2(\Omega)} \leq t^2 \|h\|_{L_2(\Omega)} \|h\|_{C^1(\Omega)},$$

$$\lim_{t \to 0} \left\| \frac{F(u_0 + th) - F(u_0)}{t} - T(u_0) h \right\|_{L_2(\Omega)} = 0,$$

i.e. $F'(u_0) h = T(u_0) h$. Hence, the operator $F$ has the Gâteaux derivative at any point $u \in D(F) = C^1(\Omega)$, $F'(u) = T(u)$. Using $F$ and $F'$, it is easy to show that the operator $F$ does not belong to set $\mathcal{D}$ and it does not have the adjoint operator in the sense of Definition 4.1. In fact, the linear skew-symmetrical part of $F$ is the operator $A$: $Av = \partial v/\partial t$. Consider the difference

$$K(u) = F(u) - Au = u\frac{\partial u}{\partial x} + au.$$

The operator $K(u)$ has the Gâteaux derivative of the form

$$K'(u)\,h = u\frac{\partial h}{\partial x} + \left(a + \frac{\partial u}{\partial x}\right)h$$

which is not a symmetrical operator. Hence, $K(u)$ is not a potential operator and, by Theorem 2.2, $F$ does not belong to $\mathcal{D}$. Note, in addition, that in this case the Gâteaux derivative $F'$ is not a continuous operator mapping $L_2$ into $L_2$, i.e.

$$\|F'(u + h) - F'(u)\|_{L_2 \to L_2} \not\to 0 \quad \text{as } \|h\|_{L_2} \to 0.$$

The next simple example shows that the definition of the adjoint operator $F^*$ given in Section 3, Chapter 1, is more general than Definition 4.1, Chapter 1 (see also the next section).

*Example* 2.2. Introduce the same spaces as in Example 4.1, Chapter 1, but instead of problem (4.5) consider a problem of the form

$$-\frac{d^2u}{dx^2} + u^2(x) + u\frac{du}{dx} = f(x), \quad x \in (0,1), \tag{2.1}$$

$$u(0) = u(1) = 0.$$

As before, we introduce relation (4.8) and equation (4.9) with the operator $F(u)$ whose derivative is defined by an equality of the type

$$(F'(u)\,v, w)_{L_2} = \int_0^1 \left(\frac{dv}{dx}\frac{dw}{dx} + 2uvw + u\frac{dv}{dx}w + \frac{du}{dx}vw\right) dx.$$

It is easy to notice that here the derivative $F'(u)$ is not a symmetrical operator. Hence, $F(u)$ is not a potential operator and, by Theorem 2.2, it does not admit the existence of the adjoint operator $F_{\mathcal{D}}^*$. On the other hand, an adjoint operator $F^*(u)$ in the sense of Definition 3.4, Chapter 1, does exist here and is given by the relation

$$(F^*(u), w)_{L_2} = \int_0^1 \left(\frac{du}{dx}\frac{dw}{dx} + u^2w - \frac{u}{2}\frac{du}{dx}w\right) dx.$$

## 3. PROPERTIES OF ADJOINT OPERATORS CONSTRUCTED WITH THE USE OF THE TAYLOR FORMULA

Consider the properties of the operators $A^*(u)$ and $F^*$ introduced by Definitions 3.1, 3.4, Chapter 1, i.e. of the operators

$$A^*(u) = \left( \int_0^1 F'(tu) \, dt \right)^*, \qquad u \in D(F),$$

$$F^*(u) = \left( \int_0^1 F'(tu) \, dt \right)^* u, \qquad u \in D(F) \subset D(A^*). \tag{3.1}$$

Some of these properties have already been formulated in Section 1. The following lemma can be useful in practical determination of the form of the operator $A^*(u)$.

**Lemma 3.1.** *Let $(F'(u))^*$ be an operator adjoint to $F'(u)$ (for a fixed element $u \in D(F)$). Then the following equality holds on $D(F)$:*

$$A^*(u) \equiv \left( \int_0^1 F'(tu) \, dt \right)^* = \int_0^1 (F'(tu))^* \, dt. \tag{3.2}$$

*Proof. Using the properties of integrals of abstract functions, we find*

$$(A^*(u) w, v)_{H_0} = \left( \left( \int_0^1 F'(tu) \, dt \right)^* w, v \right)_{H_0} = \left( w, \int_0^1 F'(tu) \, dt v \right)_{H_0}$$

$$= \int_0^1 (w, F'(tu) v)_{H_0} \, dt = \int_0^1 ((F'(tu))^* w, v)_{H_0} \, dt$$

$$= \left( \int_0^1 (F'(tu))^* \, dt w, v \right)_{H_0},$$

*that is,*

$$\left( \int_0^1 F'(tu) \, dt \right)^* = \int_0^1 (F'(tu))^* \, dt.$$

The following lemma sets up a correspondence between the operators $F^*$ and $F_{\mathcal{D}}^*$.

**Lemma 3.2.** *Let $Y = X^*$, and the operator $F$ belong to the set $\mathcal{D}$. Then the operators $F^*$ and $F_{\mathcal{D}}^*$ coincide on $D(F)$, that is, $F^*(u) = F_{\mathcal{D}}^*(u)$, $u \in D(F)$.*

*Proof. If $u \in D(F)$, then*

$$F^*(u) = \int_0^1 (F'(tu))^* \, dt u.$$

But $D(F)$ is a convex set. An element $U \equiv tu$ then belongs to $D(F)$ if $t \in [0, 1]$. Since $F \in \mathcal{D}$, we have $(F'(U))^* = (F_\mathcal{D}^*)'(U)$ and the equalities

$$F^*(u) = \int_0^1 (F_\mathcal{D}^*)'(tu) \, dtu = F_\mathcal{D}^*(u)$$

hold.

Lemma 3.2 makes it possible to extend to $F^*(u)$ a number of properties formulated above for the operator $F_\mathcal{D}^*$. Note, however, that this approach to studying the properties of the operator $F^*$ remains valid if we assume $F \in D$. But if this assumption is dropped, an additional analysis is required to find out if, for example, $F$ is symmetrical. Thus, the symmetry of the operator $\Phi = F + F^*$ is not obvious without this assumption. In order to prove this property, we have to prove the differentiability of the operator $F^*$ (this results in the differentiability of $\Phi$) and obtain the representation $\Phi(u) = B(u) u$, where $B(u) = \int_0^1 \Phi'(tu) \, dt$, and then show that $B(u) = B^*(u)$. A similar remark is true regarding the skew-symmetry of the operator $F - F^*$. In some cases, these difficulties are removed by the following theorem.

**Theorem 3.1.** *Let an operator $F$ defined in a domain $D(F)$ map $X$ into $Y = X^*$ and the following conditions be satisfied: (1) $F$ is twice Gâteaux-differentiable; (2) the set $D(F)$ belongs to the domain of each of the operators $(F'(u))^*$, $(F''(u, v))^*$, $u, v \in D(F)$; and (3) for any $u, v, w, h \in D(F)$ the equality $(F''(u, v) h, w)_{H_0} = (h, F''(u, v) w)_{H_0}$ is valid. Then the operator $F^*$ defined by (3.2) is Gâteaux-differentiable and satisfies the equalities*

$$((F^*(u))'h, w)_{H_0} = ((F'(u)^* h, w)_{H_0} + (h, F'(u) w)_{H_0}. \tag{3.3}$$

*Proof.* Condition (1) implies the Gâteaux differentiability of $F^*$, so that

$$((F^*(u))'h, w)_{H_0} = \left( h, \int_0^1 F'(tu) \, dtw \right)_{H_0} + \left( u, \int_0^1 F''(tu; th) \, dtw \right)_{H_0}$$

Taking into consideration the properties of the second Gâteaux derivative of $F$ and conditions (2) and (3), we obtain

$$\left( u, \int_0^1 F''(tu; th) \, dtw \right)_{H_0} = \left( u, \int_0^1 F''(tu, w) \, th \, dt \right)_{H_0}$$

$$= \left( u, \int_0^1 F''(tu; tw) \, dth \right)_{H_0}$$

$$= \int_0^1 (u, F''(tu, tw) \, h)_{H_0} \, dt$$

$$= \int_0^1 ((F''(tu; tw))^* u, h)_{H_0}$$

$$= \left( \int_0^1 F''(tu; tw) \, dtu, h \right)_{H_0}$$

Therefore,

$$((F^*(u))'h, w)_{H_0} = \left( h, \int_0^1 F'(tu) \, dtw \right)_{H_0} + \left( h, \int_0^1 F''(tu; tw) \, dtu \right)_{H_0}$$

$$= (h, F'(u) w)_{H_0} = ((F'(u))^* h, w)_{H_0}$$

which has to be proved.

**Corollary 1.** *If the hypotheses of Theorem 3.1 are satisfied, then the operators $F^*$ and $F^*_D$ coincide on $D(F)$, and the operator $F^*$ can be written as a sum of a potential and a linear skew-symmetrical operator.*

**Corollary 2.** *If the hypotheses of Theorem 3.1 are satisfied, then $F + F^*$ is a symmetrical operator, and $F - F^*$ a skew-symmetrical operator.*

*Remark* 3.1. It is easy to notice that the hypotheses of Theorem 3.1 are sufficient for the operator $F$ to belong to the set $\mathcal{D}$.

*Example* 3.1. Let $D(F) = X = \overset{\circ}{W^1_2}(0,1)$, $X^* = Y = \overset{\circ}{W^{-1}_2}(0,1)$, $H_0 = H^*_0 = L_2(0,1)$, and consider an operator of the type

$$F(u) = \frac{du}{dx} + u^k, \quad 2 \le k < \infty, \quad u(x) \in D(F).$$

Then we have

$$F''(v; w) \, h = k(k-1) \, v^{k-2} wh = (F''(v; w))^* h$$

and it is easily verified that all the hypotheses of Theorem 3.1 are met. Hence, the adjoint operator $F^*(u)$ which here takes the form

$$F^*(u) = -\frac{du}{dx} + u^k, \quad u(x) \in D(F)$$

satisfies relations (3.3), and the propositions of Corollaries 1 and 2 are valid for it.

# Chapter 3

# Solvability of main and adjoint equations in non-linear problems

## 1. MAIN AND ADJOINT EQUATIONS. PROBLEMS

Let $X$ and $Y$ be two spaces in a sequence $\{H^\alpha\}$ of real Hilbert spaces introduced above, among which only $H_0 \equiv H^0$ is identified with its dual space, that is, constitutes the basic space. We also assume that the space $(H^\alpha)^*$ dual to $H^\alpha$, $0 \le \alpha \le 1$, is identified with $H^{-\alpha}$. When considering the spaces $X$ and $Y$, we denote the scalar products, norms and duality relations by

$$(f,g)_X, \quad \|f\|_X = (f,f)_X^{1/2} \quad \langle f, g^* \rangle_X \equiv (f, g^*)_{H_0}, \quad f, g \in X, \quad g^* \in X^*,$$

$$(f,g)_Y, \quad \|f\|_Y = (f,f)_Y^{1/2} \quad \langle f, g^* \rangle_Y \equiv (f, g^*)_{H_0}, \quad f, g \in Y, \quad g^* \in Y^*,$$

where $X^* \equiv X^{-1}$, $Y^* \equiv Y^{-1}$.

Let us formulate the central problems we are concerned about in this chapter. Consider the non-linear equation

$$\Phi(U) = 0 \tag{1.1}$$

and study the existence problem for its solution $U \in D(\Phi)$. Take an element $U_0 \in D(\Phi)$ and turn from (1.1) to the equation

$$F(u) = f, \tag{1.2}$$

where $u = U - U_0$, $f = -\Phi(U_0)$. Equation (1.2) can be written in the form

$$A(u)\, u = f, \quad u \in D(F). \tag{1.3}$$

After having proved the solution existence for the last equation, we obtain thereby the solution existence for equation (1.1). This problem will be also solved if we show the solution existence for the equation

$$\widetilde{A}(u)\, u = f, \quad u \in D(F) \tag{1.4}$$

or, what is the same, for equation (1.2) when $F(u) = \tilde{A}(u)\,u$ with an operator $\tilde{A}(u)$ different from $A(u)$.

Along with equations (1.3) and (1.4), the solvability problem for the equations with the adjoint operators $A^*(u)$ and $\tilde{A}^*(u)$

$$A^*(u)\,u^* = p, \quad u^* \in D(A^*), \tag{1.5}$$

$$\tilde{A}^*(u)\,\tilde{u}^* = \tilde{p}, \quad \tilde{u}^* \in D(\tilde{A}^*), \tag{1.6}$$

where $u$ is an element of $D(F)$, $A^*: Y^* \rightarrow X^*$, $\tilde{A}^*: Y^* \rightarrow X^*$, is of great interest in applied studies. It is, however, well known that to study the solvability of equations (1.5) and (1.6) one should attack the solvability problem for the equations

$$A(u)\,v = g, \quad v \in D(A), \tag{1.7}$$

$$\tilde{A}(u)\,\tilde{v} = \tilde{g}, \quad \tilde{v} \in D(\tilde{A}), \tag{1.8}$$

where $u$ is an element of $D(F)$ which may be assumed to be fixed. The last equations are linear and here we can make use of standard approaches of linear analysis (which are often based on one or another property of the linear operators $A(u)$ and $\tilde{A}(u)$).

In summary, we formulate the basic problems of interest.

(1) The solvability problem for equations (1.7) and (1.8).
(2) The solvability problem for equation (1.5) with the operator $\tilde{A}(u)$ corresponding to $A(u) = \int_0^1 \Phi'(U_0 + tu)\,dt$ and for equation (1.6) with the adjoint operator $\tilde{A}^*(u)$ constructed on the basis of a different principle then $A^*(u)$.
(3) The solvability problem for equation (1.2).

It is precisely these problems that are studied in the chapter.


## 2. SOLVABILITY OF THE EQUATION $F(u) = y$

Let us turn to the solvability problem for the non-linear equation

$$F(u) \equiv A(u)\,u = y. \tag{2.1}$$

First we consider the solvability conditions for (2.1) in a neighbourhood of zero point when the Lipschitz condition (1.15), Chapter 2,

$$\|F'(u_1) - F'(u_2)\|_{X \rightarrow Y} \leq L\|u_1 - u_2\|_X \quad \forall u_1, u_2 \in \bar{S}_R(0) \tag{1.15'}$$

is satisfied and the domains of the operators $F$ and $F'(0)$ coincide. Then we will formulate the concept of 'normal solvability' and the related results.

## 2.1. Solvability in a ball

**Lemma 2.1.** *If the Lipschitz condition (1.15′) for $F'(u)$ is satisfied and $F'(0)$ is closed, then the operator $F(u)$ is also closed.*

*Proof.* Let $\{v_n\} \to v$, $\{F(v_n)\} \to f$. Then the sequence $\{F'(0)\,v_n\}$ is convergent. In fact,

$$
F'(0)\,v_n - F'(0)\,v_m = F(v_n) - F(v_m) + \int_0^1 [F'(0) - F'(tv_n)]\;dt v_n
$$

$$
- \int_0^1 [F'(0) - F'(tv_m)]\;dt v_m
$$

$$
= F(v_n) - F(v_m) + \int_0^1 [F'(0) - F'(tv_m)]\;dt(v_n - v_m)
$$

$$
+ \int_0^1 [F'(tv_m) - F'(tv_n)]\;dt v_m;
$$

hence,

$$
\|F'(0)\,v_n - F'(0)\,v_m\|_Y \leq \|F(v_n) - F(v_m)\|_Y
$$

$$
+ \int_0^1 \|F'(0) - F'(tv_n)\|_{X \to Y}\;dt\|v_n - v_m\|_X
$$

$$
+ \int_0^1 \|F'(tv_m) - F'(tv_n)\|_{X \to Y}\;dt\|v_m\|_X
$$

$$
\leq \|F(v_n) - F(v_m)\|_Y + \frac{L}{2}\|v_n\|_X\|v_n - v_m\|_X
$$

$$
+ \frac{L}{2}\|v_m\|_X\|v_m - v_n\|_X
$$

$$
\leq \|F(v_n) - F(v_m)\|_Y
$$

$$
+ L(1 + \varepsilon)\|v_n\|_X\|v_n - v_m\|_X \to 0, \quad n, m \to \infty.
$$

Since $F'(0)$ is closed, $v \in D(F'(0)) = D(F)$ and $\lim_{n\to\infty} F'(0)\,v_n = F'(0)\,v$. Then we obtain

$$
\|f - F(v)\|_Y = \lim_{n \to \infty} \|F(v_n) - F(v)\|_Y
$$

$$
= \lim_{n \to \infty} \left\| \int_0^1 [F'(0) - F'(tv)]\;dt(v_n - v) \right.
$$

$$
\left. + \int_0^1 [F'(tv) - F'(tv_n)]\;dt v_n \right\|_Y
$$

$$\leq \lim_{n \to \infty} L \int_0^1 \|tv\|_X \ dt \|v_n - v\|_X$$

$$+ \lim_{n \to \infty} L \int_0^1 t \|v_n - v\|_X \ dt \|v_n\|_X = 0.$$

**Theorem 2.1.** *Let $F'(u)$ satisfy the Lipschitz condition (1.15'), the equation*

$$F'(0) v = f \tag{2.2}$$

*be correctly solvable, and $\|F'^{-1}(0)\|_{Y \to X} \leq 1/k$, $k = $ constant $> 0$. Then any sequence $\{u_n\} \subset \bar{S}_{r_0}(0)$, $0 < r_0 < k/L$, such that $F(u_n) \to y$, converges to an element $u \in X$ which is said to be a generalized solution of (2.1). The following estimate holds:*

$$\|u_n - u\|_X \leq (k - Lr_0)^{-1} \|F(u_n) - y\|_Y. \tag{2.3}$$

*Proof.* We have

$$\|u_n - u_m\|_X \leq \frac{1}{k} \|F'(0) u_n - F'(0) u_m\|_Y$$

$$= \frac{1}{k} \left\| F(u_n) - F(u_m) + \int_0^1 [F'(0) - F'(tu_n)] \ dt u_n \right.$$

$$\left. - \int_0^1 [F'(0) - F'(tu_m)] \ dt u_m \right\|_Y$$

$$\leq \frac{1}{k} \|F(u_n) - F(u_m)\|_Y$$

$$+ \frac{1}{k} \left\| \int_0^1 [F'(0) - F'(tu_n)] \ dt \right\|_{X \to Y} \|u_n - u_m\|_X$$

$$+ \frac{1}{k} \left\| \int_0^1 [F'(tu_m) - F'(tu_n)] \ dt \right\|_{X \to Y} \|u_m\|_Y$$

$$\leq \frac{1}{k} \|F(u_n) - F(u_m)\|_Y + \frac{L}{k} \|u_n - u_m\|_X r_0;$$

hence

$$\|u_n - u_m\|_X \leq \frac{1}{k - Lr_0} \|F(u_n) - F(u_m)\|_Y \to 0, \quad n, m \to \infty, \tag{2.4}$$

i.e. the sequence $\{u_n\}$ is convergent and $\lim_{n \to \infty} u_n = u \in X$. Going to the limit in (2.4), we get (2.3).

**Theorem 2.2.** *Let the hypotheses of Lemma 2.1 be satisfied, the operator $F'(0)$ have a continuous inverse operator, $\|F'^{-1}(0)\|_{Y \to X} \leq m$, and equation (2.2) be everywhere solvable in $Y$. Then*

(1) *for any $y \in Y$, $\|y\|_Y \leq q(1-q)/m^2 L$, $0 < q < 1$, there exists $v_y \in X$ such that $F(v_y) = y$ and $\|v_y\|_X \leq q/mL$;*

(2) *if $y_n \to y$ and $\|y_n\|_Y \leq q(1-q)/m^2 L$, then $v_{y_n} \to v_y$ and*

$$\|v_{y_n} - v_y\|_X \leq \frac{m}{1-q}\|y_n - y\|_Y.$$

*Proof.* Equation (2.1) can be written in the form

$$u = F'(0)^{-1}\left[\int_0^1 [F'(0) - F'(ut)]\,dt\,u\right] + F'(0)^{-1}y \equiv T(u)\,u + f_1. \qquad (2.5)$$

Let us use the successive approximation method for solving (2.5):

$$u^0 = 0,$$

$$u^n = T(u^{n-1})\,u^{n-1} + f_1, \quad n = 1, 2, \dots$$

and find the conditions for the iterative process to be convergent. We obtain

$$
\begin{aligned}
\|u^{n+1} - u^n\|_X &= \|T(u^n)\,u^n - T(u^{n-1})\,u^{n-1}\|_X \\
&= \|T(u^n)(u^n - u^{n-1}) + (T(u^n) - T(u^{n-1}))\,u^{n-1}\|_X \\
&\leq \left\|\int_0^1 [F'(0) - F'(u^n t)]\,dt\right\|_{X \to Y} \|u^n - u^{n-1}\|_X \\
&\quad + m\left\|\int_0^1 [F'(u^{n-1}t) - F'(u^n t)]\,dt\right\|_{X \to Y} \|u^{n-1}\|_X \\
&\leq \int_0^1 L\,\|u^n t\|_X\,dt\,\|u^n - u^{n-1}\|_X \\
&\quad + m\int_0^1 L\,\|u^{n-1}t - u^n t\|_X\,dt\,\|u^{n-1}\|_X \\
&\leq \frac{mL}{2}\|u^n - u^{n-1}\|_X\,(\|u^n\|_X + \|u^{n-1}\|_X) \\
&\leq mLr\,\|u^n - u^{n-1}\|_X
\end{aligned}
$$

for $u^n \in \bar{S}_r(0)$. If $r = q/(mL)$, $0 < q < 1$, we get

$$\|u^{n+1} - u^n\|_X \leq q\|u^n - u^{n-1}\|_X,$$

i.e. the sequence $\{u^n\}$ is convergent. Let $v_y = \lim_{n\to\infty} u^n$. Find now the conditions for $y$ under which $\{u^n\} \subset \bar{S}_r(0)$. We have

$$\|u^n\|_X \le f_1(1 + q + q^2 + \ldots) \le m\|y\|_Y/(1 - q).$$

In view of $m\|y\|/(1 - q) \le q/(mL)$, we get

$$\|y\| \le q(1 - q)/(m^2 L).$$

Since

$$u^n \;=\; T(u^{n-1})\,u^{n-1} + f_1 \Rightarrow F'(0)\,u^n$$

$$=\; \int_0^1 [F'(0) - F'(u^{n-1}t)]\,dt\,u^{n-1} + y,$$

$$F(u^n) - y \;=\; \int_0^1 [F'(0) - F'(u^{n-1}t)]\,dt\,u^{n-1} - \int_0^1 [F'(0) - F'(u^n t)]\,dt\,u^n$$

$$=\; \int_0^1 [F'(0) - F'(u^{n-1}t)]\,dt\,(u^{n-1} - u^n)$$

$$+\; \int_0^1 [F'(u^n t) - F'(u^{n-1}t)]\,dt\,u^n,$$

$$\|F(u^n) - y\|_Y \;\le\; Lr\|u_n - u_{n-1}\|_X \to 0, \quad n \to \infty,$$

the sequence $\{F(u^n)\}$ converges to $y$. By Lemma 2.1, $F(u)$ is closed and, hence, $F(v_y) = y$. Proposition (1) is thus proved. Let now $y_n \to y$ as $n \to \infty$ and $\{y_n\} \subset \bar{S}_\delta(0)$, $\delta = q(1-q)/(m^2 L)$. Then we have $\{v_{y_n}\} \subset \bar{S}_r(0)$, $r = q/(mL) < 1/(mL)$. By Theorem 2.1, $\{v_{y_n}\}$ is convergent and, hence,

$$\|v_{y_n} - x\|_X \le \left(\frac{1}{m} - L\,\frac{q}{mL}\right)^{-1}\|F(v_{y_n}) - y\|_Y = \frac{m}{1-q}\|y_n - y\|_Y.$$

Since $F$ is closed, $F(x) = y$ yields $v_y = x$.

**Theorem 2.3.** *Let the hypotheses of Lemma 2.1 be satisfied and equation (2.1) is everywhere solvable in $Y$. Then $\exists r > 0$, $\delta > 0$: for any $y \in Y$, $\|y\|_Y \le \delta$, there exists $v_y \in \bar{S}_r(0)$ such that $F(v_y) = y$.*

*Proof.* Consider the iterative process

$$u_0 = 0, \quad F'(0)\Delta u_0 = y,$$

$$u_{n+1} = u_n + \Delta u_n,$$

$$F'(0)\Delta u_n = \int_0^1 [F'(0) - F'(tu_n)] \, dt u_n$$
$$- \int_0^1 [F'(0) - F'(tu_{n-1})] \, dt u_{n-1},$$

(2.6)

for solving (2.1).

The second and the fourth equations of (2.6) are not uniquely solvable. But $F'(0)$ is closed and equation (2.2) is everywhere (and, hence, normally) solvable in $Y$. Therefore, for any $y \in Y$ there exists $u_y \in D(F)$ such that $F'(0) u_y = y$ and $\|u_y\|_X \leq k \|y\|_Y$ with the constant $k > 0$ independent[105] of $y$. We consider here the only solutions possessing the property mentioned. Then for $\{u_n\} \subset \bar{S}_r(0)$ we obtain

$$\|\Delta u_n\|_X \leq k \left\| \int_0^1 [F'(0) - F'(tu_n)] \, dt \Delta u_{n-1} \right.$$

$$\left. + \int_0^1 [F'(u_{n-1}t) - F'((u_{n-1} + \Delta u_{n-1})t)] \, dt u_n \right\|_Y$$

$$\leq k \frac{L}{2} \|u_n\|_X \|\Delta u_{n-1}\|_X + k \frac{L}{2} \|u_n\|_X \|\Delta u_{n-1}\|_X$$

$$\leq kLr\|\Delta u_{n-1}\|.$$

The following inequalities

$$\|\Delta u_n\|_X \leq q\|\Delta u_{n-1}\|_X \leq q^n \|\Delta u_0\|_X \to 0, \quad n \to \infty,$$

$$\|u_n\|_X \leq \sum_{k=0}^{n-1} \|\Delta u_k\|_X \leq \|\Delta u_0\|_X(1 + q + q^2 + \ldots)$$

$$\leq \frac{\|\Delta u_0\|_X}{1 - q} \leq \frac{k\|y\|_Y}{1 - q}$$

hold for $r \leq q/(kL)$, $0 < q < 1$.

Let $v_y = \lim_{n \to \infty} u_n$. The condition $u_n \in \bar{S}_r(0)$ yields

$$\|y\|_Y \leq q(1 - q)/(Lk^2),$$

$$\|F'(0)\Delta u_n\|_Y = \left\| \int_0^1 [F'(0) - F'(tu_n)] \, dt \Delta u_{n-1} \right.$$

$$+ \int_0^1 [F'(u_{n-1}t) - F'((u_{n-1} + \Delta u_{n-1})t)] \, dt u_n \Big\|_Y$$

$$\leq Lr\|\Delta u_{n-1}\|_X \to 0, \quad n \to \infty.$$

Summing equations (2.6), we obtain

$$F(u_n) = \sum_{k=0}^{n-1} F'(0)\Delta u_k + \int_0^1 [F'(tu_n) - F'(0)] \, dt u_n = y - F'(0)\Delta u_n,$$

$$\|f(u_n) - y\|_Y = \|f'(0)\Delta u_n\|_Y \to 0, \quad n \to \infty.$$

The operator $F(u)$ is closed by Lemma 2.1 and, hence, $F(v_y) = y$. For a constant $\delta$ we may put $\delta = 1/(4k^2L)$; then $r = 1/(2kL)$.

*Remark* 2.1. The result obtained can be formulated as follows. For any $y \in Y$, $\|y\|_Y = 1$ there exists an abstract function $v_y(s)$ defined on the segment $[-\delta, \delta]$ such that $F(v_y(s)) = sy$ and $\|v_y(s)\|_X/|s| \leq 2k$.

If $F'(0)$ is closed, then $\text{Ker}\,(F'(0))$ is a linear subspace of $X$ and the Hilbert space $X$ can be represented as a sum of two orthogonal subspaces $X = E \oplus \text{Ker}\,(F'(0))$.

**Theorem 2.4.** *Let the hypotheses of Lemma 2.1 be satisfied and equation (2.2) be normally solvable. Then $u_n \to u \in E$ and $F(u) = y$ if $y_n \to y$, $n \to \infty$, $y_n = F(u_n)$, $\{u_n\} \subset E \cap \bar{S}_{r_0}(0)$ for a sufficiently small $r_0$.*

*Proof.* By $F_1$ we denote a restriction of $F$ on $E$, and by $F_1'$ a restriction of $F'(0)$ on $E$. The equation $F_1'u = f$ is uniquely and normally solvable and, hence, it is correctly solvable. By Theorem 2.1, there exists $r_0 > 0$ such that $u_n \to u$ if $\{u_n\} \subset \bar{S}_{r_0}(0)$ and $F_1(u_n) \to y$. The operator $F$ is closed by Lemma 4.1 and, hence, the operator $F_1$ is also closed as $E$ is closed. Therefore, $u \in E$ and $F_1(u) = y$.

*Remark* 2.2. The same line of reasoning can be used when proving Theorem 2.3. In fact, the equation $F_1'u = f$ is everywhere correctly solvable. Then, by Theorem 2.2, $\exists \delta > 0$, $r > 0$: for any $y \in \bar{S}_\delta(0)$ there exists $u \in D(F_1) \cap \bar{S}_r(0)$ such that $F_1(u) = y$. The operator $F$ also possesses this property, since it is a continuation of $F_1$ on $D(F)$.

Let $R_{r_0}(F) \equiv \{y \in Y \colon \exists u \in D(F) \cap \bar{S}_{r_0}(0) \colon F(u) = y\}$.

**Theorem 2.5.** *Let the hypotheses of Theorem 2.4 be satisfied and equation (2.2) be n-normal, i.e. $n(F'(0)) \equiv \dim\text{Ker}\,(F'(0)) < \infty$. Then $R_{r_0}$ is closed if $r_0$ is sufficiently small.*

*Proof.* Let $F(\tilde{u}_n) \to y$ as $n \to \infty$, $\{\tilde{u}_n\} \subset \bar{S}_{r_0}(0)$. Consider $X = E \oplus \mathrm{Ker}\,(F'(0))$. Each of the elements $\tilde{u}_n$ is uniquely represented in the form $\tilde{u}_n = \tilde{l}_n + \tilde{z}_n$, $\tilde{l}_n \in E$, $\tilde{z}_n \in \mathrm{Ker}\,(F'(0))$, $\tilde{l}_n \perp \tilde{z}_n$; hence, $\|\tilde{z}_n\|_X \leq r_0$. Since $\mathrm{Ker}\,(F'(0))$ is locally compact, the bounded sequence $\{\tilde{z}_n\}$ contains a convergent subsequence $\{z_n\}$: $z_n \to z$ as $n \to \infty$; hence, $z \in \mathrm{Ker}\,(F'(0))$. Consider the subsequence $\{u_n\}$ corresponding to $\{z_n\}$: $F(u_n) \to y$, $u_n = l_n + z_n$, $\{u_n\} \subset \bar{S}_{r_0}(0)$. Let us show that $\{l_n\}$ is convergent if $r_0$ is sufficiently small. We have

$$F(l_m + z) - F(l_n + z) = [F(u_m) - F(u_n)] - [F(l_m + z_m) - F(l_m + z)]$$

$$+[F(l_n + z_n) - F(l_n + z)].$$

But

$$F(l_m + z_m) - F(l_m + z)$$

$$= F'(0)\,l_m + \int_0^1 [F'((l_m + z_m)t) - F'(0)]\,dt(l_m + z_m)$$

$$-F'(0)\,l_m - \int_0^1 [F'((l_m + z)t) - F'(0)]\,dt(l_m + z)$$

$$= \int_0^1 [F'((l_m + z_m)t) - F'((l_m + z)t)]\,dt(l_m + z_m)$$

$$+ \int_0^1 [F'((l_m + z)t) - F'(0)]\,dt(z_m - z);$$

hence,

$$\|F(l_m + z_m) - F(l_m + z)\|_Y \leq L(r_0 + \varepsilon)\|z_m - z\|_X \to 0, \quad m \to \infty$$

($\varepsilon = $ constant $> 0$, $\varepsilon \to 0$ as $m \to \infty$: $\|l_m + z\|_X - r_0 \leq \|l_m + z\|_X - \|l_m + z_m\|_X \leq \|z - z_m\|_X$). Then (2.7) yields $F(l_m + z) - F(l_n + z) \to 0$ as $n, m \to \infty$. However,

$$F(l_m + z) - F(l_n + z)$$

$$= F'(0)\,l_m + \int_0^1 [F'((l_m + z)t) - F'(0)]\,dt(l_m + z)$$

$$-F'(0)\,l_n - \int_0^1 [F'((l_n + z)t) - F'(0)]\,dt(l_n + z)$$

$$= F'(0)(l_n - l_m) + \int_0^1 [F'((l_m + z)t) - F'(0)]\,dt(l_m - l_n)$$

$$+ \int_0^1 [F'((l_m + z)t) - F'(t(l_n + z))]\,dt(l_n + z).$$

The operator $F'(0)$ has a continuous inverse operator on $E$ (see the proof of Theorem 2.4), i.e.

$$\|F'(0)(l_n - l_m)\|_Y \geq k\|l_n - l_m\|_X;$$

hence,

$$\|F(l_m + z) - F(l_n + z)\|_Y \geq k\|l_n - l_m\|_X - L(r_0 + \varepsilon)\|l_n - l_m\|_X,$$

$\varepsilon \to 0$ as $n, m \to \infty$. If $r_0 < k/L$, then for sufficiently large $n$ and $m$ we get $k - L(r_0 + \varepsilon) > 0$. Therefore,

$$\|l_n - l_m\|_X \leq \frac{\|F(l_m + z) - F(l_n + z)\|_Y}{k - L(r_0 + \varepsilon)} \to 0, \quad n, m \to \infty.$$

Let $l = \lim_{n\to\infty} l_n$. By Lemma 2.1, $F$ is closed; hence, the convergence of $\{F(l_n + z_n)\}$ and $\{l_n + z_n\}$ gives $l + z \in D(F) \cap \bar{S}_{r_0}(0)$ and $F(u) \equiv F(l + z) = y$.

## 2.2. Normal solvability and the solvability everywhere

S.Pokhozhaev[238] introduces the concept of normal solvability for a non-linear equation and considers the conditions for its solvability everywhere. Consider equation (2.1) and assume (according to Pokhozhaev[238]) that $F \colon X \to Y$, $X$ and $Y$ are reflexive Banach spaces, $D(F) = X$, $F$ is a Frêchet-differentiable operator.

**Definition 2.1.** *The equation $F(u) = y$ is said to be normally solvable if the following conditions are satisfied:*

(1) *for any $y \in Y$ there exists $\{y_n\}$ such that $y_n \to y$ as $n \to \infty$ and for any $y_n$ there exists the minimum point $x_n$ of the functional $\|F(x) - y_n\|_Y$;*
(2) *$(F(x_n) - y_n) \in (\mathrm{Ker}\,(F'(x_n))^*)^\perp$ yields $y \in R(F)$, where $R(F)$ is the range of the operator $F$.*

When considering linear operators, Definition 2.1 gives the normal solvability in the Hausdorff sense. The following statements are valid[238].

**Lemma 2.2.** *Let $x_0$ be a point giving a (local) minimum of the functional $\|F(x) - y\|_Y$ for a fixed $y \in Y$, $v_0 = F(x_0) - y$. Then $v_0 = 0$ if $v_0 \in R(F'(x_0))$.*

**Theorem 2.6.** *Equation (2.1) is normally solvable if and only if $R(F)$ is closed in $Y$.*

**Theorem 2.7.** *Let the range $R(F)$ of the operator $F$ be closed in $Y$ and $\mathrm{Ker}\,(F'(u))^* = \{0\}$ for any $u \in X$. Then the equation $F(u) = y$ has a solution for any $y \in Y$.*

**Theorem 2.8.** *Let $R(F)$ be closed in $Y$ and $F'(u)$ be of the Fredholm type for any $u \in X$. Then (2.1) is solvable everywhere if $\mathrm{Ker}\,(F'(u)) = 0$ for any $u \in X$.*

Let now $D(F) \subseteq X$ and $F'(u)$ be the Gâteaux derivative of the operator $F$ at the point $u \in D(F)$.

**Theorem 2.9.** *If $R(F)$ is closed in $Y$ and $R(F'(u))$ is dense everywhere in $Y$ for any $u \in D(F)$, then equation (2.1) has a solution $u \in D(F)$ for any $y \in Y$.*

We have thus considered some sufficient conditions for the non-linear equation $F(u) = y$ to be solvable in a ball or in the entire space $X$. This information may turn out to be useful in the justification of a number of algorithms for computing the unknown function $u$ or a functional of $u$.

## 3. SOLVABILITY OF THE EQUATION $A(u)\,v = y$

Consider the solvability problem for the equation

$$A(u)\,v = y, \tag{3.1}$$

where $u \in D(F)$ is assumed to be given. We analyze the conditions for (3.1) to be correctly, normally and everywhere solvable. We proceed from the properties of the operators $F'(0)$ and $F'(u)$.

### 3.1. Correct solvability

From Lemma 1.2, Chapter 2, it readily follows

**Lemma 3.1.** *If the equation $F'(u)\,w = f$ is correctly solvable for any $u \in \bar{S}_r(0)$, then equation (3.1) is also correctly solvable for any $u \in \bar{S}_r(0)$.*

**Lemma 3.2.** *Let the equation $F'(0)\,w = f$ be correctly solvable, $\|F'(0)^{-1}\|_{Y \to X} \leq 1/k$ and $F'(u)$ satisfy the Lipschitz condition (1.15'). Then equation (3.1) is correctly solvable for any $u \in \bar{S}_{r_0}(0)$, $r_0 < 2k/L$.*

*Proof.* We have

$$\|A(u)\,v\|_Y = \left\| F'(0)\,v + \int_0^1 [F'(ut) - F'(0)]\,dt\,v \right\|_Y$$

$$\geq \|F'(0)\,v\|_Y - \int_0^1 \|F'(ut) - F'(0)\|_{X \to Y}\,dt\,\|v\|_X$$

$$\geq k\|v\|_X - \frac{L}{2}\|u\|_X\|v\|_X \geq \left(k - \frac{L}{2}r_0\right)\|v\|_X.$$

## 3.2. The solvability everywhere

**Lemma 3.3**[179]. *If the equation $F'(u) w = f$, $f \in Y$, is everywhere solvable in $Y$ for any $u \in \bar{S}_r(0)$, then equation (3.1) is also solvable everywhere $Y$.*

**Lemma 3.4**[179]. *Let the following hypotheses be satisfied:*

(1) *the domains of the operators $A(0) = \Phi'(U_0) = F'(0)$ and $A(u)$ coincide for any $u \in \bar{S}_r(0)$;*
(2) *the operator $\Phi'(U_0)$ is closed;*
(3) *the equation $A(0) w = f$ is everywhere solvable in $Y$;*
(4) *the following inequality is valid:*

$$\sup_{u \in \bar{S}_r(0)} \int_0^1 \|I - (\Phi'(U_0))^{-1} \Phi'(U_0 + ut)\|_{X \to X} \, dt \leq q = \text{ constant } < 1.$$

*Then equation (3.1) is solvable everywhere in $Y$ for any $u \in \bar{S}_r(0)$.*

**Lemma 3.5**[179]. *Assume that (1) the equation $A(0) w = f$ is everywhere correctly solvable in $Y$; (2) $F'(u)$ satisfies the Lipschitz condition with a constant $L$. Then equation (3.1) is everywhere correctly solvable in $Y$ for any $u \in \bar{S}_{r_0}(0)$, $r_0 < 2m/L$ if $\|A(0)^{-1}\|_{Y \to X} \leq 1/m$.*

Lemma 3.4 yields

**Lemma 3.6.** *Let (1) the operator $F'(0)$ be closed and have a continuous inverse operator; (2) $\|F'(u) - F'(0)\|_{X \to Y} \to 0$ as $\|u\|_X \to 0$; (3) the equation $F'(0) w = f$ be everywhere solvable in $Y$. Then equation (3.1) is everywhere solvable in $Y$ for any $u \in \bar{S}_r(0)$ if the value of $r$ is sufficiently small.*

**Lemma 3.7.** *Let the operator $F'(0)$ be closed, the equation $F'(0) w = f$ be everywhere solvable in $Y$ and $\|F'(u) - F'(0)\|_{X \to Y} \to 0$ as $\|u\|_X \to 0$. Then equation (3.1) is everywhere solvable in $Y$ for any $u \in \bar{S}_r(0)$ if $r$ is sufficiently small.*

*Proof.* Consider the decomposition of the Hilbert space $X$ into a sum of the orthogonal subspaces $X = E \oplus \text{Ker}\,(F'(0))$. By $F'_1$ and $A_1(u)$ we denote the restrictions on $E$ of the operators $F'(0)$ and $A(u)$, respectively. Since $E$ and $F'(0)$ are closed, the operator $F'_1$ is also closed and the equation $F'_1 w = f$ is uniquely solvable. The operator $F'^{-1}_1$ is closed and defined on the entire space $Y$; hence, it is bounded. Therefore, the hypotheses of Lemma 3.6 are satisfied and the equation $A_1(u) w = f$ has a solution $w \in E$ for any $f \in Y$ and for any $u \in \bar{S}_r(0)$ if $r$ is sufficiently small. Then equation (3.1) is also everywhere solvable in $Y$, since $A(u)$ is a continuation of $A_1(u)$ on $D(F)$.

*Remark* 3.1. In the case that $X$ is a Banach space Lemma 3.7 may be proved in the same manner as Theorem 2.3.

## 3.3. Normal solvability

**Lemma 3.8.** *If the equation $F'(u)\,w = y$ is normally solvable for any $u \in \bar{S}_r(0)$, then equation (3.1) is also normally solvable for any $u \in \bar{S}_r(0)$.*

*Proof.* Let $u \in D(F)$, $\|u\|_X \leq r$. Then

$$A(u)\,v = \int_0^1 F'(tu)\,dt\,v = F'(\tau u)\,v, \quad 0 < \tau < 1;$$

hence, $\tau u \in \bar{S}_r(0)$ and the equation $F'(\tau u)\,v = y$ is normally solvable. Since the operators $A(u)$ and $F'(\tau u)$ coincide, equation (3.1) is also normally solvable.

For an extended discussion of the conditions for normal solvability see the next section which considers the equation with an arbitrary operator $\widetilde{A}(u)$.

Note that it is important to know the solvability properties of (3.1) both when using this equation in perturbation algorithms and when studying the solvability of an equation with the adjoint operator $A^*(u)$.

# 4. SOLVABILITY OF THE EQUATION $\widetilde{A}(u)\,v = y$

This section considers in greater detail an arbitrary operator $\widetilde{A}(u)$, its form being not specified. With the assumption of an additional regularity, these operators turn to possess some general properties which, in certain cases, lead to the conclusions on the solvability of the equation $\widetilde{A}(u)\,v = y$. Our concern is in particular with the conditions of correct solvability, the solvability everywhere in $Y$ and normal solvability of this equation.

## 4.1. Correct solvability

Consider the solvability problem for the equation

$$\widetilde{A}(u)\,v = y, \tag{4.1}$$

where the element $u \in D(F)$ is assumed to be given. Using the results of Section 3, we will demonstrate that with the supposition of an additional regularity for the operator $\widetilde{A}(u)$, many of the solvability properties of the equation

$$F'(0)\,w = f \tag{4.2}$$

can be extended to equation (4.1). In particular, the following proposition is valid.

**Theorem 4.1.** *If equation (4.2) is correctly solvable and*

$$\|\widetilde{A}(u) - \widetilde{A}(0)\|_{X \to Y} \to 0 \quad \text{as } \|u\|_X \to 0,$$

*then there exists $r > 0$ such that equation (4.1) is correctly solvable for any $u \in \bar{S}_r(0)$.*

*Proof.* Lemma 1.5, Chapter 2, yields

$$\tilde{A}(0) = F'(0) \Rightarrow \|\tilde{A}(u)\,v\|_Y$$

$$= \|F'(0)\,v + (\tilde{A}(u) - \tilde{A}(0))\,v\|_Y \leq \|F'(0)\,v\|_Y - \|(\tilde{A}(u) - \tilde{A}(0))\,v\|_Y$$

$$\leq k\|v\|_X - \|\tilde{A}(u) - \tilde{A}(0)\|_{X \to Y}\|v\|_X .$$

Let $r$ be a real number such that $\|\tilde{A}(u) - \tilde{A}(0)\|_{X \to Y} \leq k/2$ for $\|u\|_X \leq r$. Then we obtain

$$\|\tilde{A}(u)\,v\|_Y \geq k\,\|v\|_X - \frac{k}{2}\|v\|_X = \frac{k}{2}\|v\|_X .$$

### 4.2. The solvability everywhere

**Theorem 4.2.** *Let the operator $F'(0)$ be closed, equation (4.2) be solvable everywhere in $Y$ and $\|\tilde{A}(u) - \tilde{A}(0)\|_{X \to Y} \to 0$ as $\|u\|_X \to 0$. Then the equation $\tilde{A}(u)\,v = y$ is solvable everywhere in $Y$ for any $u \in \bar{S}_r(0)$ if $r$ is sufficiently small.*

The proof of this proposition will be given in Section 6 (see Theorem 6.1).

### 4.3. Normal solvability

Let us consider in greater detail different types of normal solvability for equations (4.1) and (4.2). We use here the definitions and some results of S. Krein[105].

**Theorem 4.3.** *Let the equation $F'(0)\,w = f$ be n-normal, i.e.*
(1) $F'(0)$ *is closed;*
(2) *equation (4.2) is normally solvable;*
(3) $n(F'(0)) \equiv \dim(\mathrm{Ker}\,(F'(0))) < \infty$.
  *Let $\|\tilde{A}(u) - \tilde{A}(0)\|_{X \to Y} \to 0$ as $\|u\|_X \to 0$. Then there exists $r_0 > 0$ such that equation (4.1) is n-normal for any $u \in \bar{S}_{r_0}(0)$.*

*Proof.* Consider the decomposition $X = E \oplus \mathrm{Ker}\, F'(0)$. By $F_1$ and $A_1(u)$ we denote the restrictions on $E$ of the operators $F'(0)$ and $\tilde{A}(u)$, respectively. By Lemma 1.5, Chapter 2, $\tilde{A} = F'(0)$; hence, $\tilde{A}_1(0) = F_1$. Since $F'(0)$ and $E$ are closed, the operator $F_1$ is closed also and the equation $F_1 l = f$ is uniquely (normally) solvable on $E$. Therefore, it is correctly solvable, i.e. there exists $k > 0$ such that

$$\|l\| \leq k\,\|F_1 l\|_Y = k\|\tilde{A}_1(u)\,l + (\tilde{A}_1(0) - \tilde{A}_1(u))\,l\|_Y$$

$$\leq k\|\widetilde{A}_1(u)\,l\|_Y + \|\widetilde{A}_1(0) - \widetilde{A}_1(u)\|_{X \to Y}\|l\|_X.$$

Under the hypotheses, there exists $r_0 > 0$ such that $\|\widetilde{A}(0) - \widetilde{A}(u)\|_{X \to Y} \leq 1/(2k)$ for any $u \in \bar{S}_{r_0}(0)$. Then

$$\|\widetilde{A}_1(0) - \widetilde{A}_1(u))\|_{X \to Y} \leq \|\widetilde{A}(0) - \widetilde{A}(u)\|_{X \to Y} \leq 1/(2k)$$

and we find

$$\|l\|_X \leq k\|\widetilde{A}_1(u)\,l\|_Y + \frac{1}{2}\|l\|_X;$$

hence, $\|l\|_X \leq 2k\|\widetilde{A}_1(u)\,l\|_Y$. In accord with Theorem 6.1 proved below, the operators $\widetilde{A}_1(u)$ and $\widetilde{A}(u)$ are closed for any $u \in \bar{S}_{r_0}(0)$. The equation $\widetilde{A}_1(u)\,l = y$ is correctly solvable; hence, it is uniquely (normally) solvable. The operator $\widetilde{A}(u)$ is a continuation of $\widetilde{A}_1(u)$ and the normal solvability of the equation $\widetilde{A}(u)\,v = y$ holds true, with dim $(\text{Ker } \widetilde{A}(u))$ increasing, at most, by $n\,(F'(0))$.[105]

**Theorem 4.4.** *Let the equation* $F'(0)\,w = f$ *be d-normal, i.e.*
(1) $F'(0)$ *is closed;*
(2) *equation (4.2) is normally solvable;*
(3) $d\,(F'(0)) \equiv \dim R(F'(0))^{\perp} < \infty.$
*Assume that* $\|\widetilde{A}(u) - \widetilde{A}(0)\|_{X \to Y} \to 0$ *as* $\|u\|_X \to 0$. *Then there exists* $r_0 > 0$ *such that the equation* $\widetilde{A}(u)\,v = y$ *is d-normal for any* $u \in \bar{S}_{r_0}(0)$.

*Proof.* The operator $F'(0)^*$ is adjoint to a linear one; hence, it is closed and $n\,(F'(0)^*) = \dim \text{Ker}\,(F'(0)^*) = \dim R(F'(0))^{\perp} = d\,(F'(0)) < \infty$. Since $F'(0)$ is closed and equation (4.2) is normally solvable, the equation

$$F'(0)^*w = g \tag{4.3}$$

is also normally solvable.[105] Then equation (4.3) is $n$-normal. By Lemma 1.5, Chapter 2, $\widetilde{A}(0) = F'(0)$; hence, $\widetilde{A}^*(u) = [A^*(u) - \widetilde{A}^*(0)] + F'^*(0)$ and $\|\widetilde{A}^*(u) - \widetilde{A}^*(0)\|_{Y^* \to X^*} \to 0$ as $\|u\|_X \to 0$. Owing to Theorem 4.3, there exists $r_0 > 0$ such that the equation $\widetilde{A}^*(u)\,w = p$ is $n$-normal for any $u \in \bar{S}_{r_0}(0)$. It is seen from Theorem 1.1, Chapter 2, that the operator $\widetilde{A}(u)$ is closed for any $u \in \bar{S}_{r_0}(0)$. Then the normal solvability of the equation $\widetilde{A}^*(u)\,w = p$ suggests the normal solvability of (4.1)[105] and $d\,(\widetilde{A}(u)) = \dim(R(\widetilde{A}(u)))^{\perp} = \dim \text{Ker}\,(\widetilde{A}^*(u)) = n\,(\widetilde{A}^*(u)) < \infty.$

A linear equation of the form $Lx = y$ is said to be *noetherian* if it is $n$-normal and $d$-normal simultaneously. The index of this equation is defined by $\text{ind}\,(L) \equiv n(L) - d(L) < \infty.$

**Theorem 4.5.** *Let the equation $F'(0)\,w = f$ be noetherian and $\|\widetilde{A}(u) - \widetilde{A}(0)\|_{X \to Y} \to 0$ as $\|u\|_X \to 0$. Then there exists $r_0 > 0$ such that the equation $\widetilde{A}(u)\,v = y$ is noetherian for any $u \in \bar{S}_{r_0}(0)$ and ind $\widetilde{A}(u) = $ ind $F'(0)$.*

*Proof.* Consider the decomposition $X = E \oplus$ Ker $F'(0)$. Let $F_1$, $\Phi_1(u)$ and $\widetilde{A}_1(u)$ be the restrictions on $E$ of the operators $F'(0)$, $Q(u) \equiv \widetilde{A}(u) - \widetilde{A}(0)$ and $\widetilde{A}(u)$, respectively. The operator $F_1$ has a continuous inverse operator and ind $F_1 = -d(F'(0))$. By Lemma 1.5, Chapter 2, $\widetilde{A}(0) = F'(0)$; hence, $\widetilde{A}_1(0) = F_1$ and $\widetilde{A}_1(u)l = (F_1 + Q_1(u))l = (I + Q_1 \tilde{F}_1^{-1})\,F_1 l$, where $\tilde{F}_1^{-1}$ is a continuation of the bounded operator $F_1^{-1}$ on the entire space $Y$. Let $m = \|\tilde{F}_1^{-1}\|_{X \to Y}$, $B(u) \equiv I + Q_1 \tilde{F}_1^{-1}$. We find

$$\|B(u)\,y\|_Y \geq \|y\|_Y - \|Q_1(u)\|_{X \to Y}\,m\,\|y\|_Y.$$

There exists $r_0 > 0$ such that $\|Q(u)\|_{X \to Y} = \|\widetilde{A}(u) - \widetilde{A}(0)\|_{X \to Y} \leq 1/(2m)$; $\|Q_1(u)\|_{X \to Y} \leq \|Q(u)\|_{X \to Y} \leq 1/(2m)$. Therefore,

$$\|B(u)\,y\|_Y \geq \|y\|_Y - \frac{1}{2}\|y\|_Y = \frac{1}{2}\|y\|_Y.$$

The equation $B(u)\,y = g$ is correctly solvable. Since $Q_1(u)\,\tilde{F}_1^{-1}$ is bounded and $Y$ and $I$ are closed, the operator $B(u)$ is also closed. Hence, the equation $B(u)\,y = g$ is uniquely (normally) solvable, i.e. $n\,(B(u)) = 0$ for any $u \in \bar{S}_{r_0}(0)$. Then for the adjoint equation $B^*(u)\,y^* = g^*$ we have

$$\|B^*(u)\,y^*\|_{Y^*} \geq \|y^*\|_{Y^*} - \|(Q_1 \tilde{F}_1^{-1})^*\|_{Y^* \to Y^*}\|y^*\|_{Y^*}$$

$$\geq \|y^*\|_{Y^*} - \|Q_1 \tilde{F}_1^{-1}\|_{Y \to Y}\|y^*\|_{Y^*} \geq \frac{1}{2}\|y^*\|_{Y^*}$$

for any $u \in \bar{S}_{r_0}(0)$, and $n\,(B^*(u)) = 0$. Therefore, the equation

$$B\,(u)\,y = g \qquad (4.4)$$

is noetherian and ind $B(u) = n\,(B(u)) - d\,(B(u)) = n\,(B(u)) - u\,(B^*(u)) = 0$ (i.e. (4.4) is the Fredholm equation). For $\widetilde{A}_1(u)$ we find[105]

$$\text{ind } \widetilde{A}_1(u) = \text{ind } (B(u)\,F_1) = \text{ind } B(u) + \text{ind } F_1' = 0 - d\,(F'(0)).$$

The operator $\widetilde{A}(u)$ is a continuation of $\widetilde{A}_1(u)$ by the finite dimensional space Ker $(F'(0))$. The normal solvability of the equation $\widetilde{A}(u)\,v = y$ holds true[105] and

$$\text{ind } \widetilde{A}(u) = \text{ind } (\widetilde{A}_1(u) + n\,(F'(0)) = n\,(F'(0)) - d\,(F'(0)) = \text{ind } F'(0).$$

*Remark* 4.1. Since a form of the operator $\tilde{A}(u)$ has not been specified, the results of this section hold true also for $A(u)$.

## 5. SOLVABILITY OF THE EQUATION $A^*(u)\,w = p$

### 5.1. Correct solvability

Consider the equation

$$A^*(u)\,w = p, \tag{5.1}$$

where $u$ is a given element of $D(F)$, and a solution $w$ is being sought for in $D(A^*(u)) \subset Y^*$.

**Lemma 5.1.** *Let the equation $F'^*(u)\,w = g$ be correctly solvable for any $u \in \bar{S}_r(0)$. Then equation (5.1) is also correctly solvable for any $u \in \bar{S}_r(0)$.*

*Proof* is given by the equality

$$A^*(u) = F'^*(\tau u), \quad \tau \in (0,1).$$

Since the solvability everywhere of the main equation implies the correct solvability of the adjoint equation[105], the following propositions are valid.

**Lemma 5.2.** *Let the equation $F'(u)\,v = f$ be solvable everywhere in $Y$ for any $u \in \bar{S}_r(0)$. Then equation (5.1) is correctly solvable for any $u \in \bar{S}_r(0)$.*

**Lemma 5.3.** *Let the equation $F'(0)\,v = f$ be solvable everywhere in $Y$ and $\|F'(u) - F'(0)\|_{X \to Y} \to 0$ as $\|u\|_X \to 0$. Then there exists $r > 0$ such that equation (5.1) is correctly solvable for any $u \in \bar{S}_r(0)$.*

Using the conditions of the solvability everywhere for the equation $A(u)\,v = y$, $y \in Y$, stated above, we arrive at the following lemmas.

**Lemma 5.4.** *Let*
(1) *the domains of $A(0) = \Phi'(U_0)$ and $A(u)$ coincide for any $u \in \bar{S}_r(0)$;*
(2) *the operator $\Phi(U_0)$ be closed;*
(3) *the following restriction hold:*

$$\sup_{u \in \bar{S}_r(0)} \int_0^1 \|I - (\Phi'(U_0))^{-1}\Phi'(U_0 + ut)\|_{X \to Y} \; dt \le q < 1;$$

(4) *the equation $A(0)\,v = y$ be everywhere solvable in $Y$.*
*Then equation (5.1) is correctly solvable for $\|u\|_X \le r$.*

**Lemma 5.5.** *Assume that*
(1) *the equation $A(0)\,v = y$ is solvable everywhere in $Y$;*

(2) $\|A^{-1}(0)\|_{Y \to X} \le 1/m;$

(3) *the operator $A(0) = \Phi'(U_0)$ is closed;*

(4) *the following conditions are satisfied:*

$$\|\Phi'(U_0 + u) - \Phi'(U_0 + v)\|_{X \to Y} \le L \|u - v\|_X, \quad Lr/(2m) < 1.$$

*Then equation (5.1) is correctly solvable for any $u \in \bar{S}_r(0)$.*

## 5.2. The solvability everywhere

**Lemma 5.6.** *If the equation $F'^*(u) w = g$ is solvable everywhere in $X^*$ for $\|u\|_X \le r$, then equation (5.1) is also solvable everywhere for any $u \in \bar{S}_r(0)$.*

Since the solvability everywhere of the adjoint equation is equivalent to the correct solvability of the main equation, the following propositions are valid.

**Lemma 5.7.** *If the equation $F'(u) v = f$ is correctly solvable for any $u \in \bar{S}_r(0)$, then equation (5.1) is solvable everywhere in $X^*$ for any $u \in \bar{S}_r(0)$.*

**Lemma 5.8.** *If the equation $F'(u) v = f$ is correctly solvable and $\|F'(u) - F'(0)\|_{X \to Y} \to 0$ as $\|u\|_X \to 0$, then (5.1) is solvable everywhere in $X^*$ for any $u \in \bar{S}_r(0)$, the number $r$ being sufficiently small.*

**Lemma 5.9.** *Let*

(1) *the operator $A(0) = F'(0)$ have a bounded inverse operator and*

$$\|F'(0)^{-1}\|_{Y \to X} \le 1/m;$$

(2) *$F'(u)$ satisfy the Lipschitz condition*

$$\|F'(u_1) - F'(u_2)\|_{X \to Y} \le L \|u_1 - u_2\|_X;$$

(3) *$Lr/(2m) < 1$.*

*Then equation (5.1) is solvable everywhere in $X^*$ for any $u \in \bar{S}_r(0)$.*

## 5.3. Normal solvability

Equation (5.1) is normally solvable if $R(A^*(u)) = (\text{Ker } A(u))^{\perp}$. [105]

**Lemma 5.10.** *If the equation $F'(u)^* w = g$ is normally solvable for any $u \in \bar{S}_r(0)$, then the equation $A^*(u) w = p$ is also normally solvable for any $u \in \bar{S}_r(0)$.*

**Lemma 5.11.** *Let the operator $F'(u)$ be closed and the equation $F'(u) v = y$ be normally solvable for any $u \in \bar{S}_r(0)$. Then equation (5.1) is also normally solvable if $u \in \bar{S}_r(0)$.*

*Proof.* Since the normal solvability of the main equation is equivalent to the normal solvability of the adjoint equation if an operator is closed, then using Lemma 5.10, we arrive at the assertion which has to be proved.

Using the arguments of Subsection 4.3, it is easily shown that the following lemmas are valid.

**Lemma 5.12.** *Let the equation $F'(0)\, v = y$ be n-normal and $\|F'(u) - F'(0)\|_{X \to Y} \to 0$ as $\|u\|_X \to 0$. Then equation (5.1) is n-normal for any $u \in \bar{S}_r(0)$ if $r$ is sufficiently small.*

**Lemma 5.13.** *If the equation $F'(0)\, v = y$ is d-normal and*

$$\|F'(u) - F'(0)\|_{X \to Y} \to 0$$

*as $\|u\|_X \to 0$, then there exists $r_0 > 0$ such that equation (5.1) is n-normal for any $u \in \bar{S}_{r_0}(0)$.*

# 6. SOLVABILITY OF THE EQUATION $\widetilde{A}^*(u)\, w = p$

Consider the solvability problem for the equation

$$\widetilde{A}^*(u)\, w = p, \tag{6.1}$$

where $p \in X^*$, $u \in D(F) = D(\widetilde{A}) \subset X$ is a given element, and a solution $w$ is being sought for in $D(\widetilde{A}^*(u)) \subset Y^*$. In this case we can proceed from the solvability properties both of the equation

$$\widetilde{A}(u)\, v = y \tag{6.2}$$

and of the equations

$$F'^*(0)\, w = g, \tag{6.3}$$

$$F'(0)\, v = f \tag{6.4}$$

if the hypotheses of Lemma 1.13, Chapter 2, are satisfied.

## 6.1. Correct solvability

Lemma 1.14, Chapter 2, can be reformulated as follows.

**Lemma 6.1.** *Let equation (6.3) be correctly solvable and*

$$\|\widetilde{A}(u) - \widetilde{A}(0)\|_{X \to Y} \to 0 \quad \text{as } \|u\|_X \to 0.$$

*Then there exists $r_0 > 0$ such that equation (6.1) is correctly solvable for any $u \in \bar{S}_{r_0}(0)$.*

**Theorem 6.1.** *Let equation (6.4) be solvable everywhere in $Y$ and*

$$\|\widetilde{A}(u) - \widetilde{A}(0)\|_{X \to Y} \to 0$$

*as $\|u\|_X \to 0$. Then there exists $r_0 > 0$ such that equation (6.1) is correctly solvable for any $u \in \bar{S}_{r_0}(0)$. If the operator $F'(0)$ is closed, then equation (6.2) is everywhere solvable in $Y$ for any $u \in \bar{S}_{r_0}(0)$.*

*Proof.* First, following[105], we prove that the solvability everywhere of (6.4) implies the correct solvability of (6.3). In fact, we find

$$\langle y, q \rangle_Y = \langle F'(0) v, q \rangle_Y = \langle v, F'^*(0) q \rangle_X \le \|v\|_X \|F'^*(0) q\|_X.$$

for any $y \in Y$, $q \in D(F'^*(0))$.

Consider a family of functionals

$$T_q(y) = \left| \frac{\langle y, q \rangle_Y}{\|F'^*(0) q\|_{X^*}} \right| \le \|v\|_X.$$

For a fixed $q$, the functional $T_q(y)$ is continuous in $y$ and $T_q(y) \ge 0$; $T_q(\lambda) = |\lambda| T_q(y)$, $\lambda \in \mathbf{R}$; $T_q(y_1 + y_2) \le T_q(y_1) + T_q(y_2)$. We find $T_q(y) \le \|v\|_X < \infty$ for a fixed $y$ and for any $q \in D(F'^*(0))$. From the uniform boundedness principle, there exists $C > 0$ such that $T_q(y) \le C\|y\|_Y$ for any $q \in D(F'^*(0))$, $y \in Y$, i.e.

$$|\langle y, q \rangle_Y| \le C \|y\|_Y \|F'^*(0) q\|_{X^*}.$$

By definition of the norm of a linear functional, $\|q\|_{X^*} \le C \|F'^*(0) q\|_{X^*}$. Putting $k = 1/C$, we get $\|F'^*(0) q\|_{X^*} \ge k \|q\|_{Y^*}$. By Lemma 1.13, Chapter 2, $\widetilde{A}^*(0) = F'^*(0)$. There exists $r_0$ such that $\|\widetilde{A}(u) - \widetilde{A}(0)\|_{X \to Y} \le k/2$ for any $u \in \bar{S}_{r_0}(0)$. Then we obtain

$$\|\widetilde{A}^*(u) w\|_{X^*} = \|F'^*(0) w + (\widetilde{A}^*(u) - F'^*(0)) w\|_X$$

$$\ge \|F'^*(0) w\|_{X^*} - \|\widetilde{A}^*(u) - \widetilde{A}^*(0)\|_{Y^* \to X^*} \|w\|_{Y^*}$$

$$\ge k\|w\|_{Y^*} - \|\widetilde{A}(u) - \widetilde{A}(0)\|_{X \to Y} \|w\|_{Y^*} \ge \frac{k}{2}\|w\|_{Y^*}.$$

If, in addition, $F'(0)$ is closed, then, by Theorem 6.1, $\widetilde{A}(u)$ is also closed. Therefore, the correct solvability of the adjoint equation implies the solvability everywhere of the main equation[105].

**Lemma 6.2.** *Let*
*(1) the operator $F'^*(0)$ have a continuous inverse operator and*

$$\|(F'^*(0))^{-1}\|_{X^* \to Y^*} \le 1/m;$$

(2) $\widetilde{A}(u)$ satisfy the Lipschitz condition

$$\|\widetilde{A}(u_1) - \widetilde{A}(u_2)\|_{X \to Y} \leq \|u_1 - u_2\|_X$$

for any $u_1, u_2 \in \bar{S}_{r_0}(0)$;
(3) $r_0 < m/L$.
Then equation (6.1) is correctly solvable for any $u \in \bar{S}_{r_0}(0)$.

## 6.2. The solvability everywhere

**Lemma 6.3.** *If equation (6.2) is correctly solvable for any $u \in \bar{S}_{r_0}(0)$, then equation (6.1) is solvable everywhere in $X^*$ for any $u \in \bar{S}_{r_0}(0)$.*

*Proof.* Following the arguments of S.Krein[103], we prove that the correct solvability of the main equation implies the solvability everywhere of the adjoint equation. Let $f$ be an arbitrary element of $X^*$. On $R(\widetilde{A}(u))$ we define a functional $\varphi$ by the formula $\langle y, \varphi \rangle_Y = \langle x, f \rangle_X$, where $X$ is the (unique) solution of the equation $\widetilde{A}(u)\,x = y$. The functional $\varphi$ is bounded on $R(\widetilde{A}(u))$:

$$|\langle y, \varphi \rangle_Y| = |\langle x, f \rangle_X| \leq \|f\|_{X^*}\|x\|_X \leq k\,\|f\|_{X^*}\|\widetilde{A}(u)\,x\|_Y = k\,\|y\|_Y.$$

Hence, the functional $\varphi$ can be extended with continuity on $\overline{R(\widetilde{A}(u))}$ and then, by the Hahn–Banach theorem, on $Y$ (even with retention of norm). By construction, we have $\langle \widetilde{A}(u)\,x, y \rangle_Y = \langle x, f \rangle_X$ for any $x \in D(\widetilde{A})$; hence, $\varphi \in D(\widetilde{A}^*(u))$ and $\widetilde{A}^*(u)\,\varphi = f$. Since $f \in X^*$ is arbitrary, equation (6.1) is solvable everywhere in $X^*$.

Using Lemma 6.3 and Theorem 1.1, Chapter 2, we arrive at the following

**Lemma 6.4.** *Let the equation $F'(0)\,v = f$ be correctly solvable and $\|\widetilde{A}(u) - \widetilde{A}(0)\|_{X \to Y} \to 0$ as $\|u\|_X \to 0$. Then the equation $\widetilde{A}^*(u)\,w = p$ is solvable everywhere in $X^*$ for any $u \in \bar{S}_r(0)$ if $r$ is sufficiently small.*

**Lemma 6.5.** *Let*
(1) *the equation $F'(0)\,v = f$ be correctly solvable and*

$$\|F'(0)\,v\|_Y \geq k\,\|v\|_X;$$

(2) *the operator $\widetilde{A}(u)$ satisfy the Lipschitz condition*

$$\|\widetilde{A}(u_1) - \widetilde{A}(u_2)\|_{X \to Y} \leq L\,\|u_1 - u_2\|_X$$

*for any $u_1, u_2 \in \bar{S}_r(0)$;*
(3) $r < k/L$.
*Then equation (6.1) is solvable everywhere in $X^*$ for any $u \in \bar{S}_r(0)$.*

**Lemma 6.6.** *Let the equation $F'^*(0)\,w = g$ be solvable everywhere in $X^*$ and $\|\widetilde{A}(u) - \widetilde{A}(0)\|_{X \to Y} \to 0$ as $\|u\|_X \to 0$. Then equation (6.1) is solvable everywhere in $X^*$ for any $u \in \bar{S}_r(0)$ if $r$ is sufficiently small.*

*Proof.* Using the uniform boundedness principle, it is easily shown that the solvability everywhere of the adjoint equation implies the correct solvability of the main equation[105]. From Lemma 6.4, we come to the assertion which has to be proved.

### 6.3. Normal solvability

Here, as in Subsection 5.3, equation (6.1) is said to be normally solvable if $R(\widetilde{A}^*(u)) = (\text{Ker } \widetilde{A}(u))^\perp$.

**Lemma 6.7.** *If the operator $\widetilde{A}(u)$ is closed and equation (6.2) is normally solvable, then equation (6.1) is also normally solvable.*

*Proof.* This conclusion follows immediately from the fact that, for a closed linear operator, the normal solvability of the main equation is equivalent to the normal solvability of the adjoint equation[105].

Using the theorems of Subsection 4.3, we find the following propositions to be valid.

**Lemma 6.8.** *Let equation (6.4) be n-normal and $\|\widetilde{A}(u) - \widetilde{A}(0)\|_{X \to Y} \to 0$ as $\|u\|_X \to 0$. Then equation (6.1) is normally solvable for any $u \in \bar{S}_r(0)$ if $r$ is sufficiently small.*

**Lemma 6.9.** *If equation (6.4) is d-normal and $\|\widetilde{A}(u) - \widetilde{A}(0)\|_{X \to Y} \to 0$ as $\|u\|_X \to 0$, then there exists $r_0 > 0$ such that equation (6.1) is n-normal for any $u \in \bar{S}_{r_0}(0)$.*

# Chapter 4

# Transformation groups, conservation laws and constructing the adjoint operators in non-linear problems

As has been shown in Chapter 1, adjoint operators in non-linear problems may be defined in different ways. This raises a reasonable question of correlation between different adjoint operators corresponding to the same non-linear problem. At the same time, the problem of choosing the 'best' (in a sense) adjoint operator is of interest (from the numerical point of view). The solution of this problem will help to justify perturbation algorithms being applied in non-linear problems. There are a number of other questions and problems to be answered in the adjoint equation theory for a prescribed class of problems. Some of these questions are studied in this chapter.

The authors present here generalizations of some approaches developed in the previous chapters. These generalizations concern first the invocation of the results of the general theory of variational calculus, the Lie groups, the results on conservation laws, trivial and non-trivial currents and others. The authors would like to note especially that using these results of various fields of mathematics gives new possibilities for constructing and applying the adjoint equations in non-linear problems.

## 1. DEFINITIONS. NON-LINEAR EQUATIONS AND OPERATORS. CONSERVATION LAWS

**1.1.** Let $H_1 \equiv X$ and $H_0$ be Hilbert spaces, and $H_1$ be densely enclosed into $H_0$. By $H_1^*$ we denote the space dual to $H_1$. Assume that $H_0 \equiv H_0^*$, $H_1^* \equiv (H_1)^{-1}$. Hence, the following imbeddings are valid: $H_1 \subset H_0 \equiv H_0^* \subset H_1^* \equiv (H_1)^{-1}$.

Consider a non-linear operator $\Phi$ mapping $H_1$ into $H_0$ with the domain $D(\Phi) \subset H_1$ being a convex set (but not necessarily linear and including the zero element). The operator $\Phi$ is assumed to be continuously Gâteaux-differentiable (and, hence, it is Frechet-differentiable). Then the following formula holds:

$$\Phi(U) = \Phi(U_0) + \int_0^1 \Phi'(U_0 + tu) \, dt u, \tag{1.1}$$

where $u = U - U_0$, $U, U_0 \in D(\Phi)$, or

$$\Phi(U) - \Phi(U_0) = A(u) \, u \tag{1.2}$$

with the operator $A(u)$ defined by the expression

$$A(u) \, v = \int_0^1 \Phi'(U_0 + tu) \, dt v \tag{1.3}$$

and the domain

$$D(A) = \{ v \in H_1 : \; v + U_0 \in D(\Phi) \} \,. \tag{1.4}$$

We assume that $D(A)$ is dense in $H_0$. Note that the zero element belongs to $D(A)$, and by the above-introduced assumptions, the set $D(A)$ coincides with the domain $D(A(0)) = D(\Phi'(U_0))$ of the operator $A(0) = \Phi'(U_0)$, where $U_0$ is an arbitrary fixed element of $D(A)$. We assume that $D(A)$ is a linear set. If it happens that $0 \in D(\Phi)$, then $D(\Phi) = D(A)$ and $D(\Phi)$ is also linear.

**1.2.** Consider the equation

$$\Phi(U) = 0, \tag{1.5}$$

where the right-hand side is considered to be the zero element of $H_0$. We assume that a solution of this equation exists but, in general, it may be not unique. Thus, by $D_s(\Phi)$ we denote the set of solutions of equation (1.5). The solutions of (1.5) are assumed to belong to $D(\Phi)$, and, hence, $D_s(\Phi) \subseteq D(\Phi)$. Henceforward, we are interested in the case when $D_s(\Phi)$ is not empty. Note that $0 \notin D(\Phi)$ implies $0 \notin D_s(\Phi)$. If, however, $0 \in D(\Phi)$, nevertheless, it may happen that $0 \notin D_s(\Phi)$. An important class of the problems with the property $0 \in D_s(\Phi)$ is the eigenvalue problems.

Suppose that each element $U \in D_s(\Phi)$ satisfies both equation (1.5) and some additional equations

$$\Phi_j(U) = 0, \quad j = 1, \ldots, M, \tag{1.6}$$

where $\Phi_j$, $j = 1, \ldots, M$, are non-linear operators having continuous Gâteaux derivatives with the domains $D(\Phi_j) \supseteq D(\Phi)$. In view of (1.6) (and for reasons understood from the differential equation theory), the operator $\Phi_j$ is said to be an operator corresponding to the $j$-th conservation law or, simply, the $j$-th conservation law operator. Equation (1.6) may be treated as the $j$-th conservation law, although this treatment is not commonly accepted in the differential equation theory. If, however, the elements $U$ are functions of independent variables $x_i$, $i = 1, \ldots, n$, and $\Phi_j$ is defined by

$$\Phi_j = \sum_{i=1}^n \frac{\partial J_i^{(j)}}{\partial x_i} = \mathrm{div} \, \bar{J}^{(j)}, \quad \bar{J}^{(j)} = (J_1^{(j)}, \ldots, J_n^{(j)}),$$

where $J_i^{(j)} = J_i^{(j)}(x, U, U_{x_1}, \ldots)$, $x = (x_1, \ldots, x_n)$, $U_{x_1} = \partial U / \partial x_1, \ldots$, then equation (1.6) has the conventional form of the $j$-th conservation law used in many papers.

The assumption that $\Phi_j$ is continuously differentiable gives the formula

$$\Phi_j(U) = \Phi_j(U_0) + A_j(u)\, u, \tag{1.7}$$

where $U$ and $U_0$ are arbitrary elements of $D(\Phi_j) \supset D(\Phi)$, $u = U - U_0$ and the operator $A_j(u)$ is defined by expression

$$A_j(u)\, v = \int_0^1 \Phi_j'(U_0 + tu)\, dt\, v \tag{1.8}$$

and the domain $D(A)$. We assume that the operators $\Phi_j$, $j = 1, \ldots, M$, map $H_1$ into $H_0$ and the set $\{\Phi_j\}$ is at most countable.

By the above assumption, we have $D(\Phi) \subset D(\Phi_j)$, $j = 1, \ldots, M$, and, hence, the intersection of the sets $D(\Phi_j)$ is not empty and includes $D_s(\Phi)$. Furthermore, for any fixed element $U \in \bigcap_{j=1}^M D(\Phi_j)$, the values $\Phi_j(U)$, $j = 1, \ldots, M$, are elements of the Hilbert space $H_0$, and the notion of linear independence for these elements is meaningful. Hereafter, we assume that $\Phi_j$, $j = 1, \ldots, M$, are linearly independent in the following sense: the operator $\sum_{j=1}^M C_j \Phi_j$ with the constants $C_j$ is the zero operator on $\bigcap_{j=1}^M D(\Phi_j)$ ( i.e. $\sum_{j=1}^M C_j \Phi_j(U) = 0 \ \forall\, U \in \bigcap_{j=1}^M D(\Phi_j)$) if and only if all the constants $C_j$ equal zero.

In addition to operators $A$, $A_j$, $j = 1, \ldots, M$, we consider also trivial operators $A_k^{(0)}$, $k = 1, \ldots, K$, such that they (1) depend on $u \in D(A)$, i.e. $A_k^{(0)} = A_k^{(0)}(u)$; (2) have the domain $D(A^{(0)}) \equiv \bigcap_{k=1}^K D(A_k^{(0)}) \supset D(a)$; (3) map $H_1$ into $H_0$ and satisfy the equality

$$A_k^{(0)}(v)\, v = 0, \quad k = 1, \ldots, K, \tag{1.9}$$

for any element $v \in D(A^{(0)})$. Let the set $\{A_k^{(0)}\}$ be at most countable and be a minimal system. By a minimal system is meant the following. Let $\alpha_k$, $k = 1, \ldots, K$, be arbitrary numbers. By $A^{(0)}$ we denote the operator $A^{(0)} = \sum_{k=1}^K \alpha_k A_k^{(0)}$ with the domain $D(A^{(0)})$. We have in fact a set of operators $A^{(0)}$ (depending on the choice of $\alpha_k$, $k = 1, \ldots, K$). A set $\{A_k^{(0)}\}_{k=1}^K$ is said to be a minimal system if the removal of any operator $A_k^{(0)}$ from this set results in a restriction of the operator $A^{(0)}(u)$ (that is, the set of operators $A^{(0)}$) being considered on $D(A^{(0)})$ for arbitrary element $u \in D(A^{(0)})$.

**1.3.** We give below some families of operators $\{A_k^{(0)}\}$, $\{A_j\}$, $\{\Phi_j\}$ being of frequent use in non-linear problems, namely, the operators generated by the currents studied by V.Vladimirov and I.Volovich[310,311] and others. Let the elements of the spaces $H_1$ and $H_0$ be functions $U(x)$ for $x = (x_1, \ldots, x_n) \in$

$\Omega \subset R^n$. Then a set of the operators $J_i^{(j)}(U)$, $i = 1, \ldots, n$, such that

$$\sum_{i=1}^{n} \frac{\partial}{\partial x_i} J_i^{(j)}(U) \in H_0$$

is said to be the $j$-th current. Following Vladimirov-Volovich[310,311], a current is said to be conserved if the following equality (the $j$-th conservation law)

$$\sum_{i=1}^{n} \frac{\partial}{\partial x_i} J_i^{(j)}(U) = 0 \qquad (1.10)$$

holds for the solutions of equation (1.5).

If equality (1.10) holds for any $U \in D(\Phi)$ or $J_i^{(k)}(U) = 0$, $i = 1, \ldots, n$, for any $U \in D_s(\Phi)$, then the current is said to be trivial. It is known[229] that trivial currents with the property $J_i^k(U) = 0$, $i = 1, \ldots, n$, $\forall U \in D_s(\Phi)$ can be eliminated for a wide variety of equations. Thus, in what follows we do not consider the currents of this type (and the corresponding conservation laws). By a trivial current we mean the current satisfying (1.10) for any $U \in D(\Phi)$, and we denote it by $\bar{J}_0^{(k)} = (J_{0,1}^{(k)}, \ldots, J_{0,n}^{(k)})$, or $\{J_{0,i}^{(k)}(U)\}$.

Obviously, if trivial and conserved currents are found when considering the concrete equation (1.5), then for the above-introduced operators $\Phi_j$, $A_j(u)$, $j = 1, \ldots, M$, one can take the following operators (assuming that $\{J_i^{(j)}\}$, $\{J_{0,i}^{(k)}\}$ are continuously differentiable):

$$\Phi_j(U) \equiv \sum_{i=1}^{n} \frac{\partial}{\partial x_i} J_i^{(j)}(U), \qquad\qquad j = 1, \ldots, M,$$

$$A_j(u)v \equiv \sum_{i=1}^{n} \frac{\partial}{\partial x_i} \int_0^1 J_i^{(j)'}(U_0 + tu)\, dt v, \qquad j = 1, \ldots, M, \qquad (1.11)$$

$$A_k^{(0)}(u)v \equiv \sum_{i=1}^{n} \frac{\partial}{\partial x_i} \int_0^1 J_{0,i}^{(k)'}(U_0 + tu)\, dt v, \qquad k = 1, \ldots, K.$$

Thus, if the explicit form of some currents is known for equation (1.5), one can construct the corresponding operators $\{A_j\}$, $\{A_k^{(0)}\}$. Note that papers by V.Vladimirov and I.Volovich[310,311] present explicit forms of some trivial and conserved currents for a number of non-linear problems of mathematical physics. Thus, using the results of these papers, one can construct the above-mentioned operators $\{A_k^{(0)}\}$, $\{A_j\}$ when considering the equation of the form (1.5).

*Remark* 1.1. Note that the papers mentioned[310,311] do not answer the question: are the founded currents all possible for the considered equations? We can add here also that it is not clear in general whether one can construct all the operators $\{A_k^{(0)}\}$, $\{A_j\}$ if all the currents (both trivial and conserved) are known. How can one find all the currents for the concrete equation (1.5)? What is the general methodology for constructing the operators $\{A_k^{(0)}\}$, $\{A_j\}$ in the case when the currents are not found for equation (1.5) under consideration?

## 2. TRANSFORMATION OF EQUATIONS

**2.1.** Consider equation (1.5). Fixing an element $U_0 \in D(\Phi)$ and putting $u = U - U_0$, we write (1.5) as

$$\Phi(U) = A(u)\,u - f = 0, \tag{2.1}$$

where $f = -\Phi(U_0)$. Hence, if $U \in D_s(\Phi)$, then $u$ is a solution to the equation

$$F(u) \equiv A(u)\,u = f. \tag{2.2}$$

We denote by $D_f(A)$ the set of solutions of this equation for a prescribed element $f = -\Phi(U_0) \in H_0$. Obviously, $D_f(A) \subset D(A)$. Conversely, let $u \in D_f(A)$ be a solution to (2.2) for a given element $f \in H_0$. Setting $U = u + U_0$, we get $U \in D(\Phi)$ and

$$\Phi(U) = \Phi(U_0) + A(u)\,u = -f + A(u)\,u = 0, \tag{2.3}$$

that is, $U \in D_s(\Phi)$. Then, the following lemma is valid.

**Lemma 2.1.** *The problem of finding the solution to equation (1.5) is equivalent to the problem of finding the solution to equation (2.2). These solutions are related by the equality $U = u + U_0$. Moreover, $D_s(\Phi) = U_0 + D_f(A)$.*

Similarly, from equations (1.6) (where $U \in D_s(\Phi)$) we arrive at the equations:

$$\Phi_j(U) = A_j(u)\,u - f_j = 0, \quad j = 1, \ldots, M, \tag{2.4}$$

where

$$f_j = -\Phi_j(U_0), \quad u \in D_f(A),$$

or

$$F_j(u) \equiv A_j(u)\,u = f_j, \quad j = 1, \ldots, M. \tag{2.5}$$

These equations are also equivalent (in the sense of Lemma 2.1) to equations (1.6). The solutions of equations (2.2), (2.5) may be treated as corrections to the element $U_0$, and the solution of equation (1.5) may be considered as the solution of a perturbed equation (with the unperturbed equation of the form $A(0)\,u_0 = f$). If we can develop a convergent algorithm for solving the equation for the correction $u$, thereby, we will obtain a convergent algorithm for finding the solution of the original problem. Let us perform some extra transformations of equations (2.2), (2.5).

**2.2.** Let $L$, $L_j$, $j = 1, \ldots, M$, be linear operators mapping $H_0$ into $H_1^*$ with the domains $D(L)$, $D(L_j)$, $j = 1, \ldots, M$, respectively. Assume that for the ranges $R(A)$, $R(A_j)$ of the operators $A$, $A_j$, $j = 1, \ldots, M$, the following inclusions hold: $R(A) \subset D(L)$, $R(A_j) \subset D(L_j)$, $j = 1, \ldots, M$. Assume also that each of the sets $D(L)$, $D(L_j)$ is dense in $H_0$ (this assures, in particular, the uniqueness of the adjoint operators $L^*$, $L_j^*$). Considering the trivial operators $\{A_k^{(0)}\}$, we introduce also linear operators $L_k^{(0)}$ such that $R(A_k^{(0)}) \subset D(L_k^{(0)})$, $k = 1, \ldots, K$, and each domain $D(L_k^{(0)})$ is dense in $H_0$. Mapping equation (2.2) by the operator $L$ and the $j$-th equation of (2.5) by the operator $L_j$ and summing the results, we obtain the following equation:

$$\left( LA(u) + \sum_{j=1}^{M} L_j A_j(u) \right) u = Lf + \sum_{j=1}^{M} L_j f_j. \tag{2.6}$$

In view of

$$\sum_{k=1}^{K} L_k^{(0)} A_k^{(0)}(u)(u) = 0,$$

(2.6) takes the form

$$\widetilde{F}(u) \equiv \widetilde{A}(u)\, u = Lf + \sum_{j=1}^{M} L_j f_j \equiv \tilde{f}, \tag{2.7}$$

where

$$\widetilde{A}(u) = LA(u) + \sum_{j=1}^{M} L_j A_j(u) + \sum_{k=1}^{K} L_k^{(0)} A_k^{(0)}(u). \tag{2.8}$$

Thus, if $U \in D_s(\Phi)$ is a solution to equation (1.5), then the function $u = U - U_0 \in D_f(A(u))$ satisfies (2.7), where $L$, $L_j$, $L_k^{(0)}$ are arbitrary operators. On the other hand, if $D_f(\widetilde{A})$ is a solution to equation (2.7) for arbitrary $L$, $L_j$, $L_k^{(0)}$, it is easily seen that $u$ is a solution to (2.2), (2.5). However, if the set of the operators $L$, $\{L_j\}$ is fixed, then the set of the solutions of equation (2.7) may turn out to be larger than $D_f(A)$. Hence it follows that if $U = u + U_0 \in D_s(\Phi)$, then

$$\widetilde{\Phi}(U) \equiv \widetilde{A}(u)\, u - \tilde{f} = 0, \quad \tilde{f} = Lf + \sum_{j=1}^{M} L_j f_j, \tag{2.9}$$

that is, $U$ belongs to the set $D_s(\widetilde{\Phi})$ of the solutions of equation (2.9). Therefore, $D_s(\Phi) \subseteq D_s(\widetilde{\Phi})$, and if the set of solutions $D_s(\widetilde{\Phi})$ is found, then one can find the set $D_f(A)$. Thus, to solve (1.5) it is sufficient to solve equation (2.9) which is a transformed equation with the operator (2.8). Although it is evident that when transforming to (2.9), new difficulties may appear (ambiguity of the solution and so on ), our further consideration will concern the transformed equation, its adjoint equation and their specific cases.

**2.3.** Let equation (1.5) be associated with the conservation laws of the form:

$$\Phi_j(U) \equiv \sum_{i=1}^{n} \frac{\partial}{\partial x_i} J_i^{(j)}(U) = 0, \qquad (2.10)$$

where $\{J_i^{(j)}(U)\}$ are continuously differentiable operators introduced in Subsection 1.3. Then, under the above-formulated restrictions, equation (2.9) takes the form:

$$\tilde{A}(u)\,u = Lf + \sum_{j=1}^{M} L_j f_j \equiv \tilde{f}, \qquad (2.11)$$

where

$$f = -\Phi(U_0), \quad f_j = \sum_{i=1}^{n} \frac{\partial}{\partial x_i} J_i^{(j)}(U_0),$$

$$\tilde{A}(u) = LA(u) + \sum_{j=1}^{M} L_j \left( \sum_{i=1}^{n} \frac{\partial}{\partial x_i} \int_0^1 J_i^{(j)'}(U_0 + tu)\,dt \right) + \sum_{k=1}^{K} L_k^{(0)} A_k^{(0)}(u). \qquad (2.12)$$

Some of the operators $\{A_k^{(0)}\}$ may be given in the form

$$A_k^{(0)}(u) = \sum_{i=1}^{n} \frac{\partial}{\partial x_i} \int_0^1 J_{0,i}^{(k)'}(U_0 + tu)\,dt, \qquad (2.13)$$

where $\{J_{0,i}^{(k)}\}$ are trivial currents (since the set of all trivial operators may be larger than the set of the operators generated by trivial currents).

## 3. ADJOINT EQUATIONS

**3.1.** Consider equation (2.2)

$$F(u) \equiv A(u)\,u = f, \qquad (3.1)$$

where $u = U - U_0$, $U \in D_s(\Phi)$, $U_0 \in D(\Phi)$, $f = -\Phi(U_0)$. Let us fix an element $u \in D(A)$ and consider the operator $A(u)$ as a linear operator mapping $H_1$ into $H_0$ with the domain $D(A)$. Using the standard methods of linear functional analysis, we introduce the operator $A^*(u)$ adjoint to $A(u)$:

$$(A(u)\,v, w) = (v, A^*(u)\,w), \qquad (3.2)$$

mapping $H_0$ into $H_1$ with the domain $D(A^*)$. (Here $(\cdot, \cdot)$ is the scalar product in $H_0$.) The operator $A^*(u)$ may be constructed by the formula

$$A^*(u) = \int_0^1 (\Phi'(U_0 + tu))^*\,dt. \qquad (3.3)$$

Let $J$ be the imbedding operator mapping $H_0$ into $H_1^*$. Then its adjoint operator $J^*$ maps $H_1$ into $H_0$ (and it is also the identity operator). If we treat equation (3.1) as an equation in $H_1^*$, i.e. consider the equation

$$J A(u) u = J f, \tag{3.4}$$

then its adjoint equation has the form

$$A^*(u) J^* u^* = g^*, \tag{3.5}$$

where $g^*$ is an element of $H_1^*$, $u^* \in D(A^* J^*) \subset H_1$, and $A^*(u) J^*$ is a linear operator mapping $H_1$ into $H_1^*$. If $u^*$ is treated here as an element of the space $H_0$ and the operator $A^*(u)$ is adjoint to $A(u)$, mapping $H_0$ into $H_1$, then (3.5) may be written as

$$A^*(u) u^* = g^*, \tag{3.6}$$

that is, an equation adjoint to (3.1). Henceforward, we do not differentiate equations (3.1) and (3.4), (3.5) and (3.6) and omit the operators $J$, $J^*$ for simplicity (this concerns, in particular, the subsequent consideration revealing the relation between the adjoint operators).

Thus, to equation (3.1) corresponds adjoint equation (3.6), where $A^*(u)$ is the adjoint operator corresponding to the operator $F(u)$ and defined by formula (3.3). Chapter 1 formulates a number of other principles for constructing adjoint operators (we denote them by $\tilde{A}^*(u)$) satisfying (along with the operators $A^*(u)$) the Lagrange identity

$$(A(u) v, w) = (v, \tilde{A}^*(u) w), \tag{3.7}$$

where $\tilde{A}^*(u)$ is an operator distinct from $A^*(u)$ with the domain $D(\tilde{A}^*)$. If $u$ and $u^*$ are solutions to the equations

$$A(u) u = f, \tag{3.8}$$

$$A^*(u) u^* = g^*, \tag{3.9}$$

where $g^* \in H_1^* \cap R(A^*)$, $u \in D_f(A)$, then from (3.8) and (3.9) we get

$$0 = (F(u) - f, u^*) = (A(u) u - f, u^*) = (u, A^*(u) u^*) - (f, u^*) = (u, g^*) - (f, u^*).$$

Thus, we come to the adjointness relationship

$$(u, g^*) = (f, u^*), \tag{3.10}$$

where $u \in D_f(A)$, $u^* \in D_{g^*}(A^*)$, $D_{g^*}(A^*)$ is the set of the solutions to equation (3.9) for a given $g^*$.

Similarly, if $u^*$ and $\tilde{u}^*$ are solutions to the equations

$$A(u) u = f, \quad u \in D_f(A), \tag{3.11}$$

$$\tilde{A}^*(u) \tilde{u}^* = g^*, \quad \tilde{u}^* \in D_{g^*}(\tilde{A}^*), \tag{3.12}$$

we get also the relation

$$(u, g^*) = (f, \tilde{u}^*). \tag{3.13}$$

Note that the operator $\tilde{A}^*(u)$ in equation (3.12) may be different from $A^*(u)$ (it is required only to satisfy (3.7)). Equation (3.12) (as well as equation (3.7)) is said to be an adjoint equation corresponding to the equation $A(u)u = f$ or the equation $\Phi(U) = 0$.

**3.2.** Consider one of the concrete operators $\tilde{A}^*$, namely, the operator adjoint to the operator $\tilde{A}(u)$ defined by (2.8), that is, $\tilde{A}^*(u) \equiv (\tilde{A}(u))^*$, where

$$(\tilde{A}(u))^* = \left( LA(u) + \sum_{j=1}^{M} L_j A_j(u) + \sum_{j=1}^{K} L_j^{(0)} A_j^{(0)}(u) \right)^*, \quad u \in D(A). \tag{3.14}$$

The operator $(\tilde{A}(u))^* \equiv \tilde{A}(u)$ satisfies the Lagrange identity

$$(\tilde{A}(u)v, w) = \left( v, \left( LA(u) + \sum_{j=1}^{M} L_j A_j(u) + \sum_{j=1}^{K} L_j^{(0)} A_j^{(0)}(u) \right)^* w \right), \tag{3.15}$$

where $u \in D(A)$, $v \in D(A)$, $w \in D(A^*)$. Under the standard conditions (for example, the operators $LA$, $\{L_j A_j\}$, $\{L_j^{(0)} A_j^{(0)}\}$ be bounded except probably one of them), the following equality holds for $\tilde{A}^*(u)$:

$$\tilde{A}^*(u) \equiv (\tilde{A}(u))^* = A^*(u)L^* + \sum_{j=1}^{M} A_j^*(u) L_j^* + \sum_{j=1}^{K} A_j^{(0)*}(u) L_j^{(0)*}. \tag{3.16}$$

Henceforward, unless otherwise specified, we assume that the above-mentioned conditions are satisfied and equation (3.16) holds.

**Lemma 3.1.** *Let $\tilde{A}^*(u)$ be the operator defined by equalities (3.15), (3.16), where $u$ and $v$ are arbitrary elements of $D(A)$, and $w$ an arbitrary element of $D(A^*)$. Then, if $u$ is a solution to the equation $A(u)u = f$ (i.e. $u \in D_f(A) \subset D(\Phi)$) and $u^*$ is a solution to the equation $\tilde{A}^*(u)\tilde{u}^* = g^*$ (i.e. $\tilde{u}^* \in D_{g^*}(\tilde{A}^*)$), the following equality holds:*

$$(u, g^*) = \left( Lf + \sum_{j=1}^{M} L_j f_j, \tilde{u}^* \right), \tag{3.17}$$

*where*

$$f = -\Phi(U_0), \quad f_j = -\Phi_j(U_0), \quad j = 1, \ldots, M.$$

*Proof.* Let $u \in F_f(A)$. Then, (3.15) is valid. However, for $u \in D_f(A)$, the following equality holds

$$\tilde{A}(u)\, u = \left( LA(u) + \sum_{j=1}^{M} L_j A_j(u) \right) u.$$

Hence,

$$(u, g^*) = (u, \tilde{A}^*(u)\, \tilde{u}^*) = (\tilde{A}(u)\, u, \tilde{u}^*)$$

$$= \left( \left( LA(u) + \sum_{j=1}^{M} L_j A_j(u) \right) u, \tilde{u}^* \right) = \left( Lf + \sum_{j=1}^{M} L_j f_j, \tilde{u}^* \right),$$

which proves the statement.

**3.3.** Consider a specific case of equation (1.5) and equations (1.6) introduced in Subsection 1.3 (see (1.14)–(1.16)). We assume that some trivial currents $\{J_{0,i}^{(j)}\}$, $j = 1, \ldots, N_1$, and conserved currents $\{J_i^{(j)}\}$, $j = 1, \ldots, M_1$, are known for the equation $\Phi(U) = 0$. Assume that the currents are local[33], that is, each of the operators $J_{0,i}^{(j)}(U)$, $J_i^{(j)}(U)$ is a function of finite number of arguments $x$, $U$, $\partial U/\partial x_1, \ldots, x \in \Omega$. Moreover, we assume that the domain of the operator $A(u)$ is the set of functions infinitely differentiable in $\bar{\Omega}$ with a finite support in $\Omega$ (or functions defined on $\mathbf{R}^n$ and satisfying the periodicity conditions in $x_i$, $i = 1, \ldots, n$).

**Lemma 3.2.** *Given the operators*

$$A_j^{(0)}(u)\, v = \sum_{i=1}^{n} \frac{\partial}{\partial x_i} \int_0^1 J_{0,i}^{(j)'}(U_0 + tu)\, dt\, v, \quad j = 1, \ldots, N_1, \quad (3.18)$$

$$A_j(u)\, v = \sum_{i=1}^{n} \frac{\partial}{\partial x_i} \int_0^1 J_i^{(j)'}(U_0 + tu)\, dt\, v, \quad j = 1, \ldots, M_1, \quad (3.19)$$

*the operator defined by*

$$\tilde{A}^*(u)\, w = A^*(u)\, L^* w - \sum_{j=1}^{N_1} \sum_{i=1}^{n} \int_0^1 (J_{0,i}^{(j)'}(U_0 + tu))^* \, dt\, \frac{\partial}{\partial x_i} L_j^{(0)^*} w$$

$$- \sum_{j=1}^{M_1} \sum_{i=1}^{n} \int_0^1 (J_i^{(j)'}(U_0 + tu))^* \, dt\, \frac{\partial}{\partial x_i} L_j^* w \quad (3.20)$$

*with the domain $D(\tilde{A}^*) = D(A)$ is a restriction on $D(A)$ of the adjoint operator corresponding to the equation $\Phi(U) = 0$.*

*Proof* of this lemma follows from the representations (3.18), (3.19), equalities (3.15), (3.16) and the concrete form of the operators $\{A_j^*\}$, $\{A_j^{(0)^*}\}$ being constructed with the use of conventional integration by parts in the variables $x_i$, $i = 1, \ldots, n$.

**3.4.** Let us give some concrete examples for choosing the operators $L$, $\{L_j\}$, $\{L_k^{(0)}\}$.

*Example* 3.1. Let

$$L = \bar{\alpha} A^*(u), \quad L_j = \bar{\beta}_j A_j^*(u), \quad L_k^{(0)} = \bar{\gamma}_k A_k^{(0)^*}(u), \tag{3.21}$$

where $\alpha$, $\beta_j$, $\gamma_k$ are some complex numbers, and $\bar{\alpha}$, $\bar{\beta}_j$ and $\bar{\gamma}_k$ the respective conjugate values. Then the operator $\tilde{A}^*(u)$ has the form

$$\tilde{A}^*(u) = \alpha A^*(u) A(u) + \sum_{j=1}^{M} \beta_j A_j^*(u) A_j(u) + \sum_{k=1}^{K} \gamma_k A_k^{(0)^*}(u) A_k^{(0)}(u). \tag{3.22}$$

The original and adjoint equations are

$$\tilde{A}(u) u = \bar{\alpha} A^*(u) f + \sum_{j=1}^{M} \bar{\beta}_j A_j^*(u) f_j, \tag{3.23}$$

$$\tilde{A}^*(u) u^* = g^*. \tag{3.24}$$

*Example* 3.2. Let

$$L = \bar{\alpha}(\Phi'(U_0))^*, \quad L_j = \bar{\beta}_j(\Phi_j'(U_0))^*, \quad L_j^{(0)} \equiv 0. \tag{3.25}$$

Then

$$\tilde{A}^*(u) w = \alpha A^*(u) \Phi'(U_0) + \sum_{j=1}^{M} \beta_j A_j^*(u) \Phi_j'(U_0), \tag{3.26}$$

and the corresponding equations are

$$\tilde{A}(u) u = \bar{\alpha}(\Phi'(U_0))^* f + \sum_{j=1}^{M} \bar{\beta}_j(\Phi_j'(U_0))^* f_j, \tag{3.27}$$

$$\tilde{A}^*(u) u^* = g^*. \tag{3.28}$$

Note that the operator $\tilde{A}^*(0)$ is defined here by

$$\tilde{A}^*(0) w = \alpha(\Phi'(U_0))^* \Phi'(U_0) + \sum_{j=1}^{M} \beta_j(\Phi_j'(U_0))^* \Phi_j'(U_0). \tag{3.29}$$

It coincides with the operator $\widetilde{A}^*(0)$, where $\widetilde{A}^*$ is constructed in Example 3.1.

*Example* 3.3. Let all the operators $L$, $L_j$, $L_k^{(0)}$ be the operators of multiplication by the constants $\alpha$, $\beta_j$, $\gamma_k$. Then

$$\widetilde{A}^*(u) = \alpha A^*(u) + \sum_{j=1}^{M} \beta_j A_j^*(u) + \sum_{k=1}^{K} \gamma_k A_k^{(0)^*}(u). \tag{3.30}$$

If $\gamma_k \equiv 0$, then

$$\widetilde{A}^*(u) = \alpha A^*(u) + \sum_{j=1}^{M} \beta_j A_j^*(u). \tag{3.31}$$

*Example* 3.4. Let $L = \bar{\alpha} E$, $L_j = \bar{\beta}_j E$, $L_k^{(0)} \equiv 0$, where $E$ is the identity operator, and the operator $\widetilde{A}^*(u)$ has the form (3.20). Then

$$\widetilde{A}^*(u) = \alpha A^*(u) - \sum_{j=1}^{M_1} \beta_j \sum_{i=1}^{n} \int_0^1 (J_i^{(j)'}(U_0 + tu))^* \, dt \frac{\partial}{\partial x_i}. \tag{3.32}$$

If we put $\alpha = 0$ in (3.32) and assume that the currents $J_i^{(j)}$ do not depend on the derivatives of $U_0 + tu$ with respect to $x_i$, $i = 1, \ldots, n$, we get

$$\widetilde{A}^*(u) = - \sum_{j=1}^{M_1} \beta_j (\mathbf{B}_j, \nabla),$$

where

$$\mathbf{B}_j = (B_{j,1}, \ldots, B_{j,n}), \quad B_{j,i} = \int_0^1 (J_i^{(j)'}(U_0 + tu))^* \, dt, \quad \nabla = (\partial/\partial x_1, \ldots, \partial/\partial x_n),$$

that is, the operator $\widetilde{A}^*(u)$ is the sum of the advection operators $(\mathbf{B}_j, \nabla)$.

## 4. RELATIONSHIP BETWEEN DIFFERENT ADJOINT OPERATORS

**4.1.** Let us establish a relationship between different adjoint operators corresponding to the same non-linear operator $\Phi$.

It is known that if $\Phi$ is a non-linear continuously differentiable operator, then it may be represented as

$$\Phi(U) = A_i(u)\, u - f, \tag{4.1}$$

where $U = u + U_0$, $f = -\Phi(U_0)$, $U_0 \in D(\Phi)$, $(i = 1, \ldots, I)$. The integer $I$ may be greater than one, and the operators $A_i(u)$ may differ from each other. Let $I = 2$. Consider the operator $F(u)$:

$$F(u) \equiv A_i(u)\, u, \quad i = 1, 2. \tag{4.2}$$

Assume that the domains $D(A_i)$ of the operators $\{A_i(u)\}$ are dense in $H_0$ and do not depend on the element $u$ determining the operators. The operators are assumed to map $H_1$ into $H_0$. Let us fix $u \in D(A_i)$ and introduce the adjoint operator $A_i^*(u)$:

$$(A_i(u)\, v, w) = (v, A_i^*(u)\, w), \tag{4.3}$$

where $(\cdot, \cdot) = (\cdot, \cdot)_{H_0}$, $u, v \in D(A_i)$, $w \in D(A_i^*)$, $i = 1, 2$. Consider also the operator

$$\mathcal{L}(u) = A_1^*(u) - A_2^*(u). \tag{4.4}$$

From (4.2) and (4.3) we get

$$\mathcal{L}^*(u)\, u = A_1(u)\, u - A_2(u)\, u = F(u) - F(u) = 0,$$

that is, the operator $\mathcal{L}^*(u)\, v$ is trivial. The adjoint operator is always closed. Then, assuming that the above-introduced operators $\{A_j^{(0)}(u)\}$ are closed and taking into account the fact that the system $\{A_j^{(0)}(u)\}$ is supposed to be minimal, we obtain the following representation for $\mathcal{L}^*(u)$:

$$\mathcal{L}^*(u) = \sum_{j=1}^{\infty} \alpha_j A_j^{(0)}(u) \tag{4.5}$$

(here we set $N = \infty$) for some coefficients $\{\alpha_j\}$. Thus, if the system $\{A_j^{(0)}(u)\}$ constitutes a basis in the space of trivial operators with the domains coinciding with $D(A_1) \cap D(A_2)$ and independent of $u$, we come to the following statement.

**Proposition 4.1.** *Let*

(1) $A_1(u)$, $A_2(u)$: $H_1 \to H_0$ *be linear operators with the same domain $D(A)$ dense in $H_0$ such that $F(u) = A_1(u)\, u = A_2(u)\, u$;*

(2) $A_1^*(u)$, $A_2^*(u)$ *be adjoint operators satisfying the equalities $(A_i(u)\, v, w) = (v, A_i^*(u)\, w)$, $i = 1, 2$, $w \in D(A^*) \equiv D(A_1^*) \cap D(A_2^*)$ with $D(A^*)$ dense in $H_0$ and independent of $u \in D(A)$;*

(3) $\mathcal{L}^{(0)}(H_1, H_0)$ *be the space of linear trivial operators mapping $H_1$ into $H_0$ with a countable basis $\{A_j^{(0)}(u)\}$, $u \in D(A)$.*

*Then the adjoint operators $A_1^*(u)$, $A_2^*(u)$ satisfy the equality*

$$A_1^*(u) = A_2^*(u) + \sum_{j=1}^{\infty} c_j A_j^{(0)^*}(u) \tag{4.6}$$

*with some constants $\{c_j\}$.*

*Remark* 4.1. As noted above, from here on, when considering linear combinations of some operators $\{C_i\}$ and their adjoints $\{C_i^*\}$, we assume that the equality

$$\left(\sum_i \alpha_i C_i\right)^* = \sum_i \bar{\alpha}_i C_i^*$$

holds. To satisfy this equality it is sufficient that all the operators (except, possibly, one) be bounded in the corresponding spaces.

*Example* 4.1. Let $x \in \Omega \subset \mathbf{R}^1$ and the operators $F(u)$ have the form

$$F(u) = \sum_k^{|\bar{k}|<\infty} F_{\bar{k}}(x, u),$$

where $\bar{k} = (k_0, k_1, \ldots, k_n)$, $k_i$ is an integer, and

$$|\bar{k}| = k_0 + k_1 + \ldots + k_n,$$

$$F_{\bar{k}}(x, u) = a_{\bar{k}}(x)(u)^{k_0}\left(\frac{\partial u}{\partial x}\right)^{k_1} \cdots \left(\frac{\partial^n u}{\partial x^n}\right)^{k_n};$$

$a_{\bar{k}}(x) = a_{k_0 k_1 \ldots k_n}(x)$ is a regular function.

As a domain of $F_{\bar{k}}$ and $F$ we take the set of infinitely differentiable functions with a finite support in $\Omega$. Let $H_1 = H_0 = L_2(\Omega)$. Consider the operator $F_{\bar{k}}$. Let

$$F_{\bar{k}}(u) = A_i(u)\, u, \quad i = 1, 2,$$

with

$$A_1(u)\, v = \left\{a_{\bar{k}}(x)(u)^{k_0} \cdots \left(\frac{\partial^n u}{\partial x^n}\right)^{k_n}\right\} \left(\frac{\partial^p u}{\partial x^p}\right)^{k_p} \left(\frac{\partial^q u}{\partial x^q}\right)^{k_q - 1} \frac{\partial^q v}{\partial x^q},$$

$$A_2(u)\, v = \left\{a_{\bar{k}}(x)(u)^{k_0} \cdots \left(\frac{\partial^n u}{\partial x^n}\right)^{k_n}\right\} \left(\frac{\partial^p u}{\partial x^p}\right)^{k_p - 1} \left(\frac{\partial^q u}{\partial x^q}\right)^{k_q} \frac{\partial^p v}{\partial x^p},$$

where the product in braces $\theta(u) \equiv a_{\bar{k}}(x)\, u^{k_0} \ldots (\partial^n u/\partial x^n)^{k_n}$ does not include the derivatives $\partial^p u/\partial x^p$, $\partial^q u/\partial x^q$. The adjoint operators (or more precisely, their restrictions on $D(F)$) have the form:

$$A_1^*(u)\, w = (-1)^q \frac{\partial^q}{\partial x^q}\left[\theta(u)\left(\frac{\partial^p u}{\partial x^p}\right)^{k_p}\left(\frac{\partial^q u}{\partial x^q}\right)^{k_q - 1} w\right],$$

$$A_2^*(u)\, w = (-1)^p \frac{\partial^p}{\partial x^p}\left[\theta(u)\left(\frac{\partial^p u}{\partial x^p}\right)^{k_p - 1}\left(\frac{\partial^q u}{\partial x^q}\right)^{k_q} w\right].$$

For $\mathcal{L} = A_1^*(u) - A_2^*(u)$ we get

$$\mathcal{L}^*(u)\, v = \theta(u) \left(\frac{\partial^p u}{\partial x^p}\right)^{k_p-1} \left(\frac{\partial^q u}{\partial x^q}\right)^{k_q-1} \left[\frac{\partial^p u}{\partial x^p} \frac{\partial^q v}{\partial x^q} - \frac{\partial^q u}{\partial x^q} \frac{\partial^p v}{\partial x^p}\right],$$

$$\mathcal{L}^*(u)\, u = 0,$$

that is, $\mathcal{L}^*(u)$ is a trivial operator, and $A_1^*(u)$ and $A_2^*(u)$ differ by the operator $\mathcal{L}$ being adjoint to a trivial one.

It is easy to see that this example can be extended to the overall operator $F$ and to the polynomial operators $F$ mapping the functions of several variables $x_1, \ldots, x_n$.

**4.2.** Consider the operator $\Phi$ of the form

$$\Phi(U) = \sum_{i=1}^{n} \frac{\partial}{\partial x_i} \{\Phi_i(x, U)\} + \Phi_0 U + F(x), \qquad (4.7)$$

where $\partial/\partial x_i$ is the operator of differentiating in $x_i$, $\Phi_i(x, U)$ an analytical function of $x$ and $U$, and $\Phi_0$ a linear operator.

From (4.7) we come to

$$\Phi(u + U_0) = \sum_{i=1}^{n} \frac{\partial}{\partial x_i} F_i(x, u) + \Phi_0 u - f(x), \qquad (4.8)$$

where

$$u = U - U_0, \quad U, U_0 \in D(\Phi),$$

$$F_i(x, u) = A_i(x, u)\, u, \quad A_i(x, u) = \int_0^1 \Phi_i'(x, U_0 + tu)\, dt.$$

This subsection studies the relation between the adjoint operators corresponding to the operator

$$F(u) \equiv \sum_{i=1}^{n} \frac{\partial}{\partial x_i} F_i(x, u) + \Phi_0 u. \qquad (4.9)$$

We assume that the domain of this operator, and of the below-introduced operators (adjoint operators among them), is the set of infinitely differentiable functions with the supports in a bounded domain $\Omega \subset \mathbf{R}^n$. By virtue of this fact, when writing 'the adjoint operator' we mean the restriction of the original adjoint operator on $D(F)$.

Assume also that $F_i(x, u)$ and $\Phi_0 u$ are functions of $x = (x_1, \ldots, x_n)$, $u$, and of the derivatives of $u$ (up to an order $K < \infty$, inclusive).

Let us consider $k$-linear operators $Q_{il}^{(k)}(x, u_1, \ldots, u_k)$: $(H_1)^k \to H_0$ satisfying the conditions

$$Q_{il}^{(k)} = -Q_{li}^k. \qquad (4.10)$$

We assume that these operators are symmetric, i.e.

$$Q_{il}^{(k)}(x, \ldots, u_p, \ldots, u_q, \ldots) = Q_{il}^{(k)}(x, \ldots, u_q, \ldots, u_p, \ldots) \quad \forall\, p, q,$$

and $k = 1, 2, \ldots$ Using $Q_{il}^{(k)}$, we construct the operator

$$A_i^{(0)}(x, u)\, v = \sum_{k=1}^{\infty} \frac{1}{k!} \sum_{l=1}^{n} \frac{\partial}{\partial x_i} \{Q_{il}^{(k)} u^{k-1} v\}, \tag{4.11}$$

where $Q_{il}^{(k)} u^k = Q_{il}^{(k)}(x, u, \ldots, u)$ is a $k$-linear operator, and

$$Q_{il}^{(k)} u^{k-1} v = Q_{il}^{(k)}(x, u, \ldots, u, v) \equiv (Q_{il}^{(k)} u^{k-1})\, v,$$

that is, $Q_{il}^{(k)} u^{k-1}$ is a linear operator mapping $H_1$ into $H_0$. Assume that the series (4.11) is convergent absolutely and uniformly on $D(F)$.

Note that

$$\sum_{i=1}^{n} \frac{\partial}{\partial x_i} \{A_i^{(0)}(x, u)\, u\} = \sum_{i=1}^{n} \frac{\partial}{\partial x_i} \{Q_{il}(x, u)\}, \tag{4.12}$$

where

$$Q_{il}(x, u) = \sum_{k=1}^{\infty} \frac{1}{k!} Q_{il}^{(k)}(x, u, \ldots, u) = \sum_{k=1}^{\infty} \frac{1}{k!} Q_{il}^{(k)} u^k$$

$$= -\sum_{k=1}^{\infty} \frac{1}{k!} Q_{li}^{(k)}(x, u, \ldots, u) = -Q_{li}(x, u). \tag{4.13}$$

Then[229], we get

$$\begin{aligned}
&\sum_{i=1}^{n} \frac{\partial}{\partial x_i} \{A_i^{(0)}(x, u)\, u\} = \sum_{i=1}^{n} \frac{\partial}{\partial x_i} \{Q_{il}(x, u)\} = 0, \\
&\sum_{i=1}^{n} \frac{\partial}{\partial x_i} F_i(x, u) + \Phi_0 u = \sum_{i=1}^{n} \frac{\partial}{\partial x_i} \tilde{F}_i(x, u) + \Phi_0 u,
\end{aligned} \tag{4.14}$$

where

$$\tilde{F}_i(x, u) = (A_i(x, u) + A_i^{(0)}(x, u))\, u = \tilde{A}_i(x, u)\, u,$$

$$\tilde{A}_i(x, u) = A_i(x, u) + A_i^{(0)}(x, u).$$

Therefore, the operator $F$ may be associated with the following operators:

$$\begin{aligned}
&A^*(u)\, w = -\sum_{i=1}^{n} (A_i(x, u))^* \frac{\partial w}{\partial x_i} + \Phi_0^* w, \\
&\tilde{A}(u)\, w = -\sum_{i=1}^{n} (A_i^*(x, u) + (A_i^{(0)})^* (x, u)) \frac{\partial w}{\partial x_i} + \Phi_0^* w,
\end{aligned} \tag{4.15}$$

where

$$A_i^{(0)^*}(x,u)\,w = -\sum_{k=1}^{\infty}\frac{1}{k!}\sum_{l=1}^{n}(Q_{il}^{(k)}u^{k-1})^*\frac{\partial w}{\partial x_i} = -\sum_{l=1}^{n}A_{il}^{(0)^*}(x,u)\frac{\partial w}{\partial x_i},$$

$$A_{il}^{(0)^*}(x,u) = \sum_{k=1}^{\infty}\frac{1}{k!}(Q_{il}^{(k)}u^{k-1})^* = -A_{li}^{(0)^*}(x,u).$$

On the other hand, we have for regular $F_i(x,u)$:

$$\sum_{i=1}^{n}\frac{\partial}{\partial x_i}F_i(x,u) = \sum_{i=1}^{n}\frac{\partial}{\partial x_i}\widetilde{F}_i(x,u). \qquad (4.16)$$

Then

$$\widetilde{F}_i(x,u) - F_i(x,u) = \sum_{l=1}^{n}\frac{\partial}{\partial x_l}\{Q_{il}(x,u)\},$$

where $Q_{il}(x,u)$ are regular functions depending on $x$, $u$, and the derivatives of $u$, such that $Q_{il}(x,u) = Q_{li}(x,u)$, $i,l = 1,\ldots,n$, for any $x$, $u$, $\partial u/\partial x_1,\ldots,\partial^k u/\partial x_n^k$. Let us represent $Q_{il}(x,u)$ as

$$Q_{il}(x,u) = Q_{il}(x,0) + \sum_{k=1}^{\infty}\frac{1}{k!}Q_{il}^{(k)}(x,0)\,u^k,$$

where $Q_{il}^{(k)}(x,0)$ is a $k$-linear operator. Without loss of generality, we may assume that $Q_{ik}$ is generated by the $k$-linear symmetric operator $Q_{il}^{(k)}(x,u_1,\ldots,u_k)$. Note that $Q_{ik}^{(k)} = -Q_{li}^{(k)}$, $k = 1,2,\ldots$ If $F_i(x,0) = \widetilde{F}_i(x,0) = 0$, then the equality

$$\sum_{l=1}^{n}\frac{\partial}{\partial x_l}\{Q_{il}(x,0)\} = 0$$

is bound to hold. Thus, if we obtain one of the following representations for $F(u)$:

$$F(u) = \sum_{i=1}^{n}\frac{\partial}{\partial x_i}F_i(x,u) + \Phi_0 u, \qquad F_i(x,u) = A_i(x,u)\,u,$$

$$F(u) = \sum_{i=1}^{n}\frac{\partial}{\partial x_i}\widetilde{F}_i(x,u) + \Phi_0 u, \qquad \widetilde{F}_i(x,u) = \widetilde{A}_i(x,u)\,u,$$

then

$$\widetilde{F}_i(x,u) \equiv \widetilde{A}_i(x,u)\,u = F_i(x,u) + \sum_{l=1}^{n}\frac{\partial}{\partial x_l}\{A_{il}^{(0)}(x,u)\,u\},$$

where

$$A_{il}^{(0)}(x, u) \equiv \sum_{k=1}^{\infty} \frac{1}{k!} Q_{il}^{(k)} u^{k-1} = -A_{li}^{(0)}(x, u). \tag{4.17}$$

Therefore, each operator $F_i$ may be associated with the operator

$$\widetilde{A}_i(x, u) \, v = \left( A_i(x, u) + \sum_{l=1}^{n} \frac{\partial}{\partial x_l} A_{il}^{(0)}(x, u) \right) v, \tag{4.18}$$

such that

$$F(u) = \sum_{i=1}^{n} \frac{\partial}{\partial x_i} \{ A_i(x, u) \, u \} + \Phi_0 u$$

$$= \sum_{i=1}^{n} \frac{\partial}{\partial x_i} \{ \widetilde{A}_i(x, u) \, u \} + \Phi_0 u \tag{4.19}$$

and (4.17) is valid for $A_{il}^{(0)}(x, u)$.

Assume now that the operator $F(u)$ may be associated with the adjoint operator $\widetilde{A}^*(u)$ of the form:

$$\widetilde{A}^*(u) \, w = -\sum_{i=1}^{n} \widetilde{A}_i^*(x, u) \frac{\partial w}{\partial x_i} + \Phi_0^* w, \tag{4.20}$$

i.e.

$$(F(u) \, w) = \left( u, -\sum_{i=1}^{n} \widetilde{A}_i^* \frac{\partial w}{\partial x_i} + \Phi_0^* w \right), \tag{4.21}$$

with some operators $\{ \widetilde{A}_i^* \}$. Then

$$(F(u), w) = \left( \sum_{i=1}^{n} \frac{\partial}{\partial x_i} [A_i(x, u) \, u] + \Phi_0 u, w \right)$$

$$= \left( \sum_{i=1}^{n} \frac{\partial}{\partial x_i} [(\widetilde{A}_i^*(x, u))^* \, u] + \Phi_0 u, w \right).$$

Hence,

$$\sum_{i=1}^{n} \frac{\partial}{\partial x_i} \{ A_i(x, u) \, u \} = \sum_{i=1}^{n} \frac{\partial}{\partial x_i} \{ (\widetilde{A}_i^*(x, u))^* \, u \},$$

and in view of (4.18), we conclude that the operator $\widetilde{A}^*(u)$ may be associated with the operators $(\widetilde{A}_i^*(x, u))^* \equiv \widetilde{A}_i(x, u)$ (we denote them by $\widetilde{A}_i(x, u)$) such that

$$\sum_{i=1}^{n} \frac{\partial}{\partial x_i}\{\tilde{A}_i(x,u)\,u\} + \Phi_0 u \;=\; \sum_{i=1}^{n} \frac{\partial}{\partial x_i}\{A_i(x,u)\,u\} + \Phi_0 u$$

$$+ \sum_{i=1}^{n} \frac{\partial}{\partial x_i}\left\{\sum_{l=1}^{n} \frac{\partial}{\partial x_l}\{A_{il}^{(0)}(x,u)\,u\}\right\}$$

$$= \sum_{i=1}^{n} \frac{\partial}{\partial x_i}\{A_i(x,u)\,u\} + \Phi_0 u = F(u).$$

Since $A_{il}^{(0)}(x,u)\,u = -A_{li}^{(0)}(x,u)\,u$, then

$$\sum_{i=1}^{n} \frac{\partial}{\partial x_i}\left\{\sum_{l=1}^{n} \frac{\partial}{\partial x_l}\{A_{il}^{(0)}(x,u)\,u\}\right\} = 0.$$

Thus, we arrive at the following theorem.

**Theorem 4.1.** *The operator*

$$\tilde{A}^*(x,u)\,w = -\sum_{i=1}^{n} \tilde{A}_i^*(x,u)\frac{\partial w}{\partial x_i} + \Phi_0^* w$$

*is an adjoint operator corresponding to the operator $F(u)$ defined by (4.9) if and only if the operators $(\tilde{A}_i^*(x,u))^* \equiv \tilde{A}_i(x,u)$, $i = 1,\dots,n$, satisfy the equalities*

$$\tilde{A}_i(x,u)\,u = \left(A_i(x,u) + \sum_{l=1}^{n} \frac{\partial}{\partial x_l}A_{il}^{(0)}(x,u)\right)u \quad \forall\, u \in D(F), \qquad (4.22)$$

*for some operators $A_{il}^{(0)}(x,u)$ depending on $x$, $u$, and the derivatives of $u$, such that $A_{il}^{(0)} = -A_{li}^{(0)}$.*

Differentiating (4.22) in the Gâteaux sense at the point $u = 0$, we come to the following corollary.

**Corollary.** *If the operator*

$$A^*(x,u)\,w = -\sum_{i=1}^{n} \tilde{A}_i^*(x,u)\frac{\partial w}{\partial x_i} + \Phi_0^* w$$

*is an adjoint operator corresponding to the operator (4.9), then the following equalities hold:*

$$\tilde{A}_i(x,0) = \Phi_i'(x,U_0) + \sum_{l=1}^{n} \frac{\partial}{\partial x_l}Q_{il}^{(1)}(x), \qquad (4.23)$$

where $\{Q_{il}^{(1)}(x)\}$ *are linear operators regular in* $x$ *and such that* $Q_{il}^{(1)} = -Q_{li}^{(1)}$, $i, l = 1, \ldots, n,$ $\tilde{A}_i(x, 0) = \tilde{A}_i^{**}(x, 0)$. *Conversely, an operator adjoint to the operator of the form (4.18), such that (4.23) is satisfied, is an adjoint operator corresponding to (4.9) under the above restrictions for* $\{Q_{il}^{(k)}\}$, $k = 2, 3, \ldots$

**4.3.** Consider the operators $\Phi$, $\{\Phi_j\}$, equations (1.5), (1.6) and introduce the following 'transformed' operator:

$$\tilde{\Phi}(U) = \tilde{F}(u) - \tilde{f}, \qquad (4.24)$$

where

$$\tilde{F}(u) \equiv \tilde{A}(u)\, u, \quad \tilde{f} = Lf + \sum_{j=1}^{M} L_j f_j,$$

$$\tilde{A}(u) = LA(u) + \sum_{j=1}^{M} L_j A_j(u),$$

$$A(u) = \int_0^1 \Phi'(U_0 + tu)\, \mathrm{d}t, \quad f = -\Phi(U_0),$$

$$A_j(u) = \int_0^1 \Phi_j'(U_0 + tu)\, \mathrm{d}t, \quad f_j = -\Phi_j(U_0),$$

$$u = U - U_0, \quad U, U_0 \in D(\Phi).$$

When considering the operators

$$F(u) = A(u)\, u, \quad F_j(u) = A_j(u)\, u, \qquad (4.25)$$

other operators $\tilde{A}(u)$, $\tilde{A}_j(u)$ may be constructed such that

$$F(u) = A(u)\, u, = \tilde{A}(u)\, u, \qquad (4.26)$$

$$F_j(u) = A_j(u)\, u = \tilde{A}_j(u)\, u. \qquad (4.27)$$

Then we come to the operator (4.24) with

$$\tilde{F}(u) = \tilde{\tilde{A}}(u)\, u, \qquad (4.28)$$

where

$$\tilde{\tilde{A}}(u) = L\tilde{A}(u) + \sum_{j=1}^{M} L_j \tilde{A}_j(u). \qquad (4.29)$$

Introducing the adjoint operators $\tilde{A}^*$, $\approx\!\!\!A^*$, under the hypotheses of Proposition 4.1, we get the equality

$$\tilde{A}(u) L^* + \sum_{j=1}^{M} \tilde{A}_j(u) L_j^* = A^*(u) L^* + \sum_{j=1}^{M} A_j^*(u) L_j^* + \sum_{j=1}^{\infty} c_j A_j^{(0)^*}(u), \quad (4.30)$$

where $\{A_j^{(0)}(u)\}$ is a basis of trivial operators, and $\{c_j\}$ some constants. Hence, in particular, we obtain

$$\tilde{A}^*(u) = A^*(u) + \sum_{j=1}^{\infty} \alpha_j A_j^{(0)^*}(u), \quad (4.31)$$

$$\tilde{A}_j^*(u) = A_j^*(u) + \sum_{j=1}^{\infty} \gamma_i^{(j)} A_i^{(0)^*}(u), \quad j = 1, \ldots, M. \quad (4.32)$$

From (4.30)–(4.32) we come to the following conclusion. Suppose that we are required to solve the adjoint equation $\approx\!\!\!A^*(u) u^* = g^*$, where $\approx\!\!\!A^*(u)$ is the operator adjoint to the operator $\approx\!\!\!A(u)$ such that $\tilde{F}(u) = \tilde{A}(u) u = \approx\!\!\!A(u) u$ for $u \in D(\tilde{F}) = D(F)$. A reasonable question arises of a choice of the operator $\approx\!\!\!A(u)$ such that $\approx\!\!\!A^*(u)$ have the 'best' ( in a proper sense) properties and satisfy the adjointness relation $(\tilde{F}(u), w) = (u, \approx\!\!\!A^*(u) w)$. How can one construct such an operator? The relationship (4.30) gives one of the possible ways to answer this question. Adding the linear combination of trivial operators $\sum_j c_j A_j^{(0)}(u)$ to the operator

$$\tilde{A}(u) = L \int_0^1 \Phi'(u_0 + tu) \, dt + \sum_{j=1}^{M} L_j \int_0^1 \Phi_j'(U_0 + tu) \, dt$$

with properly chosen operators $L$, $\{L_j\}$, and coefficients $\{c_j\}$, one can try to construct the operators $\approx\!\!\!A(u)$, $\approx\!\!\!A^*(u)$ with the required properties.

## 5. GENERAL REMARKS ON CONSTRUCTING THE ADJOINT EQUATIONS WITH THE USE OF THE LIE GROUPS AND CONSERVATION LAWS

**5.1.** As noted above, if for a prescribed element $U_0 \in D(\Phi)$ the solution of the equation $A(u) u = f$ is found, then we can find the solution of the original equation (1.5) by $U = U_0 + u$. However, to find the element $u$ one can consider equation (2.7) with the operator (2.8). It is reasonable here to choose the operators $L$, $L_j$, $L_j^{(0)}$ such that one will be able to solve equation (2.7) in a 'simple' way (in a sense). If we put $L = I$, $L_j = 0$, $L_j^{(0)} = 0$, we obtain the equation $A(u) u = f$.

The choice of $L$, $L_j$, $L_j^{(0)}$ has an impact on constructing the adjoint operator corresponding to the original non-linear operator. This suggests the importance of finding the form of the operators $\{\Phi_j\}$, the conservation laws among them, for each concrete problem. In this context, some remarks need to be made on the ways of finding the conservation laws. One of the general procedures for constructing the conservation laws is to add the relations of the form $\sum_{i=1}^{n} J_i^{(j)}(u) = 0$ with the unknown $\{J_i^{(j)}\}$ to the main equation and study the problem of compatibility for the complete overdefined system. However, as noted in[142], this approach leads to rather cumbersome calculations which may be brought rarely to the end. Thus, this way is of little use.

On the other hand, in many cases the solution of the equations under consideration reduces to solving some variational problems. Then, for the equations obtained when considering variational problems, one can succeed in deriving the conservation laws, using transformation groups they admit, the Lie groups among them. In a sufficiently general form, this question was studied by Noether who started from the concept of the invariant functional given in the form of variational integral. This method turned out to be fruitful and gained acceptance in recent years. We point out here also the way of constructing the conservation laws, using the currents, and developed by V.Vladimirov and I.Volovich[310,311].

To summarize, we can direct some lines of investigations closely connected with the questions under consideration in relation to adjoint operators in non-linear problems. These lines are:

(1) formulation of the non-linear problem under consideration as a variational problem, and, in particular, as a conditional extremum problem;

(2) searching for the transformation groups leaving the functionals of the variational problem to be invariant;

(3) the Noether-type theorems and applications for constructing the conservation laws;

(4) constructing the operators $A(u)$, $A_j(u)$, $A_j^{(0)}$, choosing the operators $L$, $L_j$, $L_j^{(0)}$, constructing the operator $\widetilde{A}$ by formula (2.8), and solving also the problem of the optimal choice of $L$, $L_j$, $L_j^{(0)}$.

Thus, from the above discussion it appears that the solution of the problem of constructing the operators $\widetilde{A}(u)$, $\widetilde{A}^*(u)$ is closely connected with a number of fields of non-linear differential equation theory, variational calculus, and others. In its turn, the question of constructing the operators $\widetilde{A}(u)$, $\widetilde{A}^*(u)$ with 'good' properties is intimately related with perturbation algorithms and other numerical methods.

**5.2.** To illustrate the reasoning of Subsection 5.1 let us use the variational problem on finding the extremum of the functional

$$J(U) = \int_{\Omega} \mathcal{L}(x, U, U_{x_1}, \ldots, U_{x_n}) \, dx, \tag{5.1}$$

where $x = (x_1, \ldots, x_n) \in \Omega \subset \mathbf{R}^n$, $U_{x_i} = \partial U / \partial x_i$. Let equation (1.5) be the Euler equation

$$\Phi(U) \equiv \mathcal{L}_U - \sum_{i=1}^{n} \frac{\partial}{\partial x_i} \mathcal{L}_{U_{x_i}} = 0, \tag{5.2}$$

where $\mathcal{L}_U = \partial \mathcal{L} / \partial U$, $\mathcal{L}_{U_{x_i}} = \partial \mathcal{L} / \partial U_{x_i}$. Let us have the one-parameter transformation group

$$\tilde{x}_i = \tilde{\varphi}_i(x, U, U_{x_1}, \ldots, U_{x_n}, a) \sim x_i + a\varphi_i,$$
$$\tilde{U} = \tilde{\psi}(x, U, U_{x_1}, \ldots, U_{x_n}, a) \sim U + a\psi, \tag{5.3}$$

where

$$\varphi_i = \varphi_i(x, U, U_{x_1}, \ldots, U_{x_n}),$$
$$\psi = \psi(x, U, U_{x_1}, \ldots, U_{x_n}), \tag{5.4}$$

and $U$ is the solution to (5.2). Assume that the functional $J(U)$ is invariant with respect to the transformation (5.3). Then, according to the Noether theorem, there exists the conservation law of the form

$$\sum_{i=1}^{n} \frac{\partial}{\partial x_i} J_i^{(1)}(x, U, \ldots) = 0, \tag{5.5}$$

where

$$J_i^{(1)} = \mathcal{L}_{U_{x_i}} \left( \psi - \sum_{j=1}^{n} \varphi_j U_{x_j} \right) + \mathcal{L}\varphi_i. \tag{5.6}$$

The equation (2.7) for $L_j^{(0)} \equiv 0$ takes the form

$$\tilde{A}(u) \, u = Lf + L_1 f_1, \tag{5.7}$$

where

$$\tilde{A}(u) = LA(u) + L_1 A_1(u),$$

$$A(u) = \int_0^1 \left( \mathcal{L}_U - \sum_{i=1}^{n} \frac{\partial}{\partial x_i} \mathcal{L}_{U_{x_i}} \right)' (U_0 + tu) \, dt,$$

$$A_1(u) = \int_0^1 \sum_{i=1}^{n} \frac{\partial}{\partial x_i} \left( \mathcal{L}_{U_{x_i}} \left( \psi - \sum_{j=1}^{n} \varphi_j U_{x_j} \right) + \mathcal{L}\varphi_i \right) (U_0 + tu) \, dt,$$

$$f = -\left( \mathcal{L}_U - \sum_{i=1}^{n} \frac{\partial}{\partial x_i} \mathcal{L}_{U_{x_i}} \right)(U_0),$$

$$f_1 = -\sum_{i=1}^{n} \left( \mathcal{L}_{U_{x_i}} \left( \psi - \sum_{j=1}^{n} \varphi_j U_{x_j} \right) + \mathcal{L}\varphi_j \right)(U_0).$$

Introducing a concrete form of the functional $J(U)$, transformations (5.3), and the operators $L$, $\{L_j\}$, $\{L_j^{(0)}\}$ one can transform the operators (5.8) further and try to construct the adjoint operator $\widetilde{A}^*(u)$ with required properties.

## 6. CONSTRUCTION OF ADJOINT OPERATORS WITH PRESCRIBED PROPERTIES

**6.1.** As noted above, the operator $A(u)$ (as well as the operator $A^*(u)$) may be defined in different ways. This raises the question: what adjoint operator is 'best'? To answer this question one needs to give the definition of the 'best adjoint operator'. If our aim is to construct an adjoint operator with some prescribed properties and we have found such an operator, then it may be said to be best (not 'worse' than other operators with the same property). Suppose that the above-mentioned property is symmetricity. Consider some approaches that may be useful when constructing adjoint operators with the symmetricity property.

Let the operator $F(u) \equiv \widetilde{A}(u)\, u$ be given. Introduce the adjoint operator $\widetilde{A}^*(u)$ satisfying the equality

$$(\widetilde{A}(u)\, u, w) = (u, \widetilde{A}^*(u)\, w), \quad w \in D(\widetilde{A}^*),$$

and require that the relation

$$(v, \widetilde{A}^*(u)\, w) = (\widetilde{A}^*(u)\, v, w), \quad v, w \in D(\widetilde{A}^*) \tag{6.1}$$

be valid also for $\widetilde{A}^*(u)$. In the subsequent consideration we agree that the domain $D(\widetilde{A})$ of the operator $\widetilde{A}$ belongs to the domains of the adjoint operators to be introduced. All the operators are considered on $D(\widetilde{A})$ (i.e., we consider the restrictions of the operators on $D(\widetilde{A})$) and we leave the previous notations for the restrictions.

Let us formulate the condition giving the possibility to construct $\widetilde{A}^*$ with the symmetricity property. Let the operator $\widetilde{A}(u)$ have the form

$$\widetilde{A}(u) = A(u) + \sum_{j=1}^{M} \alpha_j A_j(u) \tag{6.2}$$

with some real numbers $\alpha_j$, where the operators $A(u)$, $A_j(u)$ are defined above. Note that

$$(F(u), w) = (\widetilde{A}(u)\, u, w) = \left( \left( A(u) + \sum_{j=1}^{M} \alpha_j A_j(u) + \sum_{i=1}^{N} \beta_i A_j^{(0)}(u) \right) u, w \right)$$

$$= \left( u, \left( A^*(u) + \sum_{j=1}^{M} \alpha_j A_j^*(u) + \sum_{i=1}^{N} \bar{\beta}_i (A_j^{(0)}(u))^* \right) w \right)$$

$$= \left( u, \left( \widetilde{A}_s(u) + \widetilde{A}_a(u) + \sum_{i=1}^{N} \bar{\beta}_i (A_j^{(0)}(u))^* \right) w \right), \tag{6.3}$$

where $A_j^{(0)}(u)$ are trivial operators, $\beta_j$ some complex numbers, and

$$\widetilde{A}_s = \frac{A(u) + A^*(u)}{2} + \sum_{j=1}^{M} \alpha_j \frac{A_j(u) + A_j^*(u)}{2},$$

$$\widetilde{A}_a = \frac{A^*(u) - A(u)}{2} + \sum_{j=1}^{M} \alpha_j \frac{A_j^*(u) - A_j(u)}{2}.$$

From (6.3) we conclude the following. If there exists a linear combination of trivial operators such that

$$2 \sum_{i=1}^{N} \bar{\beta}_i (A_j^{(0)}(u))^* = (A(u) - A^*(u)) + \sum_{j=1}^{M} \alpha_j (A_j(u) - A_j^*(u)), \tag{6.4}$$

then the adjoint operator defined by

$$\widetilde{A}^*(u) = A^*(u) + \sum_{j=1}^{M} \alpha_j A_j^*(u) + \sum_{i=1}^{N} \bar{\beta}_i A_i^{(0)}(u) \tag{6.5}$$

will be symmetrical (that is, it satisfies (6.1)). In fact, if a linear combination of trivial operators is chosen according to (6.4), then

$$(v, \widetilde{A}^*(u)\, w) = \left( v, \left( A^*(u) + \sum_{j=1}^{M} \alpha_j A_j^*(u) + \sum_{i=1}^{N} \bar{\beta}_j A_i^{(0)}(u) \right) w \right)$$

$$= \left( v, \left( \widetilde{A}_s(u) + \widetilde{A}_a(u) + \sum_{i=1}^{N} \bar{\beta}_i \widetilde{A}_i^{(0)}(u) \right) w \right)$$

$$= (v, (\widetilde{A}_s(u) + \widetilde{A}_a(u) - \widetilde{A}_a(u))\, w)$$

$$= (v, \widetilde{A}_s(u)\, w) = (\widetilde{A}_s(u)\, v, w) = (\widetilde{A}^*(u)\, v, w),$$

i.e., $\widetilde{A}^* = (\widetilde{A}^*)^*$ on $D(A)$. Hence, we can make the following conclusion.

In the original equation (1.5) is reduced to the equation

$$\widetilde{A}(u)\,u = f + \sum_{j=1}^{M} \alpha_j F_j \tag{6.6}$$

for the correction $u = U - U_0$ and it is required to compute the functional

$$\delta J(u) = (u, p); \tag{6.7}$$

then under the condition (6.4), we can consider the adjoint equation

$$\left( A^*(u) + \sum_{j=1}^{M} \alpha_j A_j^*(u) + \sum_{i=1}^{N} \bar{\beta}_i A_i^{(0)}(u) \right) u^* = p. \tag{6.8}$$

If equation (6.8) has a solution $u^*$, then $\delta J(u)$ may be computed by the formula

$$\delta J(u) = \left( f + \sum_{j=1}^{M} \alpha_j f_j, u^* \right). \tag{6.9}$$

Note that the operator of equation (6.8) is here symmetric. This may simplify the numerical solution of the equation (for example, by substituting the zero element for $u$ in the operator).

**6.2.** The foregoing may be obviously extended to the operators $\widetilde{A}(u)$ of the form

$$\widetilde{A}(u) = LA(u) + \sum_{j=1}^{M} L_j A_j(u) + \sum_{i=1}^{N} L_i^{(0)} A_i^{(0)}(u). \tag{6.10}$$

However, we have here a chance of choosing suitable operators $L$, $L_j$, $L_i^{(0)}$. Let us put

$$\begin{aligned} L &= \alpha A^*(0), & L_j &= \beta_j A_j^*(0), & L_i^{(0)} &= \gamma_i (A_i^{(0)}(0))^*, \\ j &= 1, \ldots, M, & i &= 1, \ldots, N, & \alpha, \beta_j, \gamma_i &> 0. \end{aligned} \tag{6.11}$$

Then equation (1.5) reduces to the equation

$$\widetilde{A}(u)\,u = A^*(0)\,f + \sum_{j=1}^{M} \beta_j A_j^*(0)\,f_j. \tag{6.12}$$

To compute (6.7) consider the adjoint equation

$$\widetilde{A}^*(u)\,u^* = p, \tag{6.13}$$

where

$$\widetilde{A}^*(u) = \alpha A^*(u)\,A(0) + \sum_{j=1}^{M} \beta_j\, A_j^*(u)\, A_j(0) + \sum_{i=1}^{N} \gamma_i\, A_j^{(0)*}(u)\, A_j^{(0)}(0).$$

Assuming that $\widetilde{A}^*(u) \cong \widetilde{A}^*(0)$, it is hoped that the solution of the equation

$$\widetilde{A}^*(0)\, u_0^* = p \tag{6.14}$$

may be close to $u^*$. Then

$$\delta J(u) \cong \left( A^*(0)\, f + \sum_{j=1}^{M} \beta_j\, A_j^*(0)\, f_j,\, u_0^* \right). \tag{6.15}$$

However, it is noteworthy that equation (6.14) possesses the symmetric operator

$$\begin{aligned}
\widetilde{A}^*(0) &= \alpha A^*(0)\,A(0) + \sum_{j=1}^{M} \beta_j\, A_j^*(0)\, A_j(0) + \sum_{i=1}^{N} \gamma_i\, A_j^{(0)*}(0)\, A_j^{(0)}(0) \\
&= \alpha(\Phi'(U_0))^*\, \Phi'(U_0) + \sum_{j=1}^{M} \beta_j\, (\Phi_j'(U_0))^*\, \Phi_j'(U_0) \\
&\quad + \sum_{i=1}^{N} \gamma_i\, (A_j^{(0)}(0))^*\, A_j^{(0)}(0),
\end{aligned} \tag{6.16}$$

which has the advantage when solving equation (6.14) numerically.

Note that Section 5 (Chapter 1) presents another approach that may be useful for constructing the adjoint operator $\widetilde{A}^*(u)$ corresponding to the operator $F(u) = A_0 u + F_1(u)$, where $A_0$ is a linear positive definite operator, and $F_1(u)$ is a non-linear operator such that $F_1(u)/u \geq 0$ for $u \in D(F)$ (see Lemma 5.1, Chapter 1).

**6.3.** Consider now the problem of constructing a positive definite adjoint operator. We reformulate here Lemma 1.12 (Chapter 2) that may be helpful.

**Lemma 6.1.** *If $\Phi(U)$ is strictly monotonic, i.e.*

$$(\Phi(U) - \Phi(V), U - V)_{H_0} \geq m\|U - V\|_{H_1}^2, \quad U, V \in D(\Phi), \tag{6.17}$$

*$m = $ constant $> 0$ and $D(A) \subset D(A^*)$, then the operator $A^*(u) \equiv (A(u))^*$ is positive definite on $D(A)$:*

$$(A^*(u)\, v, v)_{H_0} \geq m\|v\|_{H_1}^2 \tag{6.18}$$

*with the same constant $m$.*

**Corollary.** *If the hypotheses of Lemma 6.1 are satisfied and $\|v\|_{H_0} \leq \|v\|_{H_1}$, then the equation $A^*(u)\,w = g$ is correctly solvable, that is, the inequality holds:*

$$\|A^*(u)\,v\|_{H_0} \geq m\|v\|_{H_1} \quad \forall\, v \in D(A^*(u)).$$

Examine now the question of modifying the adjoint operator in such a way that it be positive definite. Consider the equation

$$\tilde{A}(u)\,u = L\left(\alpha f + \sum_{j=1}^{M} \beta_j f_j\right), \tag{6.19}$$

where $\alpha, \beta_j = $ constant $\in \mathbf{R}$ and the operators $\tilde{A}(u)$, $L$ are of the form

$$\tilde{A}(u) = L\left(\alpha A(u) + \sum_{j=1}^{M} \beta_j A_j(u)\right), \tag{6.20}$$

where

$$L = \alpha A^*(u) + \sum_{j=1}^{M} \beta_j A_j^*(u).$$

Assume that the derivatives $\Phi'$, $\Phi_j'$ satisfy the condition

$$\left\|\left(\alpha\Phi'(U) + \sum_{j=1}^{M} \beta_j \Phi_j'(U)\right) - \left(\alpha\Phi'(V) + \sum_{j=1}^{M} \beta_j \Phi_j'(V)\right)\right\|_{H_1 \to H_0}$$

$$\leq K\|U - V\|_{H_1}, \quad K = \text{constant} > 0. \tag{6.21}$$

Then

$$\|A(u) - A(0)\|_{H_1 \to H_0} \leq K \int_0^1 t\|u\|_{H_1}\, dt = \frac{K}{2}\|u\|_{H_1}, \tag{6.22}$$

and

$$(\tilde{A}(u)\,v, v) = \left\|\left(\alpha A(u) + \sum_{j=1}^{M} \beta_j A_j(u)\right)v\right\|_{H_0}^2$$

$$\geq \left(\left\|\left(\alpha A(0) + \sum_{j=1}^{M} \beta_j A_j(0)\right)U\right\|_{H_0} - \frac{K}{2}\|u\|_{H_1}\|v\|_{H_1}\right)^2.$$

If we assume that

$$\left\|\left(\alpha A(0) + \sum_{j=1}^{M} \beta_j A_j(0)\right) v\right\|_{H_0} \geq m\|v\|_{H_1}, \tag{6.23}$$

where

$$m = \text{constant} > K\|u\|_{H_1}/2, \tag{6.24}$$

then

$$(\widetilde{A}(u)\, v, v) \geq (m - K\|u\|_{H_1}/2)^2\|v\|_{H_1}^2. \tag{6.25}$$

Hence, under the conditions (6.21), (6.22), (6.23) the operator $\widetilde{A}(u)$ is positive definite and equation (6.19) is correctly solvable. (The numbers $\alpha$ and $\beta_j$ are assumed to be fixed. They may be chosen so that (6.21) is satisfied in a neighbourhood of the point $U_0 \in D(\Phi)$.) The foregoing implies the following lemma.

**Lemma 6.2.** *If the operator*

$$A = \alpha\Phi'(U_0) + \sum_{j=1}^{M} \beta_j \Phi_j'(U_0) \tag{6.26}$$

*satisfies (6.23) for some constants $\alpha$, $\{\beta_j\}$ and the derivatives $\Phi'$, $\Phi_j'$ satisfy (6.21), then the operator $\widetilde{A}(u) = (\widetilde{A}(u))^*$ of the form (6.20) is positive definite in the ball $S_{2m/K}(U_0) = \{U: \|U_0 - U\|_{H_1} < 2m/K\}$ and equation (6.19) is correctly solvable.*

Similarly to Lemma 6.1 one can prove the following statement.

**Lemma 6.3.** *If the operator*

$$\widetilde{\Phi}(U) = \alpha\Phi(U) + \sum_{j=1}^{M} \beta_j \Phi_j(U) \tag{6.27}$$

*is strictly monotonic for some constants $\alpha$, $\{\beta_j\}$, i.e.,*

$$(\widetilde{\Phi}(U) - \widetilde{\Phi}(V), U - V)_{H_0} \geq m\|U - V\|_{H_1}^2, \quad m = \text{constant} > 0, \tag{6.28}$$

*then the operator (6.20) and the operator (6.26) are positive definite with the same constant $m$.*

To close this section we make the following remark. Lemmas 6.2 and 6.3 give the following algorithm for solving the equation $\Phi(U) = 0$. If we find a set of the operators $\Phi_j$ with $D(\Phi_j) \supset D(\Phi)$, the constants $\alpha$, $\{\beta_j\}$ and the element $U_0 \in D(\Phi)$ such that the hypotheses of these lemmas are satisfied, then we can consider equation (6.19) with the operator (6.20), instead of the equation $\Phi(U) = 0$. The properties of the operator $\widetilde{A}(u) = (\widetilde{A}(u))^*$ given by Lemmas 6.2 and 6.3 may simplify the investigation and numerical solution of equation (6.19).

It may be suggested that considering equation (6.19) and solving this equation numerically will help to construct approximations to $U = u + U_0$ satisfying all the conservation laws $\Phi_j(U) = 0$, $j = 1, \ldots, M$.

## 7. THE NOETHER THEOREM, CONSERVATION LAWS AND ADJOINT OPERATORS

**7.1.** Consider a specific case of equation (1.5) being the Euler equation for the functional

$$J(U) = \int_\Omega \mathcal{L}(x, U, U') \, dx, \tag{7.1}$$

where

$$x = (x_1, \ldots, x_n) \in \Omega \subset \mathbf{R}^n, \quad U' = \left( \frac{\partial U}{\partial x_1}, \ldots, \frac{\partial U}{\partial x_n} \right).$$

We assume that the Lagrange function $\mathcal{L}$ does not depend on the derivatives of $U$ of the order greater than one, and $U$ is a scalar function. This requirement is not crucial, and the subsequent consideration may be extended to a more general case when $\mathcal{L}$ depends on the derivatives of any finite order and $U = (U_1, \ldots, U_p)$. Considering extremal values of the functional (7.1) for any domain $\Omega \subset \mathbf{R}^n$, we obtain the Euler equation

$$\Phi(U) \equiv \sum_{i=1}^n D_i \left( \frac{\partial \mathcal{L}}{\partial U_{x_i}} \right) - \frac{\partial \mathcal{L}}{\partial U} = 0, \tag{7.2}$$

where $D_i$ is the operator of differentiating with respect to $x_i$:

$$D_i = \frac{\partial}{\partial x_i} + \left( \frac{\partial u}{\partial x_i} \right) \frac{\partial}{\partial u} + \sum_{j=1}^n u_{x_i x_j} \frac{\partial}{\partial u_{x_j}} \equiv \frac{\partial}{\partial x_i} \{ \ldots \}, \quad i = 1, \ldots, n.$$

This equation is at most the second order equation for the function $U$ of independent variables $x_i$, $i = 1, \ldots, n$. A solution of the Euler equation (7.2) is said to be an extremal.

Consider continuous $M$-parameter transformation group $G_M$ (the Lie group):

$$\bar{x} = f(x, U, a), \quad \bar{U} = \varphi(x, U, a), \tag{7.3}$$

where

$$\bar{x} = (x_1, \ldots, x_n), \quad f = (f_1, \ldots, f_n), \quad a = (a_1, \ldots, a_M),$$

with the property

$$f(x, U, 0) = x, \quad \varphi(x, U, 0) = U.$$

Basis infinitesimal operators of this group are

$$X_\alpha = \sum_{i=1}^{n} \xi_\alpha^i(x, U) \frac{\partial}{\partial x_i} + \eta_\alpha(x, U) \frac{\partial}{\partial U}, \quad \alpha = 1, \ldots, M, \qquad (7.4)$$

where

$$\xi_\alpha^i(x, U) = \left. \frac{\partial f_i}{\partial a_\alpha} \right|_{a=0}, \quad \eta_\alpha(x, U) = \left. \frac{\partial \varphi}{\partial a_\alpha} \right|_{a=0} = 0, \quad \alpha = 1, \ldots, M.$$

Assume that functional (7.1) is invariant with respect to the group introduced, i.e. the equality

$$\int_\Omega \mathcal{L}(x, U, U') \, dx = \int_{\bar\Omega} \mathcal{L}(\bar x, \bar U, \bar U') \, d\bar x \qquad (7.5)$$

is satisfied for any transformation (3.3) of this group and any function $U = U(x)$, regardless of the domain $\Omega$. The domain $\bar\Omega$ is obtained here from $\Omega$ by the transformation (3.3) and, in general, depends on the function $U = U(x)$ if $f$ depends on $U$.

The following statements are valid.[77]

**Lemma 7.1.** *The functional (7.1) is invariant with respect to the group $G_M$ with infinitesimal operators (7.4) if and only if the equalities hold*

$$\sum_{i=1}^{n} D_i J_i^{(\alpha)} - Q_\alpha \left( \sum_{i=1}^{n} D_i \left( \frac{\partial \mathcal{L}}{\partial U_{x_i}} \right) - \frac{\partial \mathcal{L}}{\partial U} \right) = 0, \quad \alpha = 1, \ldots, M, \qquad (7.6)$$

*where*

$$Q_\alpha = \eta_\alpha - \sum_{i=1}^{n} \left( \frac{\partial U}{\partial x_i} \right) \xi_\alpha^i,$$

$$J_i^{(\alpha)} = \left( \eta_\alpha - \sum_{j=1}^{n} \left( \frac{\partial U}{\partial x_j} \right) \xi_\alpha^j \right) \frac{\partial \mathcal{L}}{\partial U_{x_i}} + \mathcal{L} \xi_\alpha^i, \quad \alpha = 1, \ldots, M. \qquad (7.7)$$

**Theorem 7.1.** *Let the functional (7.1) be invariant with respect to the group $G_M$ with infinitesimal operators (7.4). Then the Euler equation (7.2) has $M$ independent conservation laws. The vectors $J_\alpha = (J_1^{(\alpha)}, \ldots, J_n^{(\alpha)})$ satisfying the condition*

$$\Phi_\alpha(U) \equiv \sum_{i=1}^{n} D_i J_i^{(\alpha)} = 0, \quad \alpha = 1, \ldots, M, \qquad (7.8)$$

*are defined by (7.7).*

Theorem 7.1 is known as the Noether theorem. Thus, considering the Euler equation (7.2), we may write the equalities (7.6) as

$$\Phi_\alpha = Q_\alpha \Phi, \quad \alpha = 1, \dots, M. \tag{7.9}$$

Note that these equalities are valid not only for the extremal, but for any function $U = U(x)$ as well. We use (7.9) to obtain transformed and adjoint equations with the operators $\widetilde{A}(u)$ and $\widetilde{A}^*(u)$.

**7.2.** Consider the transformed operator $\widetilde{\Phi}$:

$$\widetilde{\Phi}(U) \equiv L\Phi(U) + \sum_{\alpha=1}^{M} L_\alpha \Phi_\alpha(U) = \widetilde{A}(u)\, u - \widetilde{f}, \tag{7.10}$$

where

$$\widetilde{f} = -\left( L\Phi(U_0) + \sum_{\alpha=1}^{M} L_\alpha \Phi_\alpha(U_0) \right), \quad U_0 \in D(\Phi),$$

$$u = U - U_0,$$

$$\widetilde{A}(u) = LA(u) + \sum_{\alpha=1}^{M} L_\alpha A_\alpha(u),$$

$$A(u) = \int_0^1 \Phi'(U_0 + tu)\, dt,$$

$$A_\alpha(u) = \int_0^1 \Phi'_\alpha(U_0 + tu)\, dt, \quad \alpha = 1, \dots, M.$$

If $U \in D_s(\Phi)$, i.e., $\Phi(U) = 0$, then

$$\widetilde{\Phi}(U) = \widetilde{\Phi}(U_0 + u) = \widetilde{A}(u)\, u - \widetilde{f} = 0 \tag{7.11}$$

for any $L$, $\{L_\alpha\}$. We assume here that the hypotheses of the Noether theorem are satisfied and the operators $\Phi$, $\{\Phi_\alpha\}$ are analytical.

Let us reveal the conditions implying that the solutions of the transformed equations ( $\widetilde{\Phi}(U) = 0$, $L_\alpha \Phi_\alpha(U) = 0$ and so on) are the solutions of the equation $\Phi(U) = 0$. If the operator $L$ is invertible, then, obviously, the equation $L\Phi(U) = 0$ is equivalent to the equation $\Phi(U) = 0$. A similar remark concerns the equations $\Phi_\alpha(U) = 0$, $L_\alpha \Phi_\alpha(U) = 0$.

Consider the equation

$$\Phi_\alpha(U) = A_\alpha(u)\, u - f_\alpha = 0, \tag{7.12}$$

where $u = U - U_0$, $f_\alpha = -Q_\alpha(U_0)\Phi(U_0)$. Let $U = U_0 + u$ be a solution to (7.12). Since $\Phi_\alpha(U) = Q_\alpha(U)\Phi(U)$, we can conclude the following. If $Q_\alpha(U) \neq 0$ almost everywhere for any $U \in S_{r_0}(U_0) = \{U \colon U = U_0 + v \in D(\Phi), \|v\| < R_0\}$ (possibly, for sufficiently small $r_0$), then $U$ is also a solution to the equation $\Phi(U) = A(u)\,u - f = 0$.

Let now $U$ be a solution to the equation

$$\widetilde{\Phi}(U) = \widetilde{A}(u)\,u - \tilde{f} = 0. \tag{7.13}$$

Since $\Phi_\alpha(U) = Q_\alpha(U)\,\Phi(U)$, then $U$ satisfies the equation

$$\left( L + \sum_{\alpha=1}^{M} L_\alpha Q_\alpha(U) \right) \Phi(U) = 0. \tag{7.14}$$

If $(L + \sum_{\alpha=1}^{M} L_\alpha Q_\alpha(U))^{-1}$ exists, then we get $\Phi(U) = 0$. This condition is satisfied if, for example,

$$L = \gamma_0 I, \quad L_\alpha = \gamma_\alpha \bar{Q}_\alpha(U_0), \tag{7.15}$$

where $\gamma_0, \gamma_\alpha = \text{constant} > 0$, $\bar{Q}_\alpha(U_0)$ is the function conjugate to $Q_\alpha(U_0)$, and $U = u + U_0 \in S_{r_0}(U_0)$ for a sufficiently small $r_0$.

Let us derive another condition implying that a solution of the equation $\widetilde{\Phi}(U) = 0$ is a solution to (7.2). We write (7.13) as

$$\widetilde{\Phi}(u + U_0) \equiv \widetilde{A}(0)\,u + R(u)\,u - \tilde{f} = 0, \tag{7.16}$$

where

$$\widetilde{A}(0) = L\Phi'(U_0) + \sum_{\alpha=1}^{M} L_\alpha \Phi'_\alpha(U_0),$$

$$R(u) = \widetilde{A}(u) - A(0)$$

$$= L \int_0^1 (\Phi'(U_0 + tu) - \Phi'(U_0))\, dt$$

$$+ \sum_{\alpha=1}^{M} L_\alpha \int_0^1 (\Phi'_\alpha(U_0 + tu) - \Phi'_\alpha(U_0))\, dt.$$

Let the operators $L$, $\{L_\alpha\}$ be bounded and $\Phi'$, $\{\Phi'_\alpha\}$ satisfy the Lipschitz condition. (The latter requirement is met in this case, because $\Phi$, $\{\Phi_\alpha\}$ are analytical, by hypotheses.) Then

$$\|R(u)\| \leq C\|u\| \leq C r_0, \quad C = \text{constant}. \tag{7.17}$$

Note that

$$\Phi_\alpha(V) \quad = \quad Q_\alpha(V)\,\Phi(V),$$

$$\Phi'_\alpha(V)\,v \quad = \quad \Phi(V)Q'_\alpha(V)\,v + Q_\alpha(V)\Phi'(V)\,v,$$

$$\Phi''_\alpha(V)\,v^2 \quad = \quad \Phi(V)Q''_\alpha(V)\,v^2 + 2(\Phi'(V)\,v)(Q'_\alpha(V)\,v) + Q_\alpha(V)\Phi''(V)\,v^2,$$

$$\ldots$$

for any $V \in D(\Phi)$. Hence,

$$\widetilde{A}(0)\,v = \left( L + \sum_{\alpha=1}^{M} L_\alpha Q_\alpha(U_0) \right) \Phi'(U_0)\,v + \sum_{\alpha=1}^{M} L_\alpha \Phi(U_0)Q'_\alpha(U_0)\,v.$$

By setting $L = -\sum_{\alpha=1}^{M} L_\alpha Q_\alpha(U_0)$, we get $\tilde{f} = 0$, and equation (7.16) is

$$(\widetilde{A}(0) + R(u))\,u = 0, \tag{7.18}$$

where

$$\widetilde{A}(0)\,v = \sum_{\alpha} L_\alpha \Phi(U_0)\,Q'_\alpha(U_0)\,v, \tag{7.19}$$

$$Q'_\alpha(U_0)\,v = \eta'_\alpha(x, U_0)\,v - \sum_{i=1}^{n} \xi_\alpha^i(x, U_0)\frac{\partial v}{\partial x_i} - \sum_{i=1}^{n} \left( \frac{\partial U_0}{\partial x_i} \right) \xi_\alpha^{i\,'}(x, U_0)\,v.$$

If the functions $\{\eta_\alpha\}$, $\{\xi_\alpha^i\}$ depend on the derivatives of $U$ of at most the first order, then

$$Q'_\alpha(U_0)\,v \quad = \quad \left( \frac{\partial \eta_\alpha}{\partial U}(x, U_0, U'_0) - \sum_{i=1}^{n} \left( \frac{\partial U_0}{\partial x_i} \right) \frac{\partial \xi_\alpha^i}{\partial U}(x, U_0, U'_0) \right) v$$

$$- \sum_{i=1}^{n} \Bigg( \xi_\alpha^i(x, U_0, U'_0) - \frac{\partial \eta_\alpha}{\partial U_{x_i}}(x, U_0, U'_0)$$

$$+ \sum_{j=1}^{n} \left( \frac{\partial U_0}{\partial x_j} \right) \frac{\partial \xi_\alpha^i}{\partial U_{x_i}}(x, U_0, U'_0) \Bigg) \frac{\partial v}{\partial x_i}, \tag{7.20}$$

$$\widetilde{A}(0)\,v \quad = \quad \sum_{i=1}^{n} V_i(x, U_0)\frac{\partial v}{\partial x_i} + W(x, U_0)\,v, \tag{7.21}$$

where

$$V_i(x, U_0) \quad = \quad \sum_{\alpha=1}^{M} L_\alpha \Phi(U_0)\left( \frac{\partial \eta_\alpha}{\partial U_{x_i}}(x, U_0, U'_0) - \xi_\alpha^i(x, U_0, U'_\alpha) \right)$$

$$-\sum_{j=1}^{n}\left(\frac{\partial U_0}{\partial x_j}\right)\frac{\partial \xi_\alpha^j}{\partial U_{x_i}}(x, U_0, U_0')\Bigg),$$

$$(7.22)$$

$$W(x, U_0) = \sum_{\alpha=1}^{M} L_\alpha \Phi(U_0)\left(\frac{\partial \eta_\alpha}{\partial U}(x, U_0, U_0')\right.$$

$$\left.-\sum_{i=1}^{n}\left(\frac{\partial U_0}{\partial x_i}\right)\frac{\partial \xi_\alpha^i}{\partial U}(x, U_0, U_0')\right).$$

Let $u$ be a solution of equation (7.18). Then $U = u + U_0$ satisfies (7.14). Notice that

$$L + \sum_{\alpha=1}^{M} L_\alpha Q_\alpha(U) = L + \sum_{\alpha=1}^{M} L_\alpha \left(Q_\alpha(U_0) + Q'_\alpha(U_0)\,u + \frac{1}{2}Q''_\alpha(U_0)\,u^2 + \ldots\right)$$

$$= \sum_{\alpha=1}^{M} L_\alpha Q'_\alpha(U_0)\,u + R_1(u)\,u,$$

where $\|R_1(v)\| \leq C r_0$ for $\|v\| \leq r_0$. If the operator

$$\widetilde{A}_\alpha v \equiv \sum_{\alpha=1}^{M} L_\alpha Q'_\alpha(U_0)\,v \equiv \sum_{i=1}^{n} V_i^{(0)}(x, U_0)\frac{\partial v}{\partial x_i} + W^{(0)}(x, U_0)\,v \qquad (7.23)$$

with

$$V_i^{(0)}(x, U_0) = \sum_{\alpha=1}^{M} L_\alpha \left(\frac{\partial \eta_\alpha}{\partial U_{x_i}} - \xi_\alpha^i - \sum_{j=1}^{n}\left(\frac{\partial U_0}{\partial x_j}\right)\frac{\partial \xi_\alpha^j}{\partial x_i}\right),$$

$$W^{(0)}(x, U_0) = \sum_{\alpha=1}^{M} L_\alpha \left(\frac{\partial \eta_\alpha}{\partial U} - \sum_{i=1}^{n}\left(\frac{\partial U_0}{\partial x_i}\frac{\partial \xi_\alpha^i}{\partial U}\right)\right),$$

satisfies the inequality

$$\|\widetilde{A}_0 v\| \geq C_0\|v\| \quad \forall\, v \in D(\widetilde{A}), \quad C_0 = \text{constant} > 0, \qquad (7.24)$$

then the operator $\widetilde{A}_0 + R_1(u)$ is invertible for $\|u\| \leq r_0$ if $r_0$ is sufficiently small. Hence, the solution of equation (7.14) satisfies the equation $u\Phi(U) = 0$, that is, $\Phi(U) = 0$ if $u \neq 0$ almost everywhere.

The foregoing proves the following lemma.

**Lemma 7.2.** *Let* $L = -\sum_{\alpha=1}^{M} L_\alpha Q_\alpha(U_0)$ *and* $\{L_\alpha\}$ *be such that (7.24) is satisfied. Then the function* $U = U_0 + u$ *(with* $u \neq 0$ *almost everywhere* *and* $\|u\| \leq r_0 \ll 1$ *) is the solution to equation (7.12) if and only if U satisfies (7.14).*

Note that if $\{L_\alpha\}$ are functions, then

$$V_i = \Phi(U_0)\, V_i^{(0)}, \quad W = \Phi(U_0)\, W^{(0)}, \quad i = 1, \ldots, n \tag{7.25}$$

and, hence,

$$\tilde{A}(0)\, v = \Phi(U_0)\, \tilde{A}_0 v. \tag{7.26}$$

If, moreover, $\{L_\alpha\}$ are defined from the equations

$$\sum_{i=1}^{n} \Phi(U_0)\, \theta_\alpha^i(U_0) \frac{\partial L_\alpha}{\partial x_i} + L_\alpha \sum_{i=1}^{n} \frac{\partial}{\partial x_i}(\Phi(U_0)\, \theta_\alpha^i(U_0)) = 0, \quad \alpha = 1, \ldots, M,$$

$$\tag{7.27}$$

where

$$\theta_\alpha^i(U_0) = \frac{\partial \eta_\alpha}{\partial U_{x_j}} - \xi_\alpha^i - \sum_{j=1}^{n} \left( \frac{\partial U_0}{\partial x_j} \right) \frac{\partial \xi_\alpha^j}{\partial U_{x_i}},$$

then the vector $\mathbf{V} = (V_1, \ldots, V_n)$ satisfies the condition

$$\operatorname{div} \mathbf{V} = \sum_{i=1}^{n} \frac{\partial}{\partial x_i} V_i = \sum_{i=1}^{n} \frac{\partial}{\partial x_i}(\Phi(U_0)\, V_i^{(0)}) = 0. \tag{7.28}$$

*Remark* 7.1. Note that if $\|\tilde{A}_0^{-1} R_1(v)\| \leq C r_0 \ll 1$ and (7.24) are satisfied, similar relations for the operators $\tilde{A}(0)$ and $R(v)$ do not have to be valid. It is easily seen, taking into account (7.26) and $\|U - U_0\| \leq r_0 \ll 1$.

Turning back to equation (7.18), we conclude that the following statement is valid.

**Lemma 7.3.** *Let*

(1) *the hypotheses of Theorem 7.1 be satisfied;*
(2) *the functions* $\{\eta_\alpha\}$, $\{\xi_\alpha^i\}$ *depend only on* $x$, $U$, $U'$;
(3) *the operators* $\Phi$, $\Phi_\alpha$, $\alpha = 1, \ldots, M$, *be analytical.*

*Then for*

$$L = -\sum_{\alpha=1}^{M} L_\alpha Q_\alpha(x, U_0)$$

*the transformed operator* $\tilde{\Phi}$ *has the form (7.16), and for the adjoint operator* $\tilde{A}^*(u)$ *corresponding to* $F(u) = \tilde{A}(u)\, u$ *one can take the operator*

$$\tilde{A}^*(u)\, w = -\sum_{i=1}^{n} \frac{\partial}{\partial x_i}(V_i(x, U_0)\, w) + W(x, U_0)\, w + R^*(u)\, w. \tag{7.29}$$

*If $\{L_\alpha\}$ are functions satisfying (7.27), then relation (7.28) holds for the vector $\mathbf{V} = (V_1, \ldots, V_n)$.*

*Remark 7.2.* The assertions of Lemma 7.3 are valid for arbitrary functions $\eta_\alpha$, $\{\xi_\alpha^i\}$ (i.e., dependent on $U''$, $U'''$ and so on). However, in this case the components of the vector $\mathbf{V}$ will be operators, and the operator (7.29) will have the form

$$\widetilde{A}^*(u)\, w = -\sum_{i=1}^{n} \frac{\partial}{\partial x_i}(V_i^*(x, U_0)\, w) + W(x, U_0)\, w + R^*(u)\, w. \qquad (7.30)$$

*Remark 7.3.* When considering specific cases of transformation (7.3) (for example, $\bar{x} = f(x, a)$, $\bar{U} = U$, and others) the form of functions $\{V_i(x, U_0)\}$, $W(x, U_0)$ may be essentially simplified. This will help to solve the problem of proper choice of $\{L_\alpha\}$.

**7.3.** To finish this section we formulate the following statement (which is of interest independently of constructing the adjoint equations).

**Lemma 7.4.** *Let the function $U_0$ and the operators $\{L_\alpha\}$ be such that (7.20) is satisfied. Then the function $u$ such that $u \neq 0$ almost everywhere, and $\|u\| \leq r_0 \ll 1$, is a solution to the equation*

$$\left(\sum_{\alpha=1}^{M} L_\alpha Q_\alpha(U_0)\right) A(u)\, u = \left(\sum_{\alpha=1}^{M} L_\alpha A_\alpha(u)\right) u \qquad (7.31)$$

*if and only if $U = u + U_0$ is a solution to equation (7.2).*

*Proof.* Let $U = u + U_0$ be a solution of (1.5). Then

$$0 = \Phi_\alpha(U) = A_\alpha(u)\, u + \Phi_\alpha(U_0) = A_\alpha(u)\, u + Q_\alpha(U_0)\, \Phi(U_0), \quad (7.32)$$

$$0 = \Phi(U) = A(u)\, u + \Phi(U_0),$$

$$0 = Q_\alpha(U_0)\, \Phi(U) = Q_\alpha(U_0)\, A(u)\, u + Q_\alpha(U_0)\, \Phi(U_0). \qquad (7.33)$$

Comparing (7.32) with (7.33), we deduce the required statement for any $\{L_\alpha\}$.

Conversely, let $u \neq 0$ be a solution of (7.31) with $\|u\| \leq r_0 \ll 1$. Then for $U = u + U_0$ we get

$$A(u)\, u = A(u)\, u + \Phi(U_0) - \Phi(U_0) = \Phi(U) - \Phi(U_0),$$

$$A_\alpha(u)\, u = \Phi_\alpha(U) - \Phi_\alpha(U_0) = Q_\alpha(U)\, \Phi(U) - Q_\alpha(U_0)\, \Phi(U_0),$$

$$0 = \sum_{\alpha=1}^{M} L_\alpha(Q_\alpha(U_0) A(u) u - A_\alpha(u) u)$$

$$= \sum_{\alpha=1}^{M} L_\alpha(Q_\alpha(U_0) \Phi(U_0) - Q_\alpha(U_0) \Phi(U_0)$$

$$-Q_\alpha(U) \Phi(U) + Q_\alpha(U_0) \Phi(U_0))$$

$$= \sum_{\alpha=1}^{M} L_\alpha(Q_\alpha(U_0) - Q_\alpha(U)) \Phi(U)$$

$$= -(\tilde{A}_0 + R_1(u)) u\Phi(U).$$

In view of (7.24) and $\|R_1(v)\| \leq C r_0$ for $\|v\| \leq r_0 \ll 1$, we conclude that $u\Phi(U) = 0$. Hence, $\Phi(U) > 0$.

## 8. ON SOME APPLICATIONS OF ADJOINT EQUATIONS

**8.1.** Consider equation (7.2). Assume that it is required to compute the functional

$$J_g(U) = \int_\Omega P(U) g \, \mathrm{d}x, \qquad (8.1)$$

where the non-linear operator $P$ has the Gâteaux derivative $P'$ satisfying the Lipschitz condition. Then

$$J_g(U) = \int_\Omega P(U_0) g \, \mathrm{d}x + \int_\Omega \left( \int_0^1 P'(U_0 + tu) \, \mathrm{d}t \right) ug \, \mathrm{d}x$$

$$= \int_\Omega P(U_0) g \, \mathrm{d}x + \int_\Omega P'(U_0) ug \, \mathrm{d}x + R_g(u), \qquad (8.2)$$

where

$$u = U - u_0, \quad |R_g| \leq C\|g\| \|u\|^2 \equiv O(\|u\|^2).$$

Let $u^*$ be a solution to the adjoint problem:

$$-\sum_{i=1}^{n} \frac{\partial}{\partial x_i}(V_i(x, U_0) u^*) + W(x, U_0) u^* = (P'(U_0))^* g, \quad u^*|_{\partial\Omega_+} = 0, \quad (8.3)$$

where $\{V_i\}$, $W$ are defined according to (7.22), $\{L_\alpha\}$ are functions, $\partial\Omega_-$ is the set of points of $\partial\Omega$ the vectors $\mathbf{V}^{(0)} = (V_1^{(0)}(x, U_0), \ldots, V_n^{(0)}(x, U_0))$ enter $\Omega$ by, $\partial\Omega_+ = \partial\Omega/ \partial\Omega_-$. Note that problem (8.3) may be written as

$$-\sum_{i=1}^{n} \frac{\partial}{\partial x_i}(V_i^0(x, U_0) U^*) + W_i^0(x, U_0) U^* = (P'(U_0))^* g, \quad U^*|_{\partial\Omega_+} = 0, \quad (8.4)$$

where $U^* = \Phi(U_0)\, u^*$. From (8.3), (8.2), (7.18), we get

$$J_g(U) = J_g(U_0) + \int_\Omega \left( \sum_{i=1}^n V_i \frac{\partial u}{\partial x_i} + Wu \right) u^*\, dx$$

$$+ R_g(u) - \int_{\partial\Omega_-} (\mathbf{n}, \mathbf{V})\, uu^*\, d\Gamma$$

$$= J_g(U_0) + R_g(u) - \int_\Omega R(u)\, uu^*\, dx - \int_{\partial\Omega_-} (\mathbf{n}, \mathbf{V})\, uu^*\, d\Gamma$$

where $\mathbf{n}$ is the outward normal vector. Assuming that $U$ is equal to a prescribed function $U_{(\Gamma)}$ on $\partial\Omega_-$, we find

$$J_g(U) = J_g(U_0) - \int_{\partial\Omega_-} (\mathbf{n}, \mathbf{V})(U_{(\Gamma)} - U_0)\, u^*\, d\Gamma + O(\|u\|^2). \qquad (8.5)$$

Thus, if the solution $u^*$ of the adjoint problem (8.3) is computed, then the value

$$\tilde{J}_g(U_0, u^*) \equiv J_g(U_0) - \int_{\partial\Omega_-} (\mathbf{n}, \mathbf{V})(U_{(\Gamma)} - U_0)\, u^*\, d\Gamma$$

approximates $J_g(U)$ with an accuracy of $O(\|u\|^2)$. If, moreover, $U|_{\partial\Omega_-} = U_0|_{\partial\Omega_-}$, then

$$|J_g(U) - J_g(U_0)| \le O(\|u\|^2) \quad \text{for } u = U - U_0, \quad \|u\| \ll 1.$$

(Thus, $J_g(U)$ is approximated by $J_g(U_0)$ with the second order, $U_0$ being an approximation to $U$ of the first order!)

**8.2.** Consider the equation

$$\Phi(U) = 0, \qquad (8.6)$$

where $U = U(x)$, $x \in \Omega \subset \mathbf{R}^n$. Assume that equation (8.6) need not be the Euler equation and has the conservation law

$$\sum_{i=1}^n \frac{\partial}{\partial x_i} J_i^{(\alpha)}(x, U) = 0. \qquad (8.7)$$

Let equations (8.6), (8.7) have a solution $U$, and we are required to compute the functional (8.1). Choosing appropriate function $U_0$, we obtain the equation

$$\sum_{i=1}^n \frac{\partial}{\partial x_i} A_i^{(\alpha)}(u)\, u = f^{(\alpha)}, \qquad (8.8)$$

where

$$f^{(\alpha)} = -\sum_{i=1}^n \frac{\partial}{\partial x_i} J_i^{(\alpha)}(x, U_0), \quad A_i^{(\alpha)} = \int_0^1 J_i^{(\alpha)\prime}(x, U_0 + tu)\, dt,$$

and also the equation

$$\sum_{i=1}^{n} \frac{\partial}{\partial x_i} \{ J_i^{(\alpha)'}(x, U_0)\, u \} + R_2(u)\, v = f^{(\alpha)}, \qquad (8.9)$$

where

$$R_2(u)\, v = \int_0^1 (J_i^{(\alpha)'}(x, U_0 + tu) - J_i^{(\alpha)'}(x, U_0))\, dt v.$$

Assume that the operators $\{J_i^{(\alpha)}\}$ are analytical. Let, for simplicity, them be independent of the derivatives of $U$. Then the estimates $\|R_2(u)\| \leq C\|u\|$, $\|R_2(u)\, u\| \leq O(\|u\|^2)$ hold.

By $\partial\Omega_-$ we denote the set of points of $\partial\Omega$ where the vector $\mathbf{J}^{(\alpha)'} = (J_1^{(\alpha)'}(x, U_0), \ldots, J_n^{(\alpha)'}(x, U_0))$ enters $\Omega$. Let $\partial\Omega_+ = \partial\Omega\backslash\partial\Omega_-$. Consider the adjoint problem

$$-\sum_{i=1}^{n} J_i^{(\alpha)}(x, U_0) \frac{\partial u^*}{\partial x_i} = (P'(U_0))^*g, \quad (\mathbf{n}, \mathbf{J}^{(\alpha)'}) u^*|_{\partial\Omega_+} = 0. \qquad (8.10)$$

Then, from (8.10), (8.2), we find

$$\begin{aligned}
J_g(U) &= J_g(U_0) + \int_\Omega u(P'(U_0))^* g\, dx + R_g(u) \\[2mm]
&= J_g(U_0) + \int_\Omega u \left( -\sum_{i=1}^{n} J_i^{(\alpha)'}(x, U_0) \frac{\partial u^*}{\partial x_i} \right) dx + R_g(u) \\[2mm]
&= J_g(U_0) - \int_{\partial\Omega_-} (\mathbf{n}, \mathbf{J}^{(\alpha)'})(U - U_0)\, u^*\, d\Gamma \\[2mm]
&\quad + \int_\Omega \left( \sum_{i=1}^{n} \frac{\partial}{\partial x_i} (J_i^{(\alpha)'}(x, U_0)\, u) \right) u^*\, dx + R_g(u) \\[2mm]
&= J_g(U_0) - \int_{\partial\Omega_-} (\mathbf{n}, \mathbf{J}^{(\alpha)'})(U - U_0)\, u^*\, d\Gamma \\[2mm]
&\quad + \int_\Omega \left( -\sum_{i=1}^{n} \frac{\partial}{\partial x_i} \{ J_i^{(\alpha)}(x, U_0) \} - R_2(u)\, u \right) u^* + R_g(u) \\[2mm]
&= J_g(U_0) - \int_{\partial\Omega_-} (\mathbf{n}, \mathbf{J}^{(\alpha)'})(U - U_0)\, u^*\, d\Gamma \\[2mm]
&\quad - \int_\Omega \sum_{i=1}^{n} \frac{\partial}{\partial x_i} \{ J_i^{(\alpha)}(x, U_0) \} u^*\, dx + O(\|u\|^2).
\end{aligned}$$

Thus, if $u^*$ is the solution to the adjoint problem (8.10), then for $\|u\| = \|U - U_0\| \ll 1$ the value

$$\tilde{J} = J_g(U_0) - \int_{\partial\Omega_-} (\mathbf{n}, \mathbf{J}^{(\alpha)'})(U_{(\Gamma)} - U_0)\, u^*\, d\Gamma$$

$$+ \int_{\Omega} \left( \sum_{i=1}^{n} \frac{\partial}{\partial x_i} \{J_i^{(\alpha)}(x, U_0)\} \right) u^*\, dx \qquad (8.11)$$

is an approximation to $J_g(U)$ with an accuracy $O(\|u\|^2)$.

*Remark* 8.1. The form of equations (8.9) and (8.10) suggests the following. If equation (8.6) and the conservation law (8.7) are considered, then for boundary conditions for the equation $\Phi(U_0)\, u = f + O(\|u\|^2)$ (as well as the equation $Q_\alpha(U_0)\, \Phi'(U_0)\, u = f Q_\alpha(U_0) + O(\|u\|^2)$) one can take the condition $(\mathbf{n}, \mathbf{J}^{(\alpha)'}(x, U_0))\, u = g$ given on $\partial\Omega$, or on a part of the boundary. To make sure that this is so indeed in many cases one can consider the Euler equation for (8.6).

**8.3.** Let us consider one of the optimal control problems that has arisen in modelling of ecological processes. It is required to find functions $U(x, t)$, $V(x)$ such that

$$\frac{\partial U}{\partial t} - \mu\Delta U + \sum_{i=1}^{n} v_i(x, t)\frac{\partial U}{\partial x_i} + \lambda_0(x, t)\, U - Q(x, t) = 0,$$

$$(x, t) \in \Omega \equiv D \times (0, T), \quad D \subset \mathbf{R}^n,$$

$$U(x, t) = U_{(\Gamma)}(x, t), \quad x \in \partial D, \quad t \in (0, T), \qquad (8.12)$$

$$U(x, 0) = V(x), \quad x \in D,$$

$$\inf_V J_\gamma(V),$$

where

$$J_\gamma(V) = \sum_{i=1}^{N} \alpha_i(U_i - C_i)^2 + \gamma \int_D (V(x))^2\, dx,$$

$$\gamma = \text{constant} > 0, \quad \mu = \text{constant} > 0, \quad \alpha_i = \text{constant} > 0,$$

$$U_i = \int_\Omega p_i(x, t)\, U(x, t)\, dx\, dt,$$

$p_i(x, t) \geq 0$ is a weight function of $L_\infty(\Omega)$; $V(x)$, $U_{(\Gamma)}(x, t)$, $Q(x, t)$ are prescribed functions, $\{C_i\}$ are constants, $N$ is an integer, $\lambda_0 \geq 0$, $\lambda_0 \in L_\infty(\Omega)$. The functions $\{v_i(x, t)\}$ are assumed to have the derivatives $\{\partial v_i/\partial x_i\}$ satisfying the condition

$$\sum_{i=1}^{n} \frac{\partial v_i}{\partial x_i} = 0, \quad (x, t) \in \Omega.$$

Assume that problem (8.12) has the solutions $U, V$. Consider (8.12) with $\tilde{V} = V + V_1$ substituted for $V$, where $V_1$ is an arbitrary function. Then the variation $\delta J_\gamma(V, V_1)$ of the functional $J_\gamma$ satisfies the relation

$$\delta J_\gamma = 2 \sum_{i=1}^{N} \alpha_i (U_i - C_i) \int_\Omega p_i(x, t) \, U_1 \, dx \, dt + 2\gamma \int_D V(x) \, V_1(x) \, dx, \quad (8.13)$$

where $U_1$ is the solution to the problem

$$\frac{\partial U_1}{\partial t} - \mu \Delta U_1 + \sum_{i=1}^{n} v_i \frac{\partial U_1}{\partial x_i} + \lambda_0 U_1 = 0, \quad (x, t) \in \Omega,$$

$$U_1 = 0, \quad (x, t) \in \partial D \times (0, T), \qquad\qquad (8.14)$$

$$U_1(x, 0) = V_1(x), \quad x \in D.$$

Consider the following adjoint problem

$$-\frac{\partial u^*}{\partial t} - \mu \Delta u^* - \sum_{i=1}^{n} v_i \frac{\partial u^*}{\partial x_i} + \lambda_0 u^* = \sum_{i=1}^{N} \alpha_i (U_i - C_i) \, p_i(x, t), \quad (x, t) \in \Omega,$$

$$u^* = 0, \quad (x, t) \in \partial D \times (0, T),$$

$$u^*(x, T) = 0.$$

$$\qquad\qquad (8.15)$$

The solution of problems (8.14) and (8.15) satisfy the Lagrange identity:

$$\int_\Omega \sum_{i=1}^{N} \alpha_i (U_i - C_i) \, p_i(x, t) \, U_1(x, t) \, dx \, dt = \int_D u^*(x, 0) \, V_1(x) \, dx. \quad (8.16)$$

From (8.16) and (8.13) we find

$$V(x) = \frac{1}{\gamma} u^*(x, 0). \qquad\qquad (8.17)$$

Thus, the solution of problem (8.12) satisfies necessarily the equations

$$\frac{\partial U}{\partial t} - \mu \Delta U + \sum_{i=1}^{n} v_i \frac{\partial U}{\partial x_i} + \lambda_0 U - Q = 0, \quad (x, t) \in \Omega,$$

$$U = U_{(\Gamma)}, \quad (x, t) \in \partial D \times (0, T), \qquad\qquad (8.18)$$

$$U(x, 0) = V(x), \quad x \in D;$$

$$-\frac{\partial u^*}{\partial t} - \mu \Delta u^* - \sum_{i=1}^{n} v_i \frac{\partial u^*}{\partial x_i} + \lambda_0 u^* = \sum_{i=1}^{N} \alpha_i (U_i - C_i) \, p_i(x, t) \, \alpha_i, \qquad (8.19)$$

$$u^* = 0, \quad (x, t) \in \partial D \times (0, T), \quad u^*(x, T) = 0, \quad x \in D.$$

It is proved[133] that if $U$ and $V$ satisfy (8.17)–(8.19), then these functions are also solutions to problem (8.6).

Having solved (8.17)–(8.19) by an appropriate method, along with the functions $U$, $V$, $u^*$, one can compute, for example, the value of the functional

$$\Phi_1(U) = \sum_{i=1}^{N} \alpha_i^{(0)} U_i = \int_{\Omega} U(x,t)\, p(x,t)\, dx\, dt, \qquad (8.20)$$

where

$$\alpha_i^{(0)} = \text{constant} \geq 0, \quad p(x,t) = \sum_{i=1}^{N} \alpha_i^{(0)} p_i(x,t),$$

or the functional

$$\Phi_2(U) = \sum_{i=1}^{N} \alpha_i^{(0)} (U_i - C_i)^2 / 2. \qquad (8.21)$$

These functionals have frequently a clear physical meaning, and to find their values is of practical importance.

**8.4.** Let $U^0$, $V^0$ be the solutions of the optimal control problem (8.12). Consider now the problem

$$\frac{\partial U}{\partial t} - \mu \Delta U + \sum_{i=1}^{n} v_i \frac{\partial U}{\partial x_i} + \lambda(x,t,U) U - Q = 0, \quad (x,t) \in \Omega,$$

$$U = U_{(\Gamma)}, \quad (x,t) \in \partial D \times (0,T), \qquad (8.22)$$

$$U(x,0) = V^0(x), \quad x \in D,$$

where the function $\lambda(x,t,U)$ depends analytically on $U$ and, possibly, on the first derivatives of $V$. Assume that

$$\lambda(x,t,U) = \lambda_0(x,t) + \varepsilon \lambda_1(x,t,U), \qquad (8.23)$$

where $\varepsilon \in [0, \varepsilon_0]$ is a small parameter, and $\lambda_1$ a prescribed function. In this case it is reasonable to assume also that $U^0$ is a 'good' approximation to the solution of problem (8.22).

Consider the equation for the correction $u = U - U^0$:

$$\frac{\partial u}{\partial t} - \mu \Delta u + \sum_{i=1}^{n} v_i \frac{\partial u}{\partial x_i} + \lambda(x,t,U^0) u + \frac{\partial \lambda}{\partial U}(x,t,U^0) U_0 u$$

$$+ \sum_{i=1}^{n} \frac{\partial \lambda}{\partial U_{x_i}}(x,t,U^0) U_0 \frac{\partial u}{\partial x_i} + R(u)\, u = f, \qquad (8.24)$$

$$u(x,0) = 0, \quad x \in D$$

$$u(x,t) = 0, \quad (x,t) \in \partial D \times (0,T),$$

or

$$\frac{\partial u}{\partial t} - \mu \Delta u + \sum_{i=1}^{n} \tilde{v}_i \frac{\partial u}{\partial x_i} + \tilde{\lambda}(x,t)\, u + R(u)\, u = f, \quad (x,t) \in \Omega,$$

$$u(x,0) = 0, \quad x \in D \tag{8.25}$$

$$u(x,t) = 0, \quad (x,t) \in \partial D \times (0,T),$$

where

$$\tilde{v}_i = v_i(x,t) + \frac{\partial \lambda}{\partial U_{x_i}}(x,t,U^0)\, U^0,$$

$$\tilde{\lambda}(x,t) = \lambda(x,t,U^0) + \frac{\partial \lambda}{\partial U}(x,t,U^0)\, U^0,$$

$$f = Q(x,t) - \frac{\partial U^0}{\partial t} + \mu \Delta U^0 - \sum_{i=1}^{n} v_i \frac{\partial U^0}{\partial x_i} - \lambda(x,t,U_0)\, U_0$$

$$= (\lambda_0(x,t) - \lambda(x,t,U^0))\, U^0,$$

$$\|R(u)\| \le C\|u\|, \quad \|R(u)\, u\| \le O(\|u\|^2)$$

( $\|\cdot\|$ is a norm of a 'properly' chosen Banach space that is not specified here for simplicity).

Assume that we are interested in computing the value of the functional $\Phi_2$. Let us represent $\Phi_2$ in the form

$$\Phi_2(U) = \Phi_2(U^0) + \int_{\Omega} \sum_{i=1}^{N}(U_i^0 - C_i)\, p_i(x,t)\, u(x,t)\, dx\, dt + O(\|u\|^2), \tag{8.26}$$

where

$$U_i^0 = \int_{\Omega} p_i(x,t)\, U^0(x,t)\, dx\, dt, \quad i = 1, \ldots, N.$$

To compute $\Phi_2$ with the accuracy $O(\|u\|^2)$ consider the adjoint problem

$$-\frac{\partial u^*}{\partial t} - \mu \Delta u^* - \sum_{i=1}^{n} \frac{\partial}{\partial x_i} \tilde{v}_i u^* + \tilde{\lambda}(x,t)\, u^* = \sum_{i=1}^{N}(U_i^0 - C_i)\, p_i(x,t),$$

$$u^* = 0, \quad (x,t) \in \partial D \times (0,T), \tag{8.27}$$

$$u^*(x,T) = 0, \quad x \in D.$$

Then, as shown above, the value

$$\tilde{\Phi}_2 = \Phi_2(U^0) - \mu \int_{\partial D} \int_0^T \frac{\partial u^*}{\partial n}(U_{(\Gamma)}(x,t) - U^0(x,t))\, d\Gamma\, dt$$

$$+ \int_{\Omega} f(x,t)\, u^*(x,t)\, dx\, dt \tag{8.28}$$

approximates $\Phi_2(U)$ with the accuracy $O(\|u\|^2)$. Since in this case the trace of the approximation $U_0$ on $\partial D$ coincides with $U|_{\partial\Omega} = U_{(\Gamma)}$, then

$$\Phi_2 = \Phi_2(U^0) + \int_\Omega f(x,t)\, u^*(x,t)\, dx\, dt. \qquad (8.29)$$

*Remark* 8.2. Using (8.28), one can solve the optimal control problem approximately. To do this it is sufficient to choose the function $U^0$ (not necessarily being a solution of (8.6)) such that $U^0 \cong U$, $U^0(x,0) = U(x,0) = V(x)$ and then, computing $\widetilde{\Phi}_2$ by formula (8.28), make the corresponding conclusion on the 'optimal' choice of $U_{(\Gamma)}$. (Note that in this case the boundary condition for $u$ in (8.24), (8.25) is not homogeneous: $u|_{\partial D} = u_{(\Gamma)} \equiv U_{(\Gamma)} - U^0|_{\partial D}$.)

**8.5.** Let us write problem (8.22) as

$$\frac{\partial U}{\partial t} + AU + f(U) = 0, \quad (x,t) \in \Omega,$$
$$U(x,0) = V^0(x), \quad x \in D, \qquad (8.30)$$

where

$$AU = -\mu\Delta U + \sum_{i=1}^n v_i \frac{\partial U}{\partial x_i}, \qquad (8.31)$$

$$f(U) = \lambda(x,t,U)U - Q(x,t), \qquad (8.32)$$

and the functions of the domain of $A$ are assumed to satisfy the condition $U = U_{(\Gamma)}$, $(x,t) \in \partial D \times (0,T)$. The functions $U_{(\Gamma)}$ and $V^0$ are supposed to be such that some optimality criteria are satisfied (see Sections 8.3 and 8.4).

Consider another optimal control problem related to (8.30). It is possible in practice that $V^0(x)$ is perturbed by $\tau V_1^0(x)$, where $V_1^0(x)$ is a prescribed function and $\tau \in \mathbf{R}$ is an unknown small parameter, that is, $\tilde{V} = V^0 + \tau V_1^0$ is substituted for $V^0$. Assume that we are interested, for example, in the value of the functional $\Phi_1(U)$ and this value is required to be insensitive to the perturbation $\tau V_1^0$ in the initial condition. The question arises: is it possible to find the control function $v(x,t)$ (that may be treated as a perturbation of the source function $Q(x,t)$) such that the functional $\Phi_1(U)$ of the solution to the problem

$$\frac{\partial U}{\partial t} + AU + f(U) = v, \quad (x,t) \in \Omega,$$
$$U(x,0) = V^0(x), \quad x \in D, \qquad (8.33)$$

be insensitive to the perturbation $\tau V_1^0$ in the initial value. (For simplicity we do not introduce here additional restrictions on $v$.) This question was answered in the papers by J.L.Lions[129,131,133] (where the strict definition of the functional to be insensitive is given). In fact, if $\Phi_1(U)$ is insensitive to perturbation $\tau V_1^0$ it is necessary that the function $U$ and $v$ be the solutions of the problem

$$\frac{\partial U}{\partial t} + AU + f(U) = v, \quad (x,t) \in \Omega,$$

$$U(x,0) = V^0(x), \quad x \in D;$$

$$-\frac{\partial q}{\partial t} - \mu \Delta q - \sum_{i=1}^{n} v_i \frac{\partial q}{\partial x_i} + (f'(U))^* q = \sum_{i=1}^{N} \alpha_i^{(0)} p_i(x,t),$$

$$q = 0, \quad (x,t) \in \partial D \times (0,T),$$   (8.34)

$$q(x,T) = 0, \quad x \in D;$$

$$\int_D q(x,0) V_1^0(x) \, dx = 0,$$

where $f'$ is the Gâteaux derivative of $f(U)$. Remembering that $f(U)$ depends on the small parameter $\varepsilon \in [0,\varepsilon_0]$, it is reasonable to suggest that one can use the perturbation algorithm to solve (8.34) numerically. Using this algorithm one can reveal some additional requirements to be satisfied when considering problem (8.34). Thus, expanding the functions in $\varepsilon \in [0,\varepsilon_0]$, it is easily seen that the following restriction is necessary to be satisfied. The function $p(x,t) = \sum_{i=1}^{N} \alpha_i^{(0)} p_i(x,t)$ needs to be orthogonal in $L_2(\Omega)$ to the solution of the problem

$$\frac{\partial U_1^0}{\partial t} + AU_1^0 + \lambda_0 U_1^0 = v, \quad (x,t) \in \Omega,$$

$$U_1^0(x,0) = V_1^0(x), \quad x \in D.$$   (8.35)

Other restrictions may be also revealed.

In conclusion note that to prove the existence theorems for problems of the form(8.34) in the case of non-linear operator $f(U)$ is not a simple problem. Sufficiently complete results concerning the existence problem were obtained by J.L.Lions[129-134] for linear problems. Some specific non-linear problems were studied by V.I.Agoshkov[5,6,8], V.M.Ipatova[11,89], and V.P.Shutyaev[266].

# Chapter 5

# Perturbation algorithms in non-linear problems

This chapter considers perturbation algorithms for non-linear equations and equations involving adjoint operators, and perturbation algorithms for linear and non-linear functionals. Along with the formulation of algorithms the questions of their justification and derivation of convergence rate estimates are considered. A separate section is concerned with a comparison of perturbation algorithms and the successive approximation method.

## 1. PERTURBATION ALGORITHMS FOR ORIGINAL NON-LINEAR EQUATIONS AND EQUATIONS INVOLVING ADJOINT OPERATORS

**1.1.** Consider a sequence of Hilbert spaces $\{H^\alpha\}$. The space $(H^\alpha)^*$, $0 \leq \alpha \leq 1$, is identified with $H^{-\alpha}$. Let $H$ and $Y$ be two spaces in the sequence $\{H^\alpha\}$. The space $X$ is assumed to be dense in $Y$.

Consider a non-linear operator $\Phi(u, \varepsilon)$ mapping $X$ into $Y$ and depending on a parameter $\varepsilon \in [0, \varepsilon_0]$. The domain $D(\Phi)$ of this operator is assumed to be a convex set. Let for any fixed $\varepsilon$ the operator $\Phi$ have the Gâteaux derivative continuous at each point $u \in D(\Phi)$. We consider $\Phi'$ as an operator mapping $X$ into $Y$. We assume also that the domain $D(\Phi')$ of the operator $\Phi'$ contains $D(\Phi)$ and is dense in $X$.

Consider the equation

$$\Phi(U, \varepsilon) = 0. \tag{1.1}$$

Fixing an element $U_0 \in D(\Phi)$ and putting $f(\varepsilon) \equiv -\Phi(U_0, \varepsilon)$, from (1.1) we come to the equation

$$A(u, \varepsilon)\, u = f(\varepsilon), \tag{1.2}$$

where

$$A(u, \varepsilon) = \int_0^1 F'(tu, \varepsilon)\, dt = \int_0^1 \Phi'(U_0 + tu, \varepsilon)\, dt,$$

$$F(u, \varepsilon) \equiv A(u, \varepsilon)\, u = \Phi(U_0 + u, \varepsilon) - \Phi(U_0, \varepsilon), \quad u = U - U_0.$$

The operator $A(u, \varepsilon)$ maps $X$ into $Y$ with the domain

$$D(A) = D(F) = \{u\colon U_0 + u \in D(\Phi)\}.$$

Its adjoint operator has the form (see Chapter 1, Section 3):

$$A^*(u, \varepsilon) = \left( \int_0^1 F'(tu, \varepsilon)\, dt \right)^* = \int_0^1 (F'(tu, \varepsilon))^*\, dt$$

$$= \int_0^1 (\Phi'(U_0 + tu, \varepsilon))^*\, dt, \quad u \in D(F). \tag{1.3}$$

This is an operator mapping $Y^*$ into $X^*$. We denote its domain by $D(A^*)$. The operator $A^*$ has been said (Chapter 1, Section 3) to be an adjoint operator corresponding to the non-linear operator $\Phi(U, \varepsilon)$. This is one of the adjoint operators that may be introduced when considering equation (1.1).

Along with (1.2), consider the adjoint equation

$$A^*(u, \varepsilon)\, u^* = g(\varepsilon), \tag{1.4}$$

where the element $g(\varepsilon) \in X^*$ is analytic in $\varepsilon$:

$$g(\varepsilon) = \sum_{i=0}^{\infty} \varepsilon^i g_i, \qquad g_i = \frac{1}{i!} \frac{d^i g}{d\varepsilon^i}\bigg|_{\varepsilon=0}, \qquad g_i \in X^*. \tag{1.5}$$

In Chapter 3 we have formulated a number of assertions on solvability of equations (1.2) and (1.4). Here we give some other similar statements which are simple and constructive.

**Lemma 1.1.** *Let the following conditions be satisfied:*

(1) *the domains of the operators $A(0, \varepsilon) = \Phi^*(U_0, \varepsilon)$ and*

$$A(u, \varepsilon) = \int_0^1 \Phi'(U_0 + tu, \varepsilon)\, dt$$

*coincide for*

$$u \in \bar{S}_X(0, R) = \{u\colon U_0 + u \in D(F),\ \|u\|_X \le R\};$$

(2) $\|A(0, \varepsilon)\, w\|_Y \ge m_\varepsilon \|w\|_X$, *where* $w \in D(A(0, \varepsilon))$, $m_\varepsilon \ge m > 0$, $m_\varepsilon, m = \text{constant}$, *and $m$ does not depend on $\varepsilon$;*

(3) *the derivative $\Phi'$ satisfies the Lipschitz condition*

$$\sup_{h \in D(\Phi')} \frac{\|(\Phi'(U_0 + u, \varepsilon) - \Phi'(U_0 + v, \varepsilon))\, h\|_Y}{\|h\|_X} \le L_\varepsilon \|u - v\|_X \tag{1.6}$$

*with the constant $L_\varepsilon$ which may depend on $\varepsilon$.*

*Then for*

$$q_\varepsilon \equiv L_\varepsilon R/m_\varepsilon \le q = \text{ constant } < 1, \qquad \|f(\varepsilon)\|_Y \le R(m_\varepsilon - L_\varepsilon) \qquad (1.7)$$

*equation (1.2) has a unique solution $u \in \bar{S}_X(0, R)$.*

Proof. Equation (1.2) is equivalent to

$$u = T(u, \varepsilon), \qquad (1.8)$$

where

$$T(u, \varepsilon) = B(u, \varepsilon)\, u + \bar{f}, \qquad \bar{f} = A^{-1}(0, \varepsilon)\, f(\varepsilon),$$

$$B(u, \varepsilon) = A^{-1}(0, \varepsilon)(A(0, \varepsilon) - A(u, \varepsilon)).$$

In view of the hypothesis (2), we get

$$\|A^{-1}(0, \varepsilon)\|_{Y \to X} \le 1/m_\varepsilon.$$

Hence,

$$\|T(u, \varepsilon) - T(v, \varepsilon)\|_X \le \|(B(u, \varepsilon) - B(v, \varepsilon))\, u\|_X + \|B(v, \varepsilon)(u - v)\|_X$$

$$\le \frac{L_\varepsilon}{m_\varepsilon} \int_0^1 \|t(u - v)\|\|u\|\, dt + \frac{L_\varepsilon}{m_\varepsilon} \int_0^1 \|tv\|_X \|u - v\|_X\, dt$$

$$\le \frac{RL_\varepsilon}{m_\varepsilon}\|u - v\|_X \le q\|u - v\|_X.$$

Therefore, $T(u, \varepsilon)$ is a contraction on $\bar{S}_X(0, R)$. Since for $u \in \bar{S}_X(0, R)$ we have

$$\|T(u, \varepsilon)\|_X \le \frac{L_\varepsilon R}{m_\varepsilon} + \frac{\|f(\varepsilon)\|_Y}{m_\varepsilon} \le \frac{L_\varepsilon R}{m_\varepsilon} + \frac{R(m_\varepsilon - L_\varepsilon)}{m_\varepsilon} = R,$$

then $T(u, \varepsilon)$ maps $\bar{S}_X(0, R)$ into $\bar{S}_X(0, R)$. Thus, in accord with the contraction principle, equation (1.8) (and, hence, equation (1.2)) has a unique solution $u$. This solution may be constructed with the use of the successive approximation method

$$u^{(n+1)} = B(u^{(n)}, \varepsilon)\, u^{(n)} + \bar{f}, \qquad n = 0, 1, \ldots, \qquad (1.9)$$

and the following convergence rate estimate is valid[290]

$$\|u - u^{(n)}\|_X \le \frac{q_\varepsilon^n}{1 - q_\varepsilon}\|u^{(0)} - B(u^{(0)}, \varepsilon)\, u^{(0)} - \bar{f}\|_X \le \frac{q_\varepsilon^n \cdot 2R}{1 - q} \equiv cq_\varepsilon^n, \qquad (1.10)$$

where $c$ is independent of $\varepsilon$ and $n$.

**Corollary.** *If the hypotheses of Lemma 1.1 are satisfied and the operator* $\Phi(U, \varepsilon)$ *is analytic for* $\varepsilon \in [0, \varepsilon_0]$, *then the solution of equation (1.2) is analytic in* $\varepsilon$ *for* $\varepsilon \in [0, \varepsilon_1]$ *with* $\varepsilon_1 \leq \varepsilon_0$.

*Proof* follows from the analyticity of $A^{-1}(0, \varepsilon)$, $B(v, \varepsilon)$ with analytic in $\varepsilon$ element $v$ and (1.9).

**1.2.** In many problems of mathematical physics, the derivative $\Phi'(U, \varepsilon)$ has the form

$$\Phi'(U, \varepsilon) = \Phi'_0 + \varepsilon \Phi'_1(U), \tag{1.11}$$

where $\Phi'_0$ is a linear operator independent of $U$, and $\Phi'_1$ an operator linearly dependent on $U$, i.e.

$$\Phi'_1(\alpha U + \beta V) = \alpha \Phi'_1(U) + \beta \Phi'_1(V), \qquad U, V \in D(\Phi), \qquad \alpha, \beta \in \mathbf{R}^1.$$

In this case the operators $A(0, \varepsilon)$ and $A(u, \varepsilon)$ are of the form

$$A(0, \varepsilon) = \Phi'_0 + \varepsilon \Phi'_1(U_0),$$

$$A(u, \varepsilon) = A(0, \varepsilon) + \frac{\varepsilon}{2} \Phi'_1(u), \qquad u \in D(F). \tag{1.12}$$

Assume that the operator $\Phi'_1$ satisfies the restriction

$$\sup_{h \in D(\Phi')} \frac{\|(\Phi'_1(U_0 + u) - \Phi'_1(U_0 + v)) h\|_Y}{\|h\|_X} \leq k_1 \|u - v\|_X, \tag{1.13}$$

where $u, v \in D(F)$, $k_1 = \text{constant} > 0$.

**Lemma 1.2.** *Let the operator* $\Phi'$ *have the form (1.11), the hypotheses (1), (2) of Lemma 1.1 be satisfied for any* $\varepsilon \in [0, \varepsilon_0]$, *and (1.13) hold. Then for*

$$q_\varepsilon \equiv \frac{\varepsilon k_1 R}{m_\varepsilon} \leq q = \text{constant} < 1, \qquad \|f(\varepsilon)\|_Y \leq R(m_\varepsilon - \varepsilon k_1) \tag{1.14}$$

*equation (1.2) has a unique solution*

$$u = \sum_{i=0}^{\infty} \varepsilon^i u_i \in \bar{S}_X(0, R), \tag{1.15}$$

*where the elements* $\{u_i\}_{i=0}^{\infty}$ *are independent of* $\varepsilon$. *For the sequence*

$$u^{(n)} = \sum_{i=0}^{\infty} \varepsilon^i u_i^{(n)} \tag{1.16}$$

*from (1.9) the following estimate holds:*

$$\|u - u^{(n)}\|_X \le c_0 \left(\frac{\varepsilon k_1 R}{m_\varepsilon}\right)^n R \le c_0 R q^n \tag{1.17}$$

*with the constant $c_0$ independent of $\varepsilon$ and $n$.*

*Proof* copies the proof of Lemma 1.2, (1.11) and (1.13) being taken into account.

**1.3.** Henceforward we shall assume that the operator $\Phi(U, \varepsilon)$ is analytic in its variables and the hypotheses of Lemma 1.1 (or Lemma 1.2) be satisfied. For $U_0$ let us take a solution of the equation

$$\Phi(U_0, 0) = 0. \tag{1.18}$$

Then $u_0 = 0$ and

$$u = \sum_{i=1}^{\infty} \varepsilon^i u_i \tag{1.19}$$

is a solution of equation (1.2). To derive the equations for $u_i$, $i = 1, 2, \dots$, let us represent $f(\varepsilon)$ as a series in powers of $\varepsilon$:

$$f(\varepsilon) = \sum_{i=0}^{\infty} \varepsilon^i f_i, \tag{1.20}$$

where

$$f_0 = f(0) = 0, \qquad f_i = \frac{1}{i!} \frac{d^i f}{d\varepsilon^i}\bigg|_{\varepsilon=0}, \qquad i = 1, 2 \dots \tag{1.21}$$

In particular,

$$f_1 = -\frac{\partial}{\partial \varepsilon} \Phi(U_0, \varepsilon)\bigg|_{\varepsilon=0}, \qquad f_2 = -\frac{1}{2} \frac{\partial^2 \Phi(U_0, \varepsilon)}{\partial \varepsilon^2}\bigg|_{\varepsilon=0}, \tag{1.22}$$

where $\partial \Phi/\partial \varepsilon$ and $\partial^2 \Phi/\partial \varepsilon^2$ are partial derivatives of $\Phi(U, \varepsilon)$ with respect to $\varepsilon$ for a fixed $U$. Substituting (1.19) and (1.20) into (1.2) and cancelling by $\varepsilon$, we get

$$A\left(\sum_{i=1}^{\infty} \varepsilon^i u_i, \varepsilon\right)\left(\sum_{i=1}^{\infty} \varepsilon^{i-1} u_i\right) = \sum_{i=1}^{\infty} \varepsilon^{i-1} f_i. \tag{1.23}$$

By putting $\varepsilon = 0$ in (1.23) we arrive at the following equation for $u_1$:

$$A_0 u_1 = f_1, \tag{1.24}$$

where

$$A_0 = A(0, 0) = \Phi'(U_0, 0). \tag{1.25}$$

To derive an equation for $u_2$, let us differentiate (1.23) in $\varepsilon$ and put $\varepsilon = 0$. As a result, we find

$$A_0 u_2 = f_2 - \int_0^1 \Phi''(U_0, t u_1, 0)\, dt u_1 - \frac{\partial \Phi'}{\partial \varepsilon}(U_0, \varepsilon), \qquad (1.26)$$

where $\Phi''$ is the second Gâteaux derivative of $\Phi(U, \varepsilon)$ with respect to $U$. In a similar manner, one can derive equations for $u_i$, $i = 3, 4, \ldots$

In a specific case when $\Phi'$ has the form (1.11), we arrive at the following equations for $\{u_i\}_{i=0}^\infty$:

$$\Phi_0' u_0 = f_0, \qquad f_0 = 0, \qquad u_0 = 0,$$

$$\Phi_0' u_1 = -\Phi_1'(U_0)\, u_0 - \frac{1}{2}\Phi_1'(u_0)\, u_0 + f_1,$$

$$\ldots \qquad\qquad\qquad\qquad\qquad\qquad\qquad\qquad (1.27)$$

$$\Phi_0' u_{k+1} = -\Phi_1'(U_0)\, u_k - \frac{1}{2}\sum_{i+j=k} \Phi_1'(u_j)\, u_i + f_{k+1}, \qquad k = 1, 2, \ldots$$

Solving successively these equations, we get

$$U = U_0 + \sum_{i=1}^\infty \varepsilon^i u_i. \qquad (1.28)$$

An element $U_{(N)}$ of the form

$$U_{(N)} = U_0 + \sum_{i=0}^N \varepsilon^i u_i \qquad (1.29)$$

is said to be the $N$-th order approximation to the element $U$.

**1.4.** The solvability of adjoint equation (1.4) is related to the solvability of the linear equation

$$A(u, \varepsilon)\, w = f, \qquad (1.30)$$

where $u \in D(F)$, $f \in Y$. We assume that the element $u$ belongs to some Banach space $E$ embedded into $X$, i.e. $\|u\|_X \le c\|u\|_E$, $c = $ constant $> 0$ (the equality $E = X$ is possible). By

$$\bar{S}_E(0, \bar{R}) = \{u : U_0 + u \in D(\Phi),\ \|u\|_E \le \bar{R}\}$$

we denote a ball in $E$.

**Lemma 1.3.** *Let for any* $\varepsilon \in [0, \bar{\varepsilon}_0]$ *the following conditions be satisfied:*

(1) *the domains of the operators $A(0,\varepsilon)$ and $A(u,\varepsilon)$ coincide for $u \in \bar{S}_E(0,\bar{R})$;*
(2) *the equation $A(0,\varepsilon)\,w = f$ is solvable everywhere in $Y$ and $\|A(0,\varepsilon)\,w\|_Y \geq m_\varepsilon \|w\|_X$, where $w \in D(A(0,\varepsilon))$, $m_\varepsilon \geq m > 0$, and $m$ is independent of $\varepsilon$;*
(3) *the derivative $\Phi'$ satisfies the restriction*

$$\sup_{h \in D(\Phi')} \frac{\|(\Phi'(U_0 + u, \varepsilon) - \Phi'(U_0 + v, \varepsilon))\,h\|_Y}{\|h\|_X} \leq \bar{L}_\varepsilon \|u - v\|_E \qquad (1.31)$$

*with the constant $\bar{L}_\varepsilon$ which may depend on $\varepsilon$;*
(4) *the operator $A(0,\varepsilon)$ is closed.*

*Then for*

$$\bar{L}_\varepsilon \bar{R}/(2m_\varepsilon) = \bar{q}_\varepsilon \leq \bar{q} = \text{ constant } < 1 \qquad (1.32)$$

*equation (1.4) is correctly solvable.*

*Proof.* From (1.30) we come to the equation

$$w = T(u, \varepsilon)\,w, \qquad (1.33)$$

where

$$T(u, \varepsilon)\,w = A^{-1}(0, \varepsilon)(A(u, \varepsilon) - A(u, \varepsilon))\,w + A^{-1}(0, \varepsilon)\,f.$$

The following relations are valid:

$$\|T(u, \varepsilon)\,w_1 - T(u, \varepsilon)\,w_2\|_X = \|A^{-1}(0, \varepsilon)(A(0, \varepsilon) - A(u, \varepsilon))(w_1 - w_2)\|_X$$

$$\leq \frac{\bar{L}_\varepsilon}{m_\varepsilon} \int_0^1 \|tu\|_E \, dt \|w_1 - w_2\|_X \leq \bar{q}_\varepsilon \|w_1 - w_2\|_X.$$

Therefore, $T$ is a contraction on $X$. Moreover, it maps $X$ into $X$. Thus, equation (1.33) has a unique solution and, hence, equation (1.30) is solvable everywhere in $Y$. Moreover, (1.30) is correctly solvable.

Furthermore, by the hypothesis, $A(0, \varepsilon)$ is closed, and the operator $A(0, \varepsilon) - A(u, \varepsilon)$ is bounded with respect to $A(0, \varepsilon)$. Hence, the operator $A(u, \varepsilon)$ is also closed. Then, according to the linear operator theory[105], the solvability everywhere of equation (1.30) with the closed operator $A(u, \varepsilon)$ implies the correct solvability of equation (1.4).

**Corollary.** *If $\Phi$ is analytic and the element $u$ is analytic in $\varepsilon$, then the solution $u^*$ of equation (1.4) is analytic in $\varepsilon$.*

**Lemma 1.4.** *Let the derivative $\Phi'$ have the form (1.11), for any $\varepsilon \in [0, \bar{\varepsilon}_0]$ the hypotheses (1), (2), (4) of Lemma 1.3 be satisfied, and*

$$\sup_{h \in D(\Phi_1')} \frac{\|(\Phi'(U_0 + u) - \Phi_1'(U_0 + v))\,h\|_Y}{\|h\|_X} \leq \bar{k}_1 \|u - v\|_E, \qquad (1.34)$$

*where $u, v \in D(F)$, $\bar{k}_1 = $ constant $> 0$. Then for*

$$\varepsilon \bar{k}_1 \bar{R}/(2m_\varepsilon) \leq \bar{q} = \text{constant} < 1 \tag{1.35}$$

*equation (1.4) is correctly solvable in $Y$. If the element $u$ therewith is analytic in $\varepsilon$, the solution $s^*$ of equation (1.4) has the form*

$$u^* = \sum_{i=0}^{\infty} \varepsilon^i u_i^*. \tag{1.36}$$

*Proof* is analogous to the proof of Lemma 1.3 with due regard for (1.11), (1.34).

**Corollary.** *If the hypotheses of Lemma 1.4 are satisfied, then the solution $u^*$ of equation (1.4) may be found with the use of the successive approximation method:*

$$A^*(0, \varepsilon)\, u^{*(n+1)} = (A^*(0, \varepsilon) - A^*(u, \varepsilon))\, u^{*(n)} + g(\varepsilon), \qquad n = 0, 1, \ldots, \tag{1.37}$$

*and*

$$\|u^* - u^{*(n)}\|_{Y^*} \leq \left(\frac{\varepsilon \bar{k}_1 \bar{R}}{2m_\varepsilon}\right)^n \frac{1}{1 - \bar{q}} \left\| A^{*-1}(0, \varepsilon) A^*(u, \varepsilon)\, u^{*(0)} - g \right\|_Y. \tag{1.38}$$

Consider the perturbation algorithm for (1.4) on the assumption that $u$ is represented in the form (1.15). Similarly, we obtain equations for $u_i^*$, $i = 0, 1, \ldots$:

$$A_0^* u_0^* = g_0, \qquad A_0^* = (\Phi'(U_0, 0))^*,$$

$$A_0^* u_1^* = g_1 - \left.\frac{d}{d\varepsilon} A^*\right|_{\varepsilon=0} u_0^*,$$

$$A_0^* u_2^* = g_2 - \left.\frac{d}{d\varepsilon} A^*\right|_{\varepsilon=0} u_1^* - \frac{1}{2} \left.\frac{d^2 A^*}{d\varepsilon^2}\right|_{\varepsilon=0} u_0^*, \tag{1.39}$$

$$\cdots$$

If $\Phi'$ has the form (1.11), then

$$A^*(u, \varepsilon) = (\Phi_0')^* + \varepsilon \left( \Phi_1' \left( U_0 + \frac{u}{2} \right) \right)^*$$

and (1.39) reads

$$(\Phi_0')^* u_0^* = g_0,$$

$$(\Phi_0')^* u_1^* = -(\Phi_1'(U_0))^* u_0^* - \frac{1}{2}(\Phi_1'(u_0))^* u_0^* + g_1, \tag{1.40}$$

$$(\Phi_0')^* u_{k+1}^* = -(\Phi_1'(U_0))^* u_k^* - \frac{1}{2} \sum_{i+j=k} (\Phi_1'(u_j))^* u_i^* + g_{k+1}, \qquad k = 1, 2, \ldots$$

Having solved the first $N + 1$ equations, one can find the $N$-th order approximation to $u^*$ by the formula

$$U^*_{(N)} = u^*_0 + \sum_{i=1}^{N} \varepsilon^i u^*_i. \tag{1.41}$$

# 2. PERTURBATION ALGORITHMS FOR NON-LINEAR FUNCTIONALS BASED ON USING MAIN AND ADJOINT EQUATIONS

**2.1.** Consider the equation

$$\Phi(U, \varepsilon) = 0 \tag{2.1}$$

depending on a parameter $\varepsilon \in [0, 1]$. Let the hypotheses of Section 1 concerning the operator $\Phi(U, \varepsilon)$ be satisfied. It is required to find the value of the functional

$$J(U) = (P(U, \varepsilon), g(\varepsilon))_{H_0}, \tag{2.2}$$

where $H_0 \equiv H^0$, and $P(U, \varepsilon)$ is a non-linear (sufficiently regular) operator mapping $X$ into $Y$ with the domain $D(P) \supset D(\Phi)$. Assume that the element $g(\varepsilon)$ belongs to $Y^*$ and is represented in the form

$$g = \sum_{i=0}^{\infty} \varepsilon^i g_i. \tag{2.3}$$

Let equation (2.1) have a unique solution $U \in D(\Phi)$. Assume that we know the solution $U_0$ of equation (2.1) for $\varepsilon = 0$:

$$\Phi_0(U_0) \equiv \Phi(U_0, 0) = 0 \tag{2.4}$$

and the value

$$J_0 = (P(U_0, \varepsilon), g(\varepsilon))_{H_0}.$$

It is required to develop a perturbation algorithm for finding $J(U)$, using $J_0$ and computing the corresponding corrections. Let us represent $J$ in the form

$$J = (P(U_0 + u, \varepsilon), g(\varepsilon))_{H_0}$$

$$= (P(U_0, \varepsilon), g(\varepsilon))_{H_0} + \left( \int_0^1 P'(U_0 + tu, \varepsilon) \, dt u, g(\varepsilon) \right)_{H_0}$$

$$\equiv J_0 + \delta J, \tag{2.5}$$

where

$$u = U - U_0 = \sum_{i=1}^{\infty} \varepsilon^i u_i, \qquad \delta J = \left( \int_0^1 P'(U_0 + tu, \varepsilon) \, dt u, g(\varepsilon) \right)_{H_0}. \tag{2.6}$$

To write $\delta J$ in the form of a series in $\varepsilon$ let us find the elements $\{Pi\}$ in the representation

$$\int_0^1 P'(U_0 + tu, \varepsilon)\, dtu = \sum_{i=1}^{\infty} \varepsilon^i P_i.$$

Cancelling by $\varepsilon$, differentiating with respect to $\varepsilon$ and putting $\varepsilon = 0$, we get

$$\int_0^1 P'(U_0 + tu, \varepsilon)\, dt \sum_{i=1}^{\infty} \varepsilon^{i-1} u_i = \sum_{i=1}^{\infty} \varepsilon^{i-1} P_i,$$

$$P'(U_0, 0)\, u_1 = P_1,$$

$$P'(U_0,0)\,u_2 + \left(\int_0^1 \frac{d}{d\varepsilon} P'\left(U_0 + t\sum_{i=1}^{\infty} \varepsilon^i u_i, \varepsilon\right) dt\bigg|_{\varepsilon=0}\right) u_1 = P_2, \qquad (2.7)$$

$$P'(U_0,0)\,u_3 + \left(\int_0^1 \frac{d}{d\varepsilon} P'\left(U_0 + t\sum_{i=1}^{\infty} \varepsilon^i u_i, \varepsilon\right) dt\bigg|_{\varepsilon=0}\right) u_2$$

$$+ \frac{1}{2}\left(\int_0^1 \frac{d^2}{d\varepsilon^2} P'\left(U_0 + t\sum_{i=1}^{\infty} \varepsilon^i u_i, \varepsilon\right) dt\bigg|_{\varepsilon=0}\right) u_1 = P_3,$$

$\ldots$

Then

$$\delta J = \sum_{i=0}^{\infty} \varepsilon^i \left(\int_0^1 P'(U_0 + tu, \varepsilon)\, dtu, g_i\right)_{H_0} = \sum_{i=0}^{\infty} \sum_{j=0}^{\infty} \varepsilon^{i+j} (P_j, g_i)_{H_0}$$

$$= \varepsilon(P_1, g_0)_{H_0} + \varepsilon^2(P_2, g_0)_{H_0} + \varepsilon^3(P_3, g_0)_{H_0}$$

$$+ \varepsilon^2(P_1, g_1)_{H_0} + \varepsilon^3(P_2, g_1)_{H_0} + \varepsilon^4(P_3, g_1)_{H_0}$$

$$+ \varepsilon^3(P_1, g_2)_{H_0} + \varepsilon^4(P_2, g_2)_{H_0} + \varepsilon^5(P_3, g_2)_{H_0} + \cdots,$$

i.e.

$$\delta J = \sum_{i=1}^{3} \varepsilon^i \delta J_i + O(\varepsilon^4), \qquad (2.8)$$

where

$$\delta J_1 = (P_1, g_0)_{H_0},$$

$$\delta J_2 = (P_2, g_0)_{H_0} + (P_1, g_1)_{H_0}, \qquad (2.9)$$

$$\delta J_3 = (P_3, g_0)_{H_0} + (P_2, g_1)_{H_0} + (P_1, g_2)_{H_0}.$$

Thus, an algorithm for finding $\delta J_i$, $i = 1, 2, 3$, consists of the following steps:

(1) Solve the equation for $u_1$:

$$A_0 u_1 = f_1, \qquad (2.10)$$

where

$$A_0 = \Phi'(U_0, 0),$$

$$f_i = - \left. \frac{\partial \Phi(U_0, \varepsilon)}{\partial \varepsilon} \right|_{\varepsilon=0}.$$

(2) Compute $\delta J_1$:

$$\delta J_1 = (P'(u_0, 0) u_1, g_0)_{H_0}. \qquad (2.11)$$

(3) Solve the equation for $u_2$:

$$A_0 u_2 = f_2 - \left( \left. \int_0^1 \frac{d}{d\varepsilon} \Phi' \left( U_0 + t \sum_{i=1}^{\infty} \varepsilon^i u_i \right) dt \right|_{\varepsilon=0} \right) u_1, \qquad (2.12)$$

where

$$f_2 = - \left. \frac{\partial^2 \Phi(U_0, \varepsilon)}{2 \partial \varepsilon^2} \right|_{\varepsilon=0}.$$

(4) Compute $\delta J_2$:

$$\delta J_2 = (P'(U_0, 0) u_1, g_1)_{H_0} + (P'(U_0, 0) u_2, g_0)_{H_0}$$

$$+ \left( \left. \int_0^1 \frac{d}{d\varepsilon} P' \left( U_0 + t \sum_{i=1}^{\infty} \varepsilon^i u_i, \varepsilon \right) dt \right|_{\varepsilon=0} u_1, g_0 \right)_{H_0}. \qquad (2.13)$$

(5) Solve the equation for $u_3$:

$$A_0 u_3 = f_3 - \left( \left. \int_0^1 \frac{d}{d\varepsilon} \Phi' \left( U_0 + t \sum_{i=1}^{\infty} \varepsilon^i u_i, \varepsilon \right) dt \right|_{\varepsilon=0} \right) u_2$$

$$- \frac{1}{2} \left( \left. \int_0^1 \frac{d^2}{d\varepsilon^2} \Phi' \left( U_0 + t \sum_{i=1}^{\infty} \varepsilon^i u_i, \varepsilon \right) dt \right|_{\varepsilon=0} \right) u_1.$$

(6) Compute $\delta J_3$:

$$\delta J_3 = (P'(U_0, 0) u_1, g_2)_{H_0} + (P'(U_0, 0) u_2, g_1)_{H_0} + (P'(U_0, 0) u_3, g)_{H_0}$$

$$+ \left( \left. \int_0^1 \frac{d}{d\varepsilon} P' \left( U_0 + t \sum_{i=1}^{\infty} \varepsilon^i u_i, \varepsilon \right) dt \right|_{\varepsilon=0} u_1, g_1 \right)_{H_0}$$

$$+ \left( \left. \int_0^1 \frac{d}{d\varepsilon} P' \left( U_0 + t \sum_{i=1}^{\infty} \varepsilon^i u_i, \varepsilon \right) dt \right|_{\varepsilon=0} u_2, g_0 \right)_{H_0}$$

$$+ \frac{1}{2} \left( \left. \int_0^1 \frac{d^2}{d\varepsilon^2} P' \left( U_0 + t \sum_{i=1}^{\infty} \varepsilon^i u_i, \varepsilon \right) dt \right|_{\varepsilon=0} u_1, g_0 \right)_{H_0}. \qquad (2.14)$$

Then we put

$$\delta J \cong \sum_{i=1}^{3} \varepsilon_i \delta J_i,$$

$$J(U) \cong (P(U_0, \varepsilon), g(\varepsilon))_{H_0} + \sum_{i=1}^{3} \varepsilon_i \delta J_i. \tag{2.15}$$

**2.2.** Let us now formulate the algorithm for finding $\{\delta J_i\}$ with the use of the solution of the adjoint equation

$$A^*(u, \varepsilon) \, u^* = G, \tag{2.16}$$

where

$$A^*(u, \varepsilon) = \left( \int_0^1 \Phi'(U_0 + tu, \varepsilon) \, dt \right)^* = (A(u, \varepsilon))^*,$$

and $G$ is an element to be prescribed. With this aim in mind let us represent $\delta J$ in the form

$$\delta J = \left( u, \left( \int_0^1 P'(U_0 + tu, \varepsilon) \, dt \right)^* g(\varepsilon) \right)_{H_0}. \tag{2.17}$$

Notice that if we put

$$G = \left( \int_0^1 P'(U_0 + tu, \varepsilon) \, dt \right)^* g(\varepsilon), \tag{2.18}$$

then

$$\delta J \cong \sum_{i=1}^{3} \varepsilon_i \delta J_i^* + O(\varepsilon^4), \tag{2.19}$$

or

$$\delta J = (u, G)_{H_0} = (u, A^*(u, \varepsilon) \, u^*)_{H_0} = (A(u, \varepsilon) \, u, u^*)_{H_0}$$

$$= (f(\varepsilon), u^*)_{H_0} = \sum_{i=1}^{\infty} \sum_{j=1}^{\infty} \varepsilon^{i+j} (f_i, u_j^*)_{H_0},$$

where

$$\delta J_1^* = (f_1, u_0^*)_{H_0},$$

$$\delta J_2^* = (f_1, u_1^*)_{H_0} + (f_2, u_0^*)_{H_0}, \tag{2.20}$$

$$\delta J_3^* = (f_1, u_2^*)_{H_0} + (f_2, u_1^*)_{H_0} + (f_3, u_0^*)_{H_0}.$$

Taking into account (2.19), (2.20), we can find the corrections $\{\delta J_i^*\}_{i=1}^{3}$ by the following algorithm:

(1) Find $u_0^*$ as the solution to

$$A_0^* u_0^* = G_0, \qquad (2.21)$$

where

$$A_0^* = (A_0)^* = (\Phi'(U_0, 0))^*, \qquad G_0 = (P'(U_0, 0))^* g_0.$$

(2) Compute $\delta J_1^*$:

$$\delta J_1^* = (f_1, u_0^*)_{H_0}. \qquad (2.22)$$

(3) Find $u_1^*$ from the equation

$$A_0^* u_1^* = G_1 - \frac{\mathrm{d}}{\mathrm{d}\varepsilon} A^* \bigg|_{\varepsilon=0} u_0^*, \qquad (2.23)$$

where

$$G_1 = \frac{\mathrm{d}G}{\mathrm{d}\varepsilon}\bigg|_{\varepsilon=0} = \left( \frac{\mathrm{d}}{\mathrm{d}\varepsilon} \int_0^1 P'\left(U_0 + t\sum_{i=1}^{\infty} \varepsilon^i u_i, \varepsilon\right) \right)^* \bigg|_{\varepsilon=0} g_0$$

$$+ (P'(U_0, 0))^* g_1,$$

$$\frac{\mathrm{d}}{\mathrm{d}\varepsilon} A^* \bigg|_{\varepsilon=0} = \left( \int_0^1 \frac{\mathrm{d}}{\mathrm{d}\varepsilon} \Phi'\left(U_0 + t\sum_{i=1}^{\infty} \varepsilon^i u_i, \varepsilon\right) \mathrm{d}t \bigg|_{\varepsilon=0} \right)^*.$$

(4) Compute $\delta J_2^*$:

$$\delta J_2^* = (f_1, u_1^*)_{H_0} + (f_2, u_0^*)_{H_0}. \qquad (2.24)$$

(5) Find $u_2^*$ as the solution to

$$A_0^* u_2^* = G_2 - \frac{\mathrm{d}}{\mathrm{d}\varepsilon} A^* \bigg|_{\varepsilon=0} u_1^* - \frac{1}{2} \frac{\mathrm{d}^2 A^*}{\mathrm{d}\varepsilon^2}\bigg|_{\varepsilon=0} u_0^*, \qquad (2.25)$$

where

$$G_2 = (P^*(U_0, 0))^* g_2 + \left( \int_0^1 \frac{\mathrm{d}}{\mathrm{d}\varepsilon} P'\left(U_0 + t\sum_{i=1}^{\infty} \varepsilon^i u_i, \varepsilon\right) \mathrm{d}t \right)^* \bigg|_{\varepsilon=0} g_1$$

$$+ \frac{1}{2} \left( \int_0^1 \frac{\mathrm{d}^2}{\mathrm{d}\varepsilon^2} P'\left(U_0 + t\sum_{i=1}^{\infty} \varepsilon^i u_i, \varepsilon\right) \mathrm{d}t \right)^* \bigg|_{\varepsilon=0} g_0.$$

(6) Compute $\delta J_3^*$:

$$\delta J_3^* = (f_1, u_2^*)_{H_0} + (f_2, u_1^*)_{H_0} + (f_3, u_0^*)_{H_0}. \qquad (2.26)$$

Then we put

$$\delta J \cong \sum_{i=1}^{3} \varepsilon_i \delta J_i^*,$$

$$(2.27)$$

$$J(U) = (P(U_0, \varepsilon), g(\varepsilon))_{H_0} + \sum_{i=1}^{3} \varepsilon_i \delta J_i^*.$$

Let us point out one of the simple approximations to $J$ — the first order approximation

$$J^{(1)} = J_0 + \varepsilon \delta J_1, \qquad (2.27a)$$

where

$$J_0 = (P(U_0, \varepsilon), g(\varepsilon))_{H_0}, \qquad \delta J_1 = (P(U_0, 0) u_1, g_0)_{H_0}.$$

The value $J^{(1)}$ can be represented also as

$$J^{(1)} = J_0 + \varepsilon \delta J_1^*, \qquad (2.28)$$

where

$$\delta J_1^* = (f, u_0^*)_{H_0}.$$

Representation (2.28) has the following property important for practical applications. If equation (2.21) is once solved, then, knowing $u_0^*$, one can compute $J^{(1)}$ for different variations $f_1$ in the right-hand side of (2.10). (It is easily seen that a similar property holds when $\delta J$ is represented in the form (2.19).)

From (2.15), (2.19), one can conclude that $\delta J_i = \delta J_i^*$, $i = 1, 2, 3$. It is not also difficult to verify these equalities immediately. The following statement is valid.

**Lemma 2.1.** *The corrections $\delta J_1$, $\delta J_2$, $\delta J_3$ may be computed using only $u_1$, $u_0^*$, $u_1^*$ (i.e. without finding $u_2$, $u_3$, $u_2^*$) by formulae (2.22), (2.24) and*

$$\delta J_3 = \delta J_3^* = \left( u_1, (P'|_{\varepsilon=0})^* g_2 + \left( \int_0^1 \frac{\mathrm{d}}{\mathrm{d}\varepsilon} P' \, \mathrm{d}t \Big|_{\varepsilon=0} \right)^* g_1 \right.$$

$$+ \frac{1}{2} \left( \int_0^1 \frac{\mathrm{d}^2}{\mathrm{d}\varepsilon^2} P' \, \mathrm{d}t \Big|_{\varepsilon=0} \right)^* g_0 \right)_{H_0} + \left( f_2 - \frac{\mathrm{d}A}{\mathrm{d}\varepsilon} \Big|_{\varepsilon=0} u_1, u_1^* \right)_{H_0}$$

$$+ \left( f_3 - \frac{1}{2} \frac{\mathrm{d}^2 A}{\mathrm{d}\varepsilon^2} \Big|_{\varepsilon=0} u_1, u_0^* \right)_{H_0}. \qquad (2.29)$$

*Proof.* The immediate verification gives

$$\delta J_1 = (P'(U_0, 0) u_1, g_0)_{H_0} = (u_1 (P'(U_0, 0))^* g_0)_{H_0}$$

$$= (u_1, A_0^* u_0^*)_{H_0} = (A_0 u_1, u_0^*)_{H_0} = (f_1, u_0^*)_{H_0} = \delta J_1^*,$$

$$\delta J_2 = (P'(U_0, 0) u_1, g_1)_{H_o} + (P'(U_0, 0) u_2, g_0)_{H_o}$$

$$+ \left( \left( \int_0^1 \frac{\mathrm{d}}{\mathrm{d}\varepsilon} P' \left( U_0 + t \sum_{i=1}^\infty \varepsilon^i u_i, \varepsilon \right) \mathrm{d}t \bigg|_{\varepsilon=0} u_1, g_0 \right)_{H_o} \right)$$

$$= (A_0 u_1, u_1^*)_{H_o} + \left( \frac{\mathrm{d}A}{\mathrm{d}\varepsilon} \bigg|_{\varepsilon=0} u_1, u_0^* \right)_{H_o} + (u_2, (P'(U_0, 0))^* g_0)_{H_o}$$

$$= (f_1, u_1^*)_{H_o} + (f_2 - A_0 u_2, u_0^*)_{H_o} + (u_2, (P'(U_0, 0))^* g_0)_{H_o}$$

$$= (f_1, u_1^*)_{H_o} + (f_2, u_0^*)_{H_o} - (u_2, A_0^* u_0^*)_{H_o} + (u_2, (P'(U_0, 0))^* g_0)_{H_o}$$

$$= (f_1, u_1^*)_{H_o} + (f_2, u_0^*)_{H_o} = \delta J_2^*.$$

(Note that to find $\delta J_2 = \delta J_2^*$ we need not know the corrections $u_2$, $u_1$. It is sufficient to compute $u_0^*$, $u_1^*$ only. If, therewith, $f_1 \equiv 0$, then to know $u_1^*$ is not also required.) Similarly

$$\delta J_3 = (u_1, (P'|_{\varepsilon=0})^* g_2)_{H_o} + (u_2, (P'|_{\varepsilon=0})^* g_1)_{H_o}$$

$$+ \left( u_1, \left( \int_0^1 \frac{\mathrm{d}}{\mathrm{d}\varepsilon} P' \, \mathrm{d}t \bigg|_{\varepsilon=0} \right)^* g_1 \right)_{H_o} + (u_3, (P'|_{\varepsilon=0})^* g_0)_{H_o}$$

$$+ \left( u_2, \left( \int_0^1 \frac{\mathrm{d}}{\mathrm{d}\varepsilon} P' \, \mathrm{d}t \bigg|_{\varepsilon=0} \right)^* g_0 \right)_{H_o}$$

$$+ \left( u_1, \frac{1}{2} \left( \int_0^1 \frac{\mathrm{d}^2}{\mathrm{d}\varepsilon^2} P' \, \mathrm{d}t \bigg|_{\varepsilon=0} \right)^* g_0 \right)_{H_o}$$

$$= (A_0 u_1, u_2^*)_{H_o} + \left( \frac{\mathrm{d}}{\mathrm{d}\varepsilon} A \bigg|_{\varepsilon=0} u_1, u_1^* \right)_{H_o} + \left( \frac{1}{2} \frac{\mathrm{d}^2 A}{\mathrm{d}\varepsilon^2} \bigg|_{\varepsilon=0} u_1, u_0^* \right)_{H_o}$$

$$+ (A_0 u_2, u_1^*)_{H_o} + \left( \frac{\mathrm{d}A}{\mathrm{d}\varepsilon} A \bigg|_{\varepsilon=0} u_2, u_0^* \right)_{H_o} + (A_0 u_3, u_0^*)_{H_o}$$

$$= (f_3, u_0^*)_{H_o} + (f_2, u_1^*)_{H_o} + (f_1, u_2^*)_{H_o} = \delta J_3^*.$$

Let us transform the expression for $\delta J_3$, eliminating $u_2$ and $u_3$. We get

$$\delta J_3 = \left( u_1, (P'|_{\varepsilon=0})^* g_2 + \left( \int_0^1 \frac{\mathrm{d}}{\mathrm{d}\varepsilon} P' \, \mathrm{d}t \bigg|_{\varepsilon=0} \right)^* g_1 \right.$$

$$+ \left. \left( \frac{1}{2} \int_0^1 \frac{\mathrm{d}^2}{\mathrm{d}\varepsilon^2} P' \, \mathrm{d}t \bigg|_{\varepsilon=0} \right)^* g_0 \right)_{H_o}$$

$$+ \left( u_2, A_0^* u_1^* + \frac{\mathrm{d}}{\mathrm{d}\varepsilon} A^* \Big|_{\varepsilon=0} u_0^* \right)_{H_0} + (u_3, A_0^* u_0^*)_{H_0}$$

$$= \ldots + (A_0 u_2, u_1^*)_{H_0} + \left( \frac{\mathrm{d}A}{\mathrm{d}\varepsilon} \Big|_{\varepsilon=0} u_2, u_0^* \right)_{H_0} + (A_0 u_3, u_0^*)_{H_0}$$

$$= \ldots + \left( f_2 - \frac{\mathrm{d}A}{\mathrm{d}\varepsilon} \Big|_{\varepsilon=0} u_1, u_1^* \right)_{H_0} + \left( \frac{\mathrm{d}A}{\mathrm{d}\varepsilon} \Big|_{\varepsilon=0} u_2, u_0^* \right)_{H_0}$$

$$+ (f_3, u_0^*)_{H_0} - \left( \frac{\mathrm{d}A}{\mathrm{d}\varepsilon} \Big|_{\varepsilon=0} u_2, u_0^* \right)_{H_0} - \left( \frac{1}{2} \frac{\mathrm{d}^2 A}{\mathrm{d}\varepsilon^2} \Big|_{\varepsilon=0} u_1, u_0^* \right)_{H_0}$$

$$= \ldots + \left( f_2 - \frac{\mathrm{d}A}{\mathrm{d}\varepsilon} \Big|_{\varepsilon=0} u_1, u_1^* \right)_{H_0} + \left( f_3 - \frac{1}{2} \frac{\mathrm{d}^2 A}{\mathrm{d}\varepsilon^2} \Big|_{\varepsilon=0} u_1, u_0^* \right)_{H_0} .$$

Hence,

$$\delta J_3 = \delta J_3^* = \left( u_1, (P'|_{\varepsilon=0})^* g_2 + \left( \int_0^1 \frac{\mathrm{d}}{\mathrm{d}\varepsilon} P' \, \mathrm{d}t \Big|_{\varepsilon=0} \right)^* g_1 \right.$$

$$+ \left( \frac{1}{2} \int_0^1 \frac{\mathrm{d}^2}{\mathrm{d}\varepsilon^2} P' \, \mathrm{d}t \Big|_{\varepsilon=0} \right)^* g_0 \Bigg)_{H_0}$$

$$+ \left( f_2 - \frac{\mathrm{d}A}{\mathrm{d}\varepsilon} \Big|_{\varepsilon=0} u_1, u_1^* \right)_{H_0} + \left( f_3 - \frac{1}{2} \frac{\mathrm{d}^2 A}{\mathrm{d}\varepsilon^2} \Big|_{\varepsilon=0} u_1, u_0^* \right)_{H_0} .$$

In conclusion, we indicate the cases where corrections $\{\delta J_i\}$ are expedient to be computed by using functions $\{u_i^*\}_{i=0}^\infty$. The computation can be efficient if we can once compute $\{u_i^*\}_{i=0}^N$ and then find the values $\delta J$ with a sufficient accuracy for different $\{f_i\}$. The computation of $\delta J$ on the basis of $\{u_i^*\}$ may appear to be also expedient in the case where equations with adjoint operators (for example, the equation $A^*(u, \varepsilon) u^* = g$) are correctly solvable. It means that for such equations the function $u$ (which is unknown) can be replaced with an approximation appropriately chosen. As a result, we obtain a stable process of computation of $\delta J$ (based, for example, on perturbation algorithms) by finding $\{\delta J_i\}$.

## 3. SPECTRAL METHOD IN PERTURBATION ALGORITHMS

As obviously follows from Sections 1 and 2, it is important that an efficient technique for inverting operators $A_0 = \Phi'(U_0, 0)$ and $A_0^* = (\Phi'(U_0, 0))^*$ be available. In some cases, for this technique we can use the spectral method. Thus, assume that $X = Y = H_0 = H_0^*$. Let the spectra of these operators be discrete and their eigenfunctions $\{\varphi_i\}$ and $\{\varphi_i\}$:

$$A_0\varphi_i = \lambda_i\varphi_i,$$

$$A_0^*\varphi_i^* = \lambda_i^*\varphi_i^* \tag{3.1}$$

form a biorthogonal system. Assume that $\bar{\lambda}_i = \lambda_i^*$ and each of the systems $\{\varphi_i\}$ and $\{\varphi_i\}$ is complete in $H_0$, i.e. for any element $u \in Y$, the following representations are valid:

$$u = \sum_{i=1}^{\infty} a_i\varphi_i = \sum_{j=1}^{\infty} b_j^*\varphi_j^*,$$

where

$$\left\| u - \sum_{i=1}^{N} a_i\varphi_i \right\|_{H_0} \to 0, \qquad \left\| u - \sum_{j=1}^{N} b_j^*\varphi_j^* \right\|_{H_0} \to 0 \quad \text{for } N \to \infty,$$

and the systems $\{\varphi_i\}$ and $\{\varphi_j^*\}$ satisfy the relationship

$$\varphi_j^*(\varphi_i) \equiv (\varphi_i, \varphi_j^*)_{H_0} = \delta_{ij},$$

where $\delta_{ij}$ is the Kronecker delta. The above formulae imply that

$$a_i = (u, \varphi_i^*)_{H_0}, \qquad b_j^* = (u, \varphi_j)_{H_0}.$$

In specific problems, it may appear expedient to preliminarily find systems $\{\varphi_i\}$ and $\{\varphi_j^*\}$, and then make use of them for finding correction functions $\{u_i\}$ and $\{u_i^*\}$ in perturbation algorithms. Thus, assume that $\{\varphi_i\}$, $\{\varphi_j^*\}$, $\{\lambda_i\}$ and $\{\lambda_j^*\}$ are found. Then the equation

$$A_0 u_1 = f_1 \tag{3.2}$$

can be easily solved by the Fourier method which results in finding $u_1$:

$$u_1 = \sum_{i=0}^{\infty} \frac{(f_1, \varphi_i^*)_{H_0}}{\lambda_i} \varphi_i. \tag{3.3}$$

The first correction $\delta J_1$ by formula (2.11) can be computed as follows:

$$\delta J_1 = \sum_{j=0}^{\infty} \frac{(f_1, \varphi_j^*)_{H_0}(\varphi_j, (P'(U_0,0))^*g_0)_{H_0}}{\lambda_j}. \tag{3.4}$$

If the solution to the adjoint equation

$$A_0^* u_0^* = (P'(U_0,0))^* g_0 \tag{3.5}$$

is found, i.e.

$$u_0^* = \sum_{i=1}^{\infty} \frac{(\varphi_i, (P'(U_0,0))^* g_0)_{H_0}}{\lambda_i^*} \varphi_i^*, \qquad (3.6)$$

then (3.6) and (2.22) will again lead to expression (3.4).

Note the advantages of the representations of $u_1$, $\delta J_1$ and $u_0^*$ in forms (3.3), (3.4) and (3.6), respectively. For example, if it is *a priori* known that the element $f_1$ and $P'(U_0,0))^* g_0$ are sufficiently accurately approximated by a set of eigenvectors, then these are the eigenvectors to be computed and stored. The other eigenvectors in this problem are actually not needed. In addition, if it is necessary to compute the value of the correction $\delta J_1$, we have to know those vectors of the systems $\{\varphi_i\}$ and $\{\varphi_j^*\}$, which satisfy the relationship $(\varphi_i, \varphi_j^*)_{H_0} = 1$, i.e. under the assumptions made, the vectors with identical indices. In practical computations, there may arise a situation where $f_1$ is expanded into a system of orthogonal vectors, into which the element $(P'(U_0,0))^* g_0$ is expanded. In this case, $\delta J_1 = 0$. Taking into account the above arguments, we can assume that there may arise a situation where only a small number of vectors from the systems $\{\varphi_j\}$ and $\{\varphi_j^*\}$ are necessary for finding the value $\delta J_1$. It is evident that the approaches considered are also efficient for solving a number of similar problems with the same operator $A$.

Let us consider the first approximation in perturbation algorithms for finding $\delta J$. By analogy with the interpretation of adjoint functions in linear problems, we can conclude that the function $u_0^*$ is a value function of the correction $f_1$ with respect to the correction of the functional $\delta J_1 = (f_1, u_0^*)_{H_0} = (P'(U_0,0) u_1, g_0)_{H_0}$, which is computed. If the elements $f_1$ and $(P'(U_0,0))^* g_0$ are put into correspondence to the sets of Fourier coefficients $\{(f_1, \varphi_j^*)_{H_0}\}$ and $\{(\varphi_j, (P'(U_0,0))^* g_0)_{H_0}\}$, then the coefficients $\{(\varphi_j, (P'(U_0,0))^* g_0)_{H_0}/\lambda_j^*\}$ can be interpreted as value coefficients (or weight coefficients) of the coefficients $\{(f_1, \varphi_j^*)_{H_0}\}$ with respect to their contribution to the value $\delta J_1$ computed by formula (3.4). It is obvious that the contribution of a coefficient $(f, \varphi_{j_0}^*)_{H_0}$ will be essential if the value of $|(\varphi_{j_0}, (P'(U_0,0))^* g_0)_{H_0}/\lambda_{j_0}^*|$ is large. Note that the annihilation of coefficients is not discussed here, as the possible impact of this effect is obvious.

Using the representation of $\delta J_1$ in form (3.4), we can realize the following: an analysis of the impact of $f_1$ and $g_0$ on the value $\delta J_1$; a formulation of the problem of the optimal choice of the functions $g_0$ on the basis of equality (3.4) with prescribed $\delta J_1$ and $f_1$; a formulation of the problem of finding the informative harmonic $\varphi_{j_0}$ by which a component in $u_1$ makes the maximal contribution into $\delta J_1$. Thus, the informative harmonic is that for which the quantity $|(f_1, \varphi_{j_0}^*)_{H_0}(\varphi_{j_0}, (P'(U_0,0))^* g_0)_{H_0}/\lambda_{j_0}|$ prevails over similar values of the other vectors. And if $f_1, g_0, P'(U_0,0), \{\varphi_j\}$ and $\{\varphi_{j_0}\}$ are given, the choice of $\varphi_{j_0}$ is easy to be made. It is obvious that this choice depends on a number of factors (the values $\lambda_{j_0}, f_1, (P'(U_0,0)^* g_0)$ and others) and is not determined only by the number of the eigenfunction depending on an increase in absolute eigenvalues $\{\lambda_j\}$.

*Example* 3.1. Let us consider one of the simplest problems of magnetohy-drodynamics

$$\tilde{\Phi}(U) = 0, \qquad U(a) = U(b) = 0, \tag{3.7}$$

where

$$\tilde{\Phi}(U) = -U'' + e^U, \qquad U'' = \frac{d^2 U}{dx^2}.$$

Set $X = Y = L_2(a, b)$ and $D(\Phi) = W_2^2 \cap \overset{\circ}{W_2^1}$. Rewrite equation (3.7) in the form

$$\Phi(U) \equiv -U'' + (e^U - 1) = -1. \tag{3.8}$$

For a certain function $U_0 \in D(\Phi)$, we have

$$A(u) u = \int_0^1 \Phi'(U + tu) \, dtu = f, \tag{3.9}$$

where

$$u = U - U_0, \qquad f = \Phi(U_0 + u) - \Phi(U_0) = U_0'' - e^{U_0},$$

$$\Phi'(U_0 + tu) v = -v'' + e^{U_0} \frac{(e^u - 1)}{u} v, \qquad \Phi(U_0) v = -v'' + e^{U_0} v.$$

For $|u| \leq$ constant, the function $(e^u - 1)/u$ is positive. Therefore, for any fixed function $u \in \overset{\circ}{W_2^1} \subset C(a, b)$, we obtain the relationship $(A(u) v, v)_{L_2} \geq c\|v\|^2_{W_2^1} \geq c\|v\|^2_{L_2}$. Hence, the equations

$$A(u) v = f,$$
$$A^*(u) u^* = g, \tag{3.10}$$

where $A^*(u) = A(u)$, are correctly solvable in $\overset{\circ}{W_2^1}$. Therefore, to find their approximate solutions, we can make use of a perturbation algorithm that will be convergent provided the choice of $U_0$ is made in a sufficiently proper manner. To apply this algorithm, we put problem (3.9) into correspondence to single-parameter families of problems of the form

$$A(\varepsilon u) u = f,$$

$$A^*(\varepsilon u) u^* = g,$$

which for $\varepsilon = 1$ coincide with the problems (3.10) we are interested in. According to the arguments contained in Section 1, we seek for $u(\varepsilon)$ in the form

$$u = \sum_{i=0}^{\infty} \varepsilon^i u_i, \tag{3.11}$$

where $\{u_i\}$ can be found from the equations

$$A_0 u_0 \; = \; f,$$

$$A_0 u_1 \; = \; -A_1(U_0, u_0)\, u_0, \qquad\qquad (3.12)$$

$$\cdots$$

where

$$A_0 \; = \; -\frac{\mathrm{d}^2}{\mathrm{d}x^2} + e^{U_0},$$

$$A_1(U_0, u_0) \; = \; \frac{\partial}{\partial \varepsilon} \int_0^1 \Phi'\left(U_0 + \varepsilon t \sum_{i=0}^{\infty} \varepsilon^i u_i\right) \mathrm{d}t \, \Bigg|_{\varepsilon=0} = \frac{1}{2} e^{U_0} u_0.$$

Equations (3.12) can be solved by the spectral method using the eigenfunctions and eigenvalues of the problem

$$-\varphi_j'' + e^{U_0} \varphi_j = \lambda_j \varphi_j, \qquad \varphi_j(0) = \varphi_j(b) = 0, \qquad j = 1, 2 \ldots \qquad (3.13)$$

which for $U_0 \equiv 0$ and $b = 1$ are of the form

$$\varphi_j = \sqrt{2} \sin j\pi x, \qquad \lambda_j = j^2 \pi^2 + 1, \qquad j = 1, 2 \ldots \qquad (3.14)$$

Making use of $\{\varphi_j\}$ and $\{\lambda_j\}$ for finding $u_0, u_1, \ldots$, we have

$$u_0 \; = \; \sum_{j=0}^{\infty} \frac{(f, \varphi_j)_{L_2(0,b)}}{\lambda_j} \, \varphi_j(x),$$

$$u_1 \; = \; -\sum_{j=0}^{\infty} \frac{\left(e^{U_0} \frac{u_0^2}{2}, \varphi_j\right)_{L_2(0,b)}}{\lambda_j} \, \varphi_j(x), \qquad\qquad (3.15)$$

$$\cdots$$

Then, for $\varepsilon = 1$ we can set

$$U \cong U_0 + u_0 + u_1.$$

Assume that we are interested in finding an approximate value of the functional $J(U) = (U, g)_{L_2(0,b)}$, where $g$ is a prescribed function. Then

$$J \cong (U_0, g)_{L_2(0,b)} + \sum_{j=1}^{\infty} \frac{(\varphi_j, g)_{L_2(0,b)}}{\lambda_j} \left\{ (f, \varphi_j)_{L_2(0,b)} - \left(e^{U_0} \frac{u_0^2}{2}, \varphi_j\right)_{L_2(0,b)} \right\}.$$

If $g = \delta(x - x_0)$ is the Dirac delta function with the support at the point $x_0$, then

$$J \cong U_0(x_0) + \sum_{j=1}^{\infty} \frac{\left[\int_0^b f(x)\,\varphi_j(x)\,dx - \left(e^{U_0}\frac{u_0^2}{2}, \varphi_j\right)_{L_2(0,b)}\right]}{\lambda_j} \varphi_j(x_0). \quad (3.16)$$

Assume that $U_0 = 0$ and $b = 1$. Then $f = -1$, and we obtain

$$u_0 = -\sum_{j=1}^{\infty} \frac{4}{(2j+1)\,\pi} \frac{\sin(2j+1)\,\pi x}{(2j+1)^2 \pi^2 + 1},$$

$$(3.17)$$

$$J \cong 2\sum_{j=1}^{\infty}\left[-\frac{1-(-1)^j}{j\pi} - \int_0^1 \frac{u_0^2(x)}{2} \sin j\pi x\,dx\right]\bigg/(j^2\pi^2 + 1)\sin j\pi x_0.$$

Note that the first addend in the last sum is

$$2\left[-\frac{2}{\pi} - \int_0^1 \frac{u_0^2(x)}{2} \sin \pi x\,dx\right]\bigg/(\pi^2 + 1)\sin \pi x_0.$$

Since $u_0 \cong 4\sin 3\pi x/(3\pi(9\pi^2 + 1))$, then neglecting the value of $\int_0^1 u_0^2 \sin \pi x\,dx/2$ for $x_0 = 1/2$, we have

$$U(x_0) \cong \frac{-4}{\pi(\pi^2 + 1)} \cong -0.117$$

with the exact value $U(x_0) = -0.1137004$. [211] Thus, even under essential approximations used at intermediate stages of the computation, the algorithms considered can produce the final result with a sufficient accuracy.

## 4. JUSTIFICATION OF THE $N$-TH ORDER PERTURBATION ALGORITHMS

In Section 1 we have considered a perturbation algorithm for non-linear equations of the general form

$$\Phi(U, \varepsilon) = 0. \quad (4.1)$$

This section is concerned with justification of perturbation algorithms for a specific class of non-linear equations.

**4.1.** Let $H_1$ and $H_0$ be the Hilbert spaces with $H_0 \equiv H_0^*$, $H_1^* \equiv (H_1)^{-1}$, $H_1$ being densely enclosed into $H_0$. Thus, the following inclusions are valid: $H_1 \subset H_0 \equiv H_0^* \equiv (H_1)^{-1}$. According to the notations of Section 1, we put $X = H_1$, $Y = H_1^*$, $H_0 \equiv H^0$.

In many problems of mathematical physics the operator $\Phi(U, \varepsilon)$ has the form

$$\Phi(U, \varepsilon) = F_0 U + \varepsilon F_{01} U + \varepsilon^{M+1} F_1(U) + F_2, \qquad (4.2)$$

where $F_0$ and $F_{01}$ are linear operators, $F_1$ is a $K$-power operator, $M \geq 0$ and $K$ are integers, $F_2$ is a prescribed element of $H_1^*$, $\varepsilon$ a small parameter. All the operators map $X = H_1$ into $Y = H_1^*$ with the same domain $D(\Phi)$. Moreover, without the loss of generality, we shall assume that $F_1(U)$ is generated by a $K$-linear symmetric operator $F_1(u_1, u_2, \ldots, u_K)$.[290] By $F_1'(U) u \equiv K F_1(u, U, \ldots, U)$ we denote the Gâteaux derivative of the operator $F_1(U)$. Then

$$\Phi'(U, \varepsilon) = F_0 + \varepsilon F_{01} + \varepsilon^{M+1} F_1'(U). \qquad (4.3)$$

Equation (4.1) reduces to equation (1.2):

$$A(u, \varepsilon) u = f(\varepsilon), \qquad (4.4)$$

where

$$A(u, \varepsilon) = \int_0^1 \Phi'(U_0 + tu, \varepsilon) \, dt, \qquad f(\varepsilon) = -\Phi(U_0, \varepsilon), \qquad u = U - U_0.$$

If the operator $\Phi$ has the form (4.2), the operators $A(0, \varepsilon)$ and $A(u, \varepsilon)$ are

$$A(0, \varepsilon) = F_0 + \varepsilon F_{01} + \varepsilon^{M+1} F_1'(U_0),$$

$$A(u, \varepsilon) = F_0 + \varepsilon F_{01} + \varepsilon^{M+1} \int_0^1 F_1'(U_0 + tu) \, dt. \qquad (4.5)$$

Assume that the operator $F_1'$ satisfies the restriction

$$\sup_{h \in D(F')} \frac{\|(F_1'(U_0 + u) - F_1'(U_0 + v)) h\|_{H_1^*}}{\|h\|_{H_1}} \leq k_1 \|u - v\|_{H_1}, \qquad (4.6)$$

where $u, v \in D(F')$, $k_1 = $ constant $> 0$.

**Lemma 4.1.** *Let the operator $\Phi$ have the form (4.2), for any $\varepsilon \in [0, \varepsilon_0]$ the hypotheses (1), (2) of Lemma 1.1 be satisfied, and (4.6) hold. Then for*

$$\frac{\varepsilon^{M+1} k_1 R}{m_\varepsilon} \leq q_1 = \text{ constant } < 1,$$

$$\|f(\varepsilon)\|_{H_1^*} \leq R(m_\varepsilon - \varepsilon^{M+1} k_1) \qquad (4.7)$$

*equation (4.4) has a unique solution*

$$u = \sum_{i=0}^{\infty} \varepsilon^i u_i, \qquad (4.8)$$

*in $\bar{S}_{H_1}(0, R) = \{u: (U_0+u) \in D(\Phi), \|u\|_{H_1} \leq R\}$, where the elements $\{u_i\}_{i=0}^{\infty}$ are independent of $\varepsilon$, and for the successive approximations in (1.9):*

$$u^{(n)} = \sum_{i=0}^{\infty} \varepsilon^i u_i^{(n)} \tag{4.9}$$

*the following estimate is valid:*

$$\|u - u^{(n)}\|_{H_1} \leq \left(\frac{\varepsilon^{M+1} k_1 R}{m_\varepsilon}\right)^n \frac{\|u^{(0)} - B(u^{(0)}, \varepsilon) u^{(0)} - \bar{f}\|_{H_1}}{1 - q_1}$$

$$\leq C_0 R q_1^n. \tag{4.10}$$

*Proof is similar to the proof of Lemma 1.1 with due regard for (4.3), (4.6).*

**4.2.** Thus, if the operator $\Phi(U, \varepsilon)$ is analytical in its variables and the hypotheses of Lemma 4.1 are satisfied, then, in view of the representations

$$f(\varepsilon) = \sum_{i=0}^{\infty} \varepsilon^i f_i, \qquad f_i \in H_1^*, \tag{4.11}$$

$$u(\varepsilon) = \sum_{i=0}^{\infty} \varepsilon^i u_i, \qquad u_i \in H_1, \tag{4.12}$$

it is not difficult to derive the equations for finding the elements $\{u_i\}$. One can do this, for example, by using the general scheme of regular perturbation algorithms given in Section 1. In the specific cases when $K = 1, 2, 3$ the operator $A(u, \varepsilon)$ has the form

$$K = 1: \quad A(u, \varepsilon) v = F_0 v + \varepsilon F_{01} v + \varepsilon^{M+1} F_1(v),$$

$$K = 2: \quad A(u, \varepsilon) v = F_0 v + \varepsilon F_{01} v + \varepsilon^{M+1}(2F_1(v, U_0) + F_1(v, u)), \tag{4.13}$$

$$K = 3: \quad A(u, \varepsilon) v = F_0 v + \varepsilon F_{01} v + \varepsilon^{M+1}(3F_1(v, U_0, U_0)$$

$$+ 3F_1(v, u, U_0) + F_1(v, u, u)),$$

and equations of perturbation algorithm are

$$K = 1: \quad F_0 u_0 = f_0,$$

$$F_0 u_k = f_k - F_{01} u_{k-1} - F_1(u_{k-M-1}), \quad k = 1, 2, \ldots \tag{4.14}$$

$$(F_1(u_{k-M-1}) \equiv 0 \text{ if } k - M - 1 < 0);$$

$$K = 2: \quad F_0 u_0 = f_0,$$

$$F_0 u_k = f_k - F_{01} u_{k-1} - 2F_1(u_{k-M-1}, U_0)$$

$$- \sum_{i+j+M+1=k} F_1(u_i, u_j), \quad k = 1, 2, \ldots \quad (4.15)$$

$(F_1(u_{k-M-1}, U_0) \equiv 0$ if $k - M - 1 < 0$; and if $i + j > k - M - 1$, the corresponding addend in the last sum is omitted);

$$K = 3: \quad F_0 u_0 = f_0,$$

$$F_0 u_k = f_k - F_{01} u_{k-1} - 3F_1(u_{k-M-1}, U_0, U_0)$$

$$-3 \sum_{i+j+M+1=k} F_1(u_i, u_j, U_0)$$

$$- \sum_{i+j+l+M+1=k} F_1(u_i, u_j, U_l), \quad k = 1, 2, \ldots \quad (4.16)$$

(with respective modifications of the sums when $k - M - 1 < 0$, $i + j + l + M + 1 > k$).

Solving successively these equations, we obtain

$$U = U_0 + \sum_{i=0}^{\infty} \varepsilon^i u_i. \quad (4.17)$$

An element of the form

$$U_{(N)} = U_0 + \sum_{i=0}^{N} \varepsilon^i u_i \quad (4.18)$$

is said to be the $N$-th order approximation to $U$. To find $U_{(N)}$ it is sufficient to solve the first $N + 1$ equations of (4.16) (or (4.14), (4.15)).

*Remark* 4.1. If $U_0$ is a solution of the equation

$$F(U_0, 0) = 0, \quad (4.19)$$

then

$$f(\varepsilon)|_{\varepsilon=0} = -F(U_0, \varepsilon)|_{\varepsilon=0} = 0.$$

Hence, $f_0 \equiv 0$, $u_0 \equiv 0$ and the solution of equation (4.4) has the form

$$u = \sum_{i=1}^{\infty} \varepsilon^i u_i, \quad (4.20)$$

i.e. $u|_{\varepsilon=0} \equiv 0$.

**4.3.** Along with equation (4.4), consider the adjoint equation

$$A^*(u, \varepsilon) u^* = g(\varepsilon), \quad (4.21)$$

where $g(\varepsilon)$ is defined by (1.5), $A^*(u, \varepsilon)$ is given by (1.3), and $u \in D(A) = D(F)$.

The following lemma is valid.

**Lemma 4.2.** *Let the operator $\Phi$ have the form (4.2), for any $\varepsilon \in [0, \bar{\varepsilon}_0]$ the hypotheses (1), (2), (4) of Lemma 1.3 be satisfied, and*

$$\sup_{h \in D(F')} \frac{\|(F_1'(U_0 + u) - F_1'(U_0 + v)) h\|_{H_1^*}}{\|h\|_{H_1}} \le k_1^{(1)} \|u - v\|_E, \qquad (4.22)$$

*where $k_1^{(1)} = $ constant $> 0$. Then for*

$$\frac{\varepsilon^{M+1} k_1^{(1)} \bar{R}}{2m_\varepsilon} \le q^{(1)} = \text{constant} < 1 \qquad (4.23)$$

*equation (4.21) is correctly solvable in $H_1^*$. If, moreover, the element $u$ is analytic in $\varepsilon$, then the solution $u^*$ of equation (4.21) has the form*

$$u^* = \sum_{i=1}^{\infty} \varepsilon^i u_i. \qquad (4.24)$$

*Proof* is similar to the proof of Lemma 1.3, with due regard for (4.2), (4.22).

**Corollary.** *If the hypotheses of Lemma 4.2 are satisfied, the solution $u^*$ of equation (4.21) may be constructed by the successive approximation method*

$$A^*(0, \varepsilon) u^{*(n+1)} = (A^*(0, \varepsilon) - A^*(u, \varepsilon)) u^{*(n)} + g(\varepsilon), \qquad n = 0, 1, \ldots,$$

*and*

$$\|u^* - u^{*(n)}\|_{H_1} \le \left( \frac{\varepsilon^{M+1} k_1^{(1)} \bar{R}}{2m_\varepsilon} \right)^n \frac{1}{1 - q^{(1)}} \|A^{*-1}(0, \varepsilon)(A^*(u, \varepsilon) u^{*(0)} - g)\|_{H_1}.$$

**4.4** Let $\Phi$ be an analytical operator, and the element $u$ determining the form of $A(u, \varepsilon)$ according to (4.5) is also analytic in $\varepsilon$. Then to find $u^* = \sum_{i=0}^{\infty} \varepsilon^i u_i^*$, as for main equations, we can use the regular perturbation algorithm. Let $\Phi$ have the form (4.2) and the equality hold:

$$A^*(u, \varepsilon) = F_0^* + \varepsilon F_{01}^* + \varepsilon^{M+1} \int_0^1 (F_1'(U_0 + tu))^* \, dt.$$

Then a set of equations for finding $\{u_i^*\}$ have the form

$$K = 1: \quad F_0^* u_0^* = p_0,$$

$$F_0^* u_k^* = p_k - F_{01}^* u_{k-1}^* - F_1^*(u_{k-M-1}^*), \quad k = 1, 2, \ldots; \quad (4.25)$$

$$K = 2: \quad F_0^* u_0^* = p_0,$$

$$F_0^* u_k^* = p_k - F_{01}^* u_{k-1}^* - 2F_1^*(U_0, \cdot) u_{k-M-1}^*$$

$$- \sum_{i+j+M+1=k} F_1^*(u_i, \cdot) u_j^*, \quad k = 1, 2, \ldots; \quad (4.26)$$

$$K = 3: \quad F_0^* u_0^* = p_0,$$

$$F_0^* u_k^* = p_k - F_{01}^* u_{k-1}^* - 3F_1^*(U_0, U_0, \cdot) u_{k-M-1}^*$$

$$-3 \sum_{i+j+M+1=k} F_1^*(U_0, u_i, \cdot) u_j^*$$

$$- \sum_{i+j+l+M+1=k} F_1^*(u_i, u_j, \cdot) u_l^*, \quad k = 1, 2, \ldots \quad (4.27)$$

Here $F_1^*(U_0, \cdot)$ is the operator adjoint to $F_1(U_0, \cdot)$ linear with respect to the argument denoted by the symbol '$\cdot$':

$$F_1(U_0, \lambda u + \mu v) = F_1(U_0, \cdot)(\lambda u + \mu v)$$

$$= \lambda F_1(U_0, \cdot) u + \mu F_1(U_0, \cdot) v$$

$$= \lambda F_1(U_0, u) + \mu F_1(U_0, v), \quad \lambda, \mu \in \mathbf{R}.$$

The operators $F_1^*(U_0, u_i, \cdot)$ and $F_1^*(u_i, u_j, \cdot)$ are defined in a similar way. Comparing (4.14)–(4.16) and (4.25)–(4.27), it is easy to notice that we have taken into account the symmetricity of the operator $F_1$, i.e. the property $F_1(\ldots, u_i, \ldots, u_j, \ldots) = F_1(\ldots, u_j, \ldots, u_i, \ldots)$.

Having solved the first $N + 1$ equations, we can find the $N$-th order approximation to $u^*$ by the formula

$$U_{(N)}^* = u_0^* + \sum_{i=1}^N \varepsilon^i u_i^*. \quad (4.28)$$

**4.5.** Assume now that it is required to find the value of the functional

$$J(U) = (p, U) \quad ((\cdot, \cdot) \equiv (\cdot, \cdot)_{H_0}), \quad (4.29)$$

where $U$ is the element satisfying (4.1), and $p = p(\varepsilon)$ is an element of $H_1^*$, analytically depending on $\varepsilon \in [0, \varepsilon_0]$:

$$p(\varepsilon) = \sum_{i=0}^\infty \varepsilon^i p_i, \quad p_i = \frac{1}{i!} \left. \frac{d^i p(\varepsilon)}{d\varepsilon^i} \right|_{\varepsilon=0}, \quad p_i \in H_1^*.$$

The following representation holds:

$$J(U) = (p, U_0) + (p, u),$$

where $u$ is the solution of equation (4.4). Using the above-given perturbation algorithm for finding $U_{(N)}$ by (4.18), we obtain the following formula for computing $J(U)$ approximately:

$$J(U) \cong J(U_{(N)}) = (p, U_0) + \sum_{i=0}^{N} \varepsilon^i (p, u_i), \tag{4.30}$$

where $J_{(N)} \equiv J(U_{(N)})$ is the $N$-th order approximation to $J$.

*Remark* 4.2. Since $p = \sum_{i=0}^{\infty} \varepsilon_i p_i$, then, instead of formula (4.30), one can consider the value

$$J_{(N)} \equiv (p_{(N)}, U_0) + \sum_{i+j \leq N} \varepsilon^{i+j} (p_j, u_i),$$

where $p_{(N)} = \sum_{i=0}^{N} \varepsilon^i p_i$, which is also the $N$-th order approximation to $J(U)$.

If we find the solution $u^*$ of adjoint problem (4.21) for $g(\varepsilon) = p(\varepsilon)$, then $J(U)$ may be also computed. In fact, in view of the adjointness relation

$$(f(\varepsilon), u^*) = (A(u, \varepsilon) u, u^*) = (u, A^*(u, \varepsilon) u^*) = (u, p),$$

we get the following representation for $J(U)$:

$$J(U) = (p, U_0) + (u^*, f(\varepsilon)).$$

If $U_{(N)}^*$ is computed by (4.28), we come to the $N$-th order approximation formula for $J(U)$:

$$J_{(N)} = (p, U_0) + (U_{(N)}^*, f(\varepsilon)). \tag{4.31}$$

As for main (or adjoint) equations, Lemma 4.1 (or Lemma 4.2) gives the justification of regular perturbation algorithms for computing $J(U)$.

**4.6.** Furthermore, we assume that all the below-considered operators are analytic, and the elements $u, u^*, \ldots$ are analytic in $\varepsilon$, except as otherwise noted.

Let it be required to compute $J(U)$ with an accuracy of $O(\varepsilon^{N_0+N_1+1})$, where $N_0, N_1 \geq 0$ are integers. One can do this by the perturbation algorithm using the solutions $u_0, u_1, \ldots, u_{N_0+N_1}$ of the main equations by the formula

$$J_{(N_0+N_1)} = (p, U_0) + \sum_{i=0}^{N_0+N_1} \varepsilon^i (p, u_i) \tag{4.32}$$

or, using the solutions $u_0^*, u_1^*, \ldots, u_{N_0+N_1}^*$ of the adjoint equations, by the formula

$$J_{(N_0+N_1)} = (p, U_0) + \sum_{i=0}^{N_0+N_1} \varepsilon^i (u_i^*, f(\varepsilon)), \qquad (4.33)$$

where to compute $u_0^*, u_1^*, \ldots, u_{N_0+N_1}^*$ one needs to find also $u_0, u_1, \ldots, u_{N_0+N_1-1}$.

Along with formulae (4.32),(4.33), other representations for approximate values of the sought-for functional may be useful. Let us derive some of them. Write $\delta J = (p, u)$ in the form:

$$\delta J = \sum_{i=0}^{N_0} \varepsilon^i (p, u_i) + \sum_{i=N_0+1}^{\infty} \varepsilon^i (p, u_i)$$

$$= \sum_{i=0}^{N_0} \varepsilon^i (p, u_i) + \sum_{i=N_0+1}^{\infty} \varepsilon^i (u^*, A(u, \varepsilon) u_i)$$

$$= \sum_{i=0}^{N_0} \varepsilon^i (p, u_i) + \sum_{i=N_0+1}^{\infty} \varepsilon^i \left( \sum_{j=0}^{N_1-1} \varepsilon^j u_j^*, A(u, \varepsilon) u_i \right) + O(\varepsilon^{N_0+N_1+1})$$

$$= \sum_{i=0}^{N_0} \varepsilon^i (p, u_i) + \sum_{j=0}^{N_1-1} \varepsilon^j (u_j^*, f) + O(\varepsilon^{N_0+N_1+1})$$

$$- \left( \sum_{j=0}^{N_1-1} \varepsilon^j u_j^*, (A(u, \varepsilon) - A(u_{(N_0)}, \varepsilon)) \sum_{i=0}^{N_0} \varepsilon^i u_i \right)$$

$$- \left( \sum_{j=0}^{N_1-1} \varepsilon^j u_j^*, A(u_{(N_0)}, \varepsilon) \sum_{i=0}^{N_0} \varepsilon^i u_i \right).$$

Here $u_{(N_0)} = \sum_{i=0}^{N_0} \varepsilon^i u_i$. Let us estimate the quantity

$$R_{N_0, N_1} \equiv \left( \sum_{j=0}^{N_1-1} \varepsilon^j u_j^*, (A(u, \varepsilon) - A(u_{(N_0)}, \varepsilon)) \sum_{i=0}^{N_0} \varepsilon^i u_i \right).$$

Assume that $\Phi$ has the form (4.2), (1.6) is satisfied and

$$\sum_{i=0}^{\infty} \varepsilon^i \|u_i\|_{H_1} \leq d, \quad \sum_{i=0}^{\infty} \varepsilon^j \|u_j^*\|_{H_1} \leq d_1, \quad d, d_1 = \text{constant} < \infty.$$

Then

$$A(u, \varepsilon) - A(u_{(N_0)}, \varepsilon) = \varepsilon^{M+1} \int_0^1 (F_1'(U_0 + tu) - F_1'(U_0 + tu_{(N_0)})) \, dt,$$

$$\|A(u,\varepsilon) - A(u_{(N_0)},\varepsilon)\|_{H_1 \to H_1^*} \leq k_1 \varepsilon^{M+1} \int_0^1 t\|u - u_{(N_0)}\|_{H_1}\, dt$$

$$\leq \frac{1}{2} k_1 \varepsilon^{M+1} \varepsilon^{N_0+1} \left\| \sum_{i=0}^{\infty} \varepsilon^i u_{i+N_0+1} \right\|_{H_1}$$

$$\leq \frac{1}{2} k_1 \varepsilon^{N_0+M+2}\, d.$$

Hence,

$$|R_{N_0,N_1}| \leq d_1 d \cdot \frac{1}{2} k_1 \varepsilon^{N_0+M+2}. \tag{4.34}$$

Note also that if we take the solution of equation (4.19) for $U_0$, then, in view of $f_0 \equiv 0$, $u_0 \equiv 0$, we have $\|\sum_{i=0}^{\infty} \varepsilon^i u_i\|_{H_1} \leq \varepsilon d$, and, hence,

$$|R_{N_0,N_1}| \leq d d_1 \cdot \frac{1}{2} k_1 \varepsilon^{N_0+M+3}.$$

Therefore, if we make use of estimate (4.34), then we get for $\delta J$:

$$\delta J = \sum_{i=0}^{N_0} \varepsilon^i (p, u_i) + \sum_{j=0}^{N_1-1} \varepsilon^j (u_j^*, f(\varepsilon))$$

$$- \sum_{j=0}^{N_1-1} \sum_{i=0}^{N_0} \varepsilon^{i+j} (u_j^*, A(u_{(N_0)},\varepsilon) u_i) + O(\varepsilon^{N_0+N_1+1}) + O(\varepsilon^{N_0+M+2}).$$

(Here $O(\varepsilon^{N_0+M+2})$ should be replaced by $O(\varepsilon^{N_0+M+3})$ if $U_0$ satisfies (4.19).) Neglecting the last addends in $\delta J$, we obtain the following approximation formula for computing $J(U)$:

$$J(U) \cong J_{N_0,N_1} = (p, U_0) + \sum_{i=0}^{N_0} \varepsilon^i (p, u_i) + \sum_{j=0}^{N_1-1} \varepsilon^j (u_j^*, f(\varepsilon))$$

$$- \sum_{j=0}^{N_1-1} \sum_{i=0}^{N_0} \varepsilon^{i+j} (u_j^*, A(u_{(N_0)},\varepsilon) u_i) \tag{4.35}$$

with the accuracy $O(\varepsilon^{N_0+N_1+1} + \varepsilon^{N_0+M+2})$. In particular, if $N_0 = 0$, $N_1 = 1$, we get the formula

$$J_{0,1} = (p, U_0) + (p, u_0) + (u_0^*, f(\varepsilon)) - (u_0^*, A(u_0,\varepsilon) u_0),$$

with

$$|J(U) - J_{0,1}| \leq O(\varepsilon^2).$$

In the case when $U_0$ satisfies (4.19) this formula takes the following simple form:

$$J_{0,1} = (p, U_0) + (u_0^*, f(\varepsilon)). \tag{4.36}$$

This is a so-called 'small-perturbation formula'. It is of frequent use in practical computations. Thus, if the solution $U_0$ of the equation $\Phi(U_0, 0) = 0$ is known, then, having once solved the equation $F_0^* u_0^* = p_0$, one can compute by (4.36) the necessary values $\{J_{0,1}^{(i)}\}$ corresponding to different right-hand sides $f = f^{(i)}$.

*Remark* 4.3. We indicate a special case of linear operator $\Phi$. In this case, $\Phi(U, \varepsilon) \equiv \Phi(\varepsilon) U$, $A(u, \varepsilon) u \equiv A(\varepsilon) u$, $A(u, \varepsilon) - A(u_{(N)}, \varepsilon) \equiv 0$, and, hence, the formula (4.35) is of the accuracy $O(\varepsilon^{N_0 + N_1 + 1})$.

**4.7.** Let us give one more approximate formula for computing $J(U)$. Represent $J(U)$ in the form

$$J(U) = (p, U_0) + \sum_{i=0}^{3} \varepsilon^i \delta J_i + O(\varepsilon^4), \tag{4.37}$$

where

$$\delta J_0 = (f_0, u_0^*),$$

$$\delta J_1 = (f_0, u_1^*) + (f_1, u_0^*),$$

$$\delta J_2 = (f_0, u_2^*) + (f_1, u_1^*) + (f_2, u_0^*),$$

$$\delta J_3 = (f_0, u_3^*) + (f_1, u_2^*) + (f_2, u_1^*) + (f_3, u_0^*).$$

Let $U_0$ satisfy (4.19). Then $f_0 \equiv 0$, $u_0 \equiv 0$, and, hence,

$$\delta J_0 \equiv 0,$$

$$\delta J_1 = (f_1, u_0^*),$$

$$\delta J_2 = (f_1, u_1^*) + (f_2, u_0^*),$$

$$\delta J_3 = (f_1, u_2^*) + (f_2, u_1^*) + (f_3, u_0^*).$$

Let, for definiteness, $K = 3$. Then, since $f_1 = F_0 u_1$, we get

$$(f_1, u_2^*) = (F_0 u_1, u_2^*) = (u_1, F_0^* u_2^*)$$

$$= \left( u_1, p_2 - F_{01}^* u_1^* - 3F_1^*(U_0, U_0, \cdot) u_{1-M}^* \right.$$

$$\left. -3 \sum_{i+j=1-M} F_1^*(U_0, u_i, \cdot) u_j^* - \sum_{i+j+l=1-M} F_1^*(u_i, u_j, \cdot) u_l^* \right),$$

where the last addend vanishes. Hence, for any $M \geq 0$ the values $\delta J_1$, $\delta J_2$, $\delta J_3$ depend only on $u_0^*$, $u_1^*$, $u_1$. Thus, the following lemma is valid.

**Lemma 4.3.** *Let $U_0$ be a solution of the equation $\Phi(U_0, 0) = 0$, and $K = 3$. Then an approximate formula for $J(U)$ of the accuracy $O(\varepsilon^4)$ may be constructed on the basis of only $u_1$, $u_0^*$, $u_1^*$ (i.e. without finding $u_2$, $u_2^*$, $u_3$, $u_3^*$). One of these formulae is*

$$J(U) \cong \tilde{J} = (p, U_0) + \sum_{i=1}^{3} \varepsilon^i \delta J_i,$$

*where*

$$\delta J_1 = (f_1, u_0^*),$$

$$\delta J_2 = (f_1, u_1^*) + (f_2, u_0^*),$$

$$\delta J_3 = (f_3, u_0^*) + (f_2, u_1^*)$$

$$+ \left( u_1, p_2 - F_{01}^* u_1^* - 3 F_1^*(U_0, U_0, \cdot) u_{1-M}^* \right.$$

$$\left. -3 \sum_{i+j=1-M} F_1^*(U_0, u_i, \cdot) u_j^* \right).$$

*Remark* 4.4. The value $\delta J_3$ has the form depending on $M$:

$M = 0$: $\delta J_3 = (f_3, u_0^*) + (f_2, u_1^*) + (u_1, p_2 - F_{01}^* u_1^*$
$\qquad\qquad -3 F_1^*(U_0, U_0, \cdot) u_1^* - 3 F_1^*(U_0, u_1, \cdot) u_0^*)$;

$M = 1$: $\delta J_3 = (f_3, u_0^*) + (f_2, u_1^*) + \left( u_1, p_2 - F_{01}^* u_1^* - 3 F_1^*(U_0, U_0, \cdot) u_0^* \right)$;

$M > 1$: $\delta J_3 = (f_3, u_0^*) + (f_2, u_1^*) + (u_1, p_2 - F_{01}^* u_1^*)$.

*Remark* 4.5. What actually happens is that the assertions of Lemma 4.3 are valid for any $K$. In fact, if $U_0$ satisfies (4.19) and $M \geq 0$, formula (4.35) takes the form

$$J(U) = J_{N_0, N_1} + O(\varepsilon^{N_0 + N_1 + 1} + \varepsilon^{N_0 + M + 3}).$$

By putting $N_0 = 1$, $N_1 = 2$, we get

$$J(U) = J_{1,2} + O(\varepsilon^4)$$

or

$$J(U) = (p, U_0) + \sum_{i=0}^{1} \varepsilon^i (p, u_i) + \sum_{j=0}^{1} \varepsilon^j (u_j^*, f)$$

$$- \sum_{j=0}^{1} \sum_{i=0}^{1} \varepsilon^{i+j} (u_j^*, A(u_{(1)}, \varepsilon) u_i) + O(\varepsilon^4).$$

Hence, to compute $J(U)$ with the accuracy $O(\varepsilon^4)$ it is required to know $u_0^*$, $u_1$, $u_1^*$ only.

## 5. CONVERGENCE RATE ESTIMATES FOR PERTURBATION ALGORITHMS. COMPARISON WITH THE SUCCESSIVE APPROXIMATIOM METHOD

**5.1.** The estimation of the convergence rate of perturbation algorithms has been actually initiated in Section 4. Here we explore this question further under the following restrictions: the operator $\Phi$ has the form (4.2); $\|F_0 v\|_{H_1^*} \geq m\|v\|_{H_1}$, $m = $ constant $> 0$; $\|F_{01}\| \equiv \|F_{01}\|_{H_1 \to H_1^*} < \infty$; the operator $F_1$ is bounded, i.e.

$$\|F_1(u_1, \ldots, u_K)\|_{H_1^*} \leq \|F_1\|\|u_1\|_{H_1} \cdots \|u_K\|_{H_1},$$

where $\|F_1\| = $ constant $< \infty$; $\varepsilon \in [0, \varepsilon_0]$.

Let us introduce the following notations:

$$\xi_N = \sum_{k=0}^{N} \varepsilon_0^k \|u_k\|_{H_1}, \qquad \eta_N = \sum_{k=0}^{N} \varepsilon_0^k \|f_k\|_{H_1^*}.$$

From the $k$-th equation of the perturbation algorithm system, we find

$$\|F_0 u_k\|_{H_1^*} \leq \|f_k\|_{H_1^*} + \|F_{01} u_{k-1}\|_{H_1^*} + \left\|\varepsilon^{M+1} \sum_{\mathbf{i}} \int_0^1 F_1'(U_0 + tu)\, dt u\right\|_{H_1^*},$$

$$\|\varepsilon^k F_0 u_k\|_{H_1^*} \leq \|\varepsilon^k f_k\|_{H_1^*} + \|\varepsilon^k F_{01}\|\|u_{k-1}\|_{H_1^*}$$

$$+ \left\|\varepsilon^{M+1} \sum_{\mathbf{i}, |\mathbf{i}|=k-M-1} \int_0^1 F_1'(U_0 + tu)\, dt u\right\|_{H_1^*}, \tag{5.1}$$

$$m\|\varepsilon^k u_k\|_{H_1^*} \leq \varepsilon^k \|f_k\|_{H_1^*} + \varepsilon^k \|F_{01}\|\|u_{k-1}\|_{H_1^*}$$

$$+ \left\|\varepsilon^{M+1} \sum_{\mathbf{i}, |\mathbf{i}|=k-M-1} \int_0^1 F_1'(U_0 + tu)\, dt u\right\|_{H_1^*},$$

where $\mathbf{i} = (i_1, \ldots, i_K)$ ($i_j \geq 0$ are integers), $|\mathbf{i}| = \sum_{j=1}^{K} i_j$,

$$F_1'(U_0 + tu)\, u = K F_1(U_0 + tu, \ldots, U_0 + tu, u),$$

$U_0 + tu = U_0 + t \sum_{i_l=0}^{\infty} \varepsilon^{i_l} u_{i_l}$ is the $l$-th argument of $F_1$.

Let us estimate the last addend in the right-hand side of (5.1):

$$\left\| \varepsilon^{M+1} \sum_{|i|=k-M-1} \int_0^1 F_1'(U_0+tu)\,dtu \right\|_{H_1^*}$$

$$\leq K\|F_1\|\varepsilon_0^{M+1} \sum_{|i|=k-M-1} \int_0^1 (\|U_0\|_{H_1}+t\|u\|_{H_1})$$

$$\times \ldots \times (\|U_0\|_{H_1}+t\|u\|_{H_1})\|u\|_{H_1}\,dt.$$

Hence,

$$\sum_{k=0}^N \left\| \varepsilon^{M+1} \sum_{|i|=k-M-1} \int_0^1 F_1'(U_0+tu)\,u\,dt \right\|_{H_1^*}$$

$$\leq K\|F_1\|\varepsilon_0^{M+1} \sum_{k=0}^N \sum_{|i|=k-M-1} \int_0^1 (\|U_0\|_{H_1}+t\xi_{N-M-1})$$

$$\times \ldots \times (\|U_0\|_{H_1}+t\xi_{N-M-1})\xi_{N-M-1}\,dt$$

$$= \|F_1\|\varepsilon_0^{M+1}((\|U_0\|_{H_1}+\xi_{N-M-1})^K - \|U_0\|_{H_1}^K).$$

Taking into account these relations, from (5.1) we find

$$m\sum_{k=0}^N \varepsilon^k\|u_k\|_{H_1} \leq \eta_N + \varepsilon_0\|F_{01}\|\xi_{N-1}$$

$$+\|F_1\|\varepsilon_0^{M+1}((\|U_0\|_{H_1}+\xi_{N-M-1})^K - \|U_0\|_{H_1}^K).$$

By setting $\varepsilon = \varepsilon_0$ in the left-hand side, we get

$$m\eta_N \leq \eta_N + \varepsilon_0\|F_{01}\|\xi_{N-1}$$

$$+\|F_1\|\varepsilon_0^{M+1}((\|U_0\|_{H_1}+\xi_{N-M-1})^K - \|U_0\|_{H_1}^K). \qquad (5.2)$$

This implies that if $\xi_i \leq d$, $i = 0, 1, \ldots, N-1$, and

$$\eta_\infty + \varepsilon_0\|F_{01}\|\,d + \|F_1\|\,\varepsilon_0^{M+1}((\|U_0\|_{H_1}+d)^K - \|U_0\|_{H_1}^K) \leq md, \qquad (5.3)$$

then $\xi_N \leq d$. Therefore, if $\xi_0 \leq d$ and (5.3) is satisfied, then

$$\xi_\infty = \sum_{i=0}^\infty \varepsilon_0^i\|u_i\|_{H_1} \leq d.$$

Now, satisfied (5.3) and $\xi_0 \leq d$, it is not difficult to estimate the convergence rate of the perturbation algorithm. In fact, if $u_{(N)} = \sum_{i=0}^{N} \varepsilon^i u_i$ and $\varepsilon < \varepsilon_0$, then

$$\|u - u_{(N)}\|_{H_1} = \left\| \sum_{i=N+1}^{\infty} \varepsilon^i u_i \right\|_{H_1} \leq (\varepsilon/\varepsilon_0)^{N+1} \sum_{i=N+1}^{\infty} \varepsilon_0^i \|u_i\|_{H_1}$$

$$\leq (\varepsilon/\varepsilon_0)^{N+1} \sum_{i=0}^{\infty} \varepsilon_0^i \|u_i\|_{H_1}$$

$$\leq (\varepsilon/\varepsilon_0)^{N+1} d. \tag{5.4}$$

If $\varepsilon = \varepsilon_0$, then from the above relations we can state the convergence $u_{(N)} \to u$:

$$\|u - u_{(N)}\|_{H_1} \leq \sum_{i=N+1}^{\infty} \varepsilon_0^i \|u_i\|_{H_1} \to 0, \qquad N \to \infty. \tag{5.5}$$

(To derive a convergence rate estimate one appears to need a further study of the decrease rate of the values $\|u_i\|_H$.) Thus, we come to the following theorem.

**Theorem 5.1.** *Let $\Phi$ have the form (4.2) with $\|F_{01}\| < \infty$, $\|F_1\| < \infty$, and the initial approximation $U_0$ and $\varepsilon_0$ be such that*

$$\sum_{i=0}^{\infty} \frac{\varepsilon_0^i}{i!} \left\| \frac{d^i \Phi(U_0, \varepsilon)}{d\varepsilon^i} \right|_{\varepsilon=0} \right\|_{H_1^*} \leq (m - \varepsilon_0 \|F_{01}\|) d - \|F_1\| \varepsilon_0^{M+1}$$

$$\times ((\|U_0\|_{H_1} + d)^K - \|U_0\|_{H_1}^K) \tag{5.6}$$

*for a constant $d \leq \|u_0\|_{H_1}$. Then $u_{(N)} \to u$ as $N \to \infty$, and*

$$\|u - u_{(N)}\|_{H_1} \leq (\varepsilon/\varepsilon_0)^{N+1} d, \tag{5.7}$$

*if $0 < \varepsilon < \varepsilon_0$.*

**Corollary 1.** *If $d \equiv \|U_0\|_{H_1} \neq 0$ and inequality*

$$\sum_{i=0}^{\infty} \frac{\varepsilon_0^i}{i!} \left\| \frac{d^i \Phi(U_0, \varepsilon)}{d\varepsilon^i} \right|_{\varepsilon=0} \right\|_{H_1^*} \leq (m - \varepsilon_0 \|F_{01}\|) \|U_0\|_{H_1}$$

$$-\|F_1\| \varepsilon_0^{M+1} (2^K - 1) \|U_0\|_{H_1}^K \tag{5.8}$$

*is satisfied, then the estimate*

$$\|u - u_{(N)}\|_{H_1} \leq (\varepsilon/\varepsilon_0)^{N+1} \|U_0\|_{H_1} \tag{5.9}$$

*holds if $0 < \varepsilon < \varepsilon_0$.*

**Corollary 2.** *Let $M = 0$ and $U_0$ satisfy the equation $F(U_0, 0) = 0$. Then, under the hypotheses of Theorem 5.1, the estimate (5.9) holds if*

$$\|F_{01}U_0 + F_1(U_0)\|_{H_1^*}$$

$$\leq \|U_0\|_{H_1}(m/\varepsilon_0 - \|F_{01}\| - \|F_1\|(2^K - 1)\|U_0\|_{H_1}^{K-1}) \qquad (5.10)$$

*and $0 < \varepsilon < \varepsilon_0$.*

*Remark* 5.1. It is easily seen that, under the hypotheses of Corollary 2, (5.10) is always satisfied if $\varepsilon_0$ is sufficiently small.

*Remark* 5.2. In Section 4 we have derived estimates under the restriction $d = \sum_{i=0}^{\infty} \|u_i\|_{H_1} < \infty$. This condition will be satisfied if we put $\varepsilon_0 = 1$ in Theorem 5.1. Otherwise, under the hypotheses of Theorem 5.1, all the estimates of Section 4 will be valid if we substitute $\varepsilon$ with $\varepsilon/\varepsilon_0$, and $d$ with $d = \sum_{i=0}^{\infty} \varepsilon_0^i \|u_i\|_{H_1}$.

**5.2.** We have derived the convergence rate estimate for the perturbation algorithm under some additional requirements. Though these conditions are sufficient, nevertheless, a comparison of the results of Section 4 and the above-formulated results makes it possible to suppose that there may appear the case when the application of the perturbation algorithm is not preferable (from computational point of view) over other approximate algorithms for solving the problem. One of these algorithms may be the successive approximation method (1.9). In the case when the operator $\Phi$ has the form (4.2) for $K = 3$, (1.9) may be represented as

$$(F_0 + \varepsilon F_{01} + \varepsilon^{M+1} \cdot 3F_1(U_0, U_0)) u^{(n+1)}$$
$$= f(\varepsilon) - \varepsilon^{M+1}(3F_1(u^{(n)}, u^{(n)}, U_0) + F_1(u^{(n)}, u^{(n)}, u^{(n)})). \quad (5.11)$$

Let $n = -1, 0, 1, \dots$ and $u^{(-1)} \equiv 0$. Then the functions $u^{(n)}$, $n = 0, 1, \dots$, will be analytic in $\varepsilon$. Let $u^{(n)} = \sum_{i=0}^{\infty} \varepsilon^i u_i^{(n)}$. Substituting these representations into (5.11), we derive the equations for $u_i^{(n)}$:

$$F_0 u_0^{(n+1)} = f_0,$$

$$(5.12)$$

$$F_0 u_k^{(n+1)} = f_k - F_{01} u_{k-1}^{(n+1)} - 3F_1(U_0, U_0, u_{k-M-1}^{(n+1)})$$

$$-3 \sum_{i+j=k-M-1} F_1(U_0, u_i^{(n)}, u_j^{(n)})$$

$$- \sum_{i+j+l=k-M-1} F_1(u_i^{(n)}, u_j^{(n)}, u_l^{(n)}).$$

Comparing (5.12) and (4.16), we find

$$n = -1: \quad u_0^{(0)} = u_0, \quad u_1^{(0)} = u_1, \ldots, u_M^{(0)} = u_M;$$

$$n = 0: \quad u_0^{(1)} = u_0, \quad u_1^{(1)} = u_1, \ldots, u_{2M+1}^{(1)} = u_{2M+1};$$

$$n = 1: \quad u_0^{(2)} = u_0, \quad u_1^{(2)} = u_1, \ldots, u_{3M+2}^{(2)} = u_{3M+2}; \qquad (5.13)$$

$$\ldots$$

$$n = N: \quad u_0^{(N)} = u_0, \quad \ldots, u_{(N+1)M+N}^{(N)} = u_{(N+1)M+N}.$$

Hence, after the $N$-th step of the successive approximation method the element $u^{(N)}$ may be represented in the form:

$$u^{(N)} = \sum_{i=0}^{(N+1)M+N} \varepsilon^i u_i + \sum_{i=(N+1)M+N+1}^{\infty} \varepsilon^i u_i^{(N)}$$

$$= u_{((N+1)M+N)} + \sum_{i=(N+1)M+N+1}^{\infty} \varepsilon^i u_i^{(N)}. \qquad (5.14)$$

Taking into account (4.10) and the fact that the process starts with $u^{(-1)} \equiv 0$, we get

$$\|u - u^{(N)}\|_{H_1} \leq \left( \frac{\varepsilon^{M+1} k_1 R}{m_\varepsilon} \right)^{N+1} \frac{\|\bar{f}\|_{H_1}}{1-q} \leq O(\varepsilon^{(M+1)(N+1)}), \qquad (5.15)$$

where the quantity $(k_1 R/m_\varepsilon)^{N+1}$ is assumed to be bounded. Under the hypotheses of Theorem 5.1, we obtain the estimate:

$$\|u - u_{((N+1)M+N)}\|_{H_1} = \left\| \sum_{i=(N+1)M+N+1}^{\infty} \varepsilon^i u_i \right\|_{H_1}$$

$$\leq (\varepsilon/\varepsilon_0)^{(M+1)(N+1)} d. \qquad (5.16)$$

Note that (for example, for $\varepsilon_0 = 1$) estimates (5.15), (5.16) are of the same order of accuracy.

Thus, the following theorem is proved.

**Theorem 5.2.** *Let the hypotheses of Lemma 4.1 be satisfied and the process (5.11) start with $n = -1$ and $u^{(-1)} \equiv 0$. Then the approximation $u^{(N)}$ may be represented as*

$$u^{(N)} = u_{((N+1)M+N)} + \sum_{i=(M+1)(N+1)}^{\infty} \varepsilon^i u_i^{(N)}, \qquad (5.17)$$

*where* $u_{((N+1)M+N)}$ *is the approximation by the perturbation algorithm, and estimate (5.15) holds.*

From what was stated above, the following suggestion may be made. If the computational costs for solving the equations $F_0 v = g$ and $(F_0 + \varepsilon F_{01} + \varepsilon^{M+1} 3 F_1(U_0, U_0, \cdot)) v = g$ are of the same order, the successive approximation method (5.11) will be preferable over the perturbation algorithm if $M > 0$.

*Remark* 5.3. The above-drawn conclusions hold true for any finite $K$.

**5.3.** Let us return to computing the functional $J(U)$. Assume that after the $(N + 1)$-th iteration we have obtained, according to process (5.11), the approximation $u^{(N)}$. Define $\bar{u}^*$ as the solution to the equation

$$A^*(u^{(N)}, \varepsilon)\, \bar{u}^* = p. \tag{5.18}$$

Assume also that the equation with the operator $A^*(u, \varepsilon)$ is correctly solvable and (4.6) is satisfied. Then, it is not difficult to show that $\|u^* - \bar{u}^*\|_{H_1} \leq O(\varepsilon^{(M+1)(N+2)})$ if estimate (5.15) holds. In fact, for any $v \in D(A)$ we get

$$A^*(u, \varepsilon)(u^* - \bar{u}^*) = (A^*(u^{(N)}, \varepsilon) - A^*(u, \varepsilon))\, \bar{u}^*,$$

$$(A^*(u, \varepsilon)(u^* - \bar{u}^*), v) = \left( \bar{u}^*, \varepsilon^{M+1} \int_0^1 (F_1'(U_0 + tu^{(N)}) - F_1'(U_0 + tu))\, dt v \right),$$

$$|(A^*(u, \varepsilon)(u^* - \bar{u}^*), v)| \leq \|\bar{u}^*\|_{H_1} \|v\|_{H_1} C \varepsilon^{M+1} \|u - u^{(N)}\|_{H_1}$$

$$\leq C \varepsilon^{(N+1)(N+2)} \|\bar{u}^*\|_{H_1} \|v\|_{H_1},$$

$$m\|u^* - \bar{u}^*\|_{H_1} \leq \|A^*(u, \varepsilon)(u^* - \bar{u}^*)\|_{H_1^*}$$

$$= \sup_{v \in H_1} \frac{|(A^*(u, \varepsilon)(u^* - \bar{u}^*), v)|}{\|v\|_{H_1}}$$

$$\leq C \varepsilon^{(M+1)(N+2)} \|\bar{u}^*\|_{H_1} \leq C \varepsilon^{(M+1)(N+2)} \|p\|_{H_1^*},$$

i.e.

$$\|u^* - \bar{u}^*\|_{H_1} \leq O(\varepsilon^{(M+1)(N+2)}). \tag{5.19}$$

Hence, we can conclude that the formula

$$J(U) \cong (p, U_0) + (f, \bar{u}^*) \tag{5.20}$$

is of the accuracy $O(\varepsilon^{(M+1)(N+2)})$. Then, the computing of $J(U)$ by (5.20) may be very efficient if $u^{(N)}$ converges to $u$ fast. For example, if we

(1) solve the equation

$$(F_0 + \varepsilon F_{01} + \varepsilon^{M+1} \cdot 3F_1(U_0, U_0, \cdot)) \, u^{(0)} = f(\varepsilon) \qquad (5.21)$$

(that is, one iteration of the process);
(2) find $\bar{u}^*$ as the solution to

$$(F_0^* + \varepsilon F_{01}^* + \varepsilon^{M+1}(3F_1^*(U_0, U_0, \cdot)$$
$$+ \; 3F_1^*(U_0, u^{(0)}, \cdot) + F_1^*(u^{(0)}, u^{(0)}, \cdot))) \, \bar{u}^* = p; \qquad (5.22)$$

(3) compute

$$J_{(0)} = (p, U_0) + (f, \bar{u}^*), \qquad (5.23)$$

then, according to (5.20), it is valid to say that $|J(U) - J_{(0)}| \leq O(\varepsilon^{2(M+1)})$.
To obtain the same order of accuracy for $M = 2$ with the use of the formula

$$J(U) \cong (p, U_0) + \sum_{i=0}^{5} \varepsilon^i (p, u_i),$$

one needs to solve six equations. The algorithm (5.21)–(5.23) uses the solutions of only two equations (but with more complicated operators to be
inverted!).

**5.4.** In conclusion we consider an algorithm being a superposition of the
above-discussed methods. The principle of the algorithm is that first we find
some approximation $u_{(N)}$ by the perturbation algorithm. Then, using $u_{(N)}$,
we consider the iterative process

$$A(u_{(N)}, \varepsilon) \, u^{(n+1)} = (A(u_{(N)}, \varepsilon) - A(u^{(n)}, \varepsilon)) \, u^{(n)} + f(\varepsilon),$$
$$n = -1, 0, 1, 2, \ldots, \qquad u^{(-1)} = u_{(N)}. \qquad (5.24)$$

Note that since the operators to be inverted in the perturbation algorithm
are simple, the computational cost for constructing $u_{(N)}$ (for example, for
small $N$) is not high. As a result, we obtain the element $u_{(N)}$ being conceivably
of not necessary accuracy but being a sufficiently good initial approximation in
process (5.24). (As is well-known, the algorithms of the form (5.24) converge
very fast if the initial approximation is judiciously chosen.) Then, after few
iterations, one can construct the element $u^{(n)}$ to needed accuracy.

We indicate also the iterative algorithm

$$A(u^{(n)}, \varepsilon) \, u^{(n+1)} = f(\varepsilon), \qquad n = 0, 1, \ldots \qquad (5.25)$$

If $\|A(v, \varepsilon) \, w\|_{H_1^*} \geq m\|w\|_{H_1}$ ($m = $ constant $> 0$, $v, w \in D(A)$), then for
$u^{(n+1)}$ we get

$$\|u^{(n+1)}\|_{H_1} \leq \|f(\varepsilon)\|_{H_1^*}/m. \qquad (5.26)$$

Let us consider the equation for the difference $u^{(n+1)} - u$:

$$A(u, \varepsilon)(u^{(n+1)} - u) = \left( A(u, \varepsilon) - A(u^{(n)}, \varepsilon) \right) u^{(n+1)}.$$

If the operator $\Phi$ has the form (4.2) and (4.6) is satisfied, then

$$m\|u^{(n+1)} - u\|_{H_1} \leq \varepsilon^{M+1} k_1 \cdot \frac{1}{2} \|u - u^{(n)}\|_{H_1} \|u^{(n+1)}\|_{H_1}$$

$$\leq \frac{\varepsilon^{M+1} k_1 \|f(\varepsilon)\|_{H_1^*}}{2m} \|u - u^{(n)}\|_{H_1}.$$

Hence,

$$\|u^{(n+1)} - u\|_{H_1} \leq \left( \frac{\varepsilon^{M+1} k_1 \|f(\varepsilon)\|_{H_1^*}}{2m^2} \right)^{n+1} \|u - u^{(0)}\|_{H_1}. \qquad (5.27)$$

If $F(U_0, 0) = 0$, $u^{(0)} \equiv 0$, and $\sum_{i=1}^{\infty} \varepsilon_0^i \|u_i\|_{H_1} \leq d < \infty$, then, from (5.27), we obtain the estimate

$$\|u^{(n+1)} - u\|_{H_1} \leq \left( \frac{\varepsilon^{M+2} k_1 \|F_{01} U_0 + \varepsilon^M F_1(U_0)\|_{H_1^*}}{2m^2} \right)^{n+1} \frac{\varepsilon}{\varepsilon_0} d.$$

Therefore

$$\|u^{(n+1)} - u\|_{H_1} \sim O\left( \varepsilon^{(M+2)(N+1)+1} \right)$$

if $\varepsilon$ is sufficiently small.

Thus, the algorithm (5.25) may converge very fast (however, at each step of the process, one needs to construct a new operator $A(u^{(n)}, \varepsilon)$, or its approximation).

## 6. JUSTIFICATION OF PERTURBATION ALGORITHMS IN QUASI-LINEAR ELLIPTIC PROBLEMS

This section considers a perturbed quasi-linear second-order elliptic problem. The solution of the problem is shown to exist in the Sobolev space $\overset{\mathrm{o}}{W_2^1}(\Omega)$ and be represented in the form of a power series in a perturbation parameter $\varepsilon$. The $N$-th order perturbation algorithm is used to find an approximate solution. The convergence of the algorithm is proved, and a convergence rate estimate is derived.

The presentation considers a specific elliptic problem as an example. However, as will readily be observed, the formulated statements remain valid for more general elliptic problems.

## 6.1. Statement of the problem

Let $\Omega$ be a bounded domain of $\mathbf{R}^n$, $n = 2, 3$, with a piecewise smooth boundary of the class $C^2$. Introduce the spaces

$$H_0 = H_0^* = L_2(\Omega), \quad H_1 = \overset{\circ}{W_2^1}(\Omega) \subset H_0 \equiv H_0^* \subset H_1^* = \overset{\circ}{W_2^{-1}}(\Omega).$$

Consider $U_0 \in H_1$ which is the solution to the linear elliptic problem of the form

$$-\nabla\mu\nabla U_0 + \sum_{i=1}^{n} b_i(x)\frac{\partial U_0}{\partial x_i} + a(x)U_0 = G, \quad G \in H_1^*, \tag{6.1}$$

$$U_0|_{\partial\Omega} = 0,$$

where

$$0 < \mu_0 \le \mu = \mu(x, y) \le \mu_1, \quad \mu_i = \text{constant}, \quad i = 1, 2, \quad x = (x_1, \ldots, x_n) \in \Omega,$$

the functions $\mu(x, y)$, $a(x)$, $b_i(x)$, $i = 1, \ldots, n$, are assumed to be real-valued and sufficiently regular, and

$$\sum_{i=1}^{n} \frac{\partial b_i}{\partial x_i} = 0, \quad a(x) \ge 0.$$

Along with the main problem (6.1), consider a perturbed problem which is non-linear and of the form

$$-\nabla\mu\nabla U + \sum_{i=1}^{n} v_i(x, U)\frac{\partial U}{\partial x_i} + l(x, U)U = G, \quad G \in H_1^*, \tag{6.2}$$

$$U|_{\partial\Omega} = 0,$$

where

$$v_i(x, U) = b_i(x) + \varepsilon c_i(x) + \varepsilon^{M+1}q_i(x)U,$$

$$l(x, U) = a(x) + \varepsilon d(x) + \varepsilon^{M+1}p(x)U,$$

$c_i(x)$, $q_i(x)$ $(i = 1, \ldots, n)$, $d(x)$, $p(x)$ are sufficiently regular real-valued functions,

$$\sum_{i=1}^{n} \frac{\partial c_i}{\partial x_i} = 0, \quad d(x) \ge 0, \quad \varepsilon \in [0, 1], \quad M \ge 0.$$

Suppose that the solution $U_0$ of problem (6.1) is known and we need to find $U$. To do this, let us use the perturbation algorithm. There arises a number of questions to be answered. First, does the solution $U$ of problem (6.2) exists? Second, is $U$ represented in the form of a series in powers $\varepsilon$? An important question concerns the convergence rate estimation. These are the questions to be considered in this section.

We shall consider weak formulations of problems (6.1) and (6.2). A weak formulation of problem (6.2) is as follows: find a function $U \in H_1$ such that

$$(\mu \nabla U, \nabla V) + \left( \sum_{i=1}^{n} b_i(x) \frac{\partial U}{\partial x_i}, V \right) + (a(x) U, V)$$

$$+ \varepsilon \left( \sum_{i=1}^{n} c_i(x) \frac{\partial U}{\partial x_i} + d(x) U, V \right)$$

$$+ \varepsilon^{M+1} \left( \sum_{i=1}^{n} q_i(x) U \frac{\partial U}{\partial x_i} + p(x) U^2, V \right) = (G, V) \quad (6.3)$$

for any $V \in H_1$.

## 6.2. Operator formulation of the problem

Let us rewrite (6.3) in the form

$$[U, V]_\varepsilon = (G, V), \tag{6.4}$$

where

$$[U, V]_\varepsilon = (\mu \nabla U, \nabla V) + \left( \sum_{i=1}^{n} b_i(x) \frac{\partial U}{\partial x_i}, V \right) + (a(x) U, V)$$

$$+ \varepsilon \left( \sum_{i=1}^{n} c_i(x) \frac{\partial U}{\partial x_i} + d(x) U, V \right)$$

$$+ \varepsilon^{M+1} \left( \sum_{i=1}^{n} q_i(x) U \frac{\partial U}{\partial x_i} + p(x) U^2, V \right).$$

To the form $[U, V]_\varepsilon$ there corresponds a non-linear operator $\Phi(U, \varepsilon)$ mapping $H_1 \equiv \overset{o}{W_2^1}$ into $H_1^*$ with the domain $D(\Phi) = H_1$, given by the equality

$$(\Phi(U, \varepsilon), V) = [U, V]_\varepsilon. \tag{6.5}$$

From (6.4) and (6.5) we conclude that problem (6.3) has a weak solution if and only if the equation

$$(\Phi(U, \varepsilon), V) = (G, V) \quad \forall V \in H_1 \tag{6.6}$$

has a solution $U \in D(\Phi) = H_1$. These solutions coincide.

Henceforward, we write equation (6.6) in the form

$$\Phi(U, \varepsilon) = G. \tag{6.7}$$

This is an operator formulation of the perturbed problem (6.2). We consider (6.7) as an operator equation in $H_1^*$ with the operator $\Phi$. Setting $\varepsilon = 0$, we get the unperturbed problem

$$\Phi(U, 0) = G \tag{6.8}$$

which is equivalent to the equality

$$(\mu \nabla U, \nabla V) + \left( \sum_{i=1}^{n} b_i(x) \frac{\partial U}{\partial x_i}, V \right) + (a(x) U, V) = (G, V) \quad \forall V \in \overset{\circ}{W}_2^1(\Omega). \tag{6.9}$$

## 6.3. Transformation of the problem.
## Properties of the non-linear operator

Subtracting $\Phi(U_0, \varepsilon)$ from the left and the right-hand sides of (6.8), we come to the equality

$$F(u, \varepsilon) = f(\varepsilon), \tag{6.10}$$

where

$$F(u, \varepsilon) = \Phi(U, \varepsilon) - \Phi(U_0, \varepsilon) = \Phi(U_0 + u, \varepsilon) - \Phi(U_0, \varepsilon),$$

$$u = U - U_0, \qquad f(\varepsilon) = G - \Phi(U_0, \varepsilon).$$

The operator $F(u, \varepsilon)$ is defined by

$$(F(u, \varepsilon), w) = (\Phi(U_0 + u, \varepsilon) - \Phi(u_0, \varepsilon), w) \quad \forall w \in \overset{\circ}{W}_2^1(\Omega), \tag{6.11}$$

mapping $H_1$ into $H_1^*$ with the domain $D(F) = H_1$. Since $U_0$ is the solution to problem (6.1), we have for the element $f(\varepsilon)$:

$$(f(\varepsilon), w) = -\varepsilon \left( \sum_{i=1}^{n} c_i(x) \frac{\partial U_0}{\partial x_i} + d(x) U_0, w \right)$$

$$-\varepsilon^{M+1} \left( \sum_{i=1}^{n} q_i(x) U_0 \frac{\partial U_0}{\partial x_i} + p(x) U_0^2, w \right), \quad w \in H_1. \tag{6.12}$$

Using the definition of the operator $\Phi$ from (6.5), let us write (6.11) in the explicit form:

$$(F(u, \varepsilon), w) = (\mu \nabla u, \nabla w) + \left( \sum_{i=1}^{n} b_i(x) \frac{\partial u}{\partial x_i}, w \right) + (a(x) u, w)$$

$$+ \varepsilon \left( \sum_{i=1}^{n} c_i(x) \frac{\partial u}{\partial x_i} + d(x) u, w \right)$$

$$+\varepsilon^{M+1}\left(\sum_{i=1}^{n}q_i(x)(U_0+u)\frac{\partial(U_0+u)}{\partial x_i}\right.$$

$$\left.-\sum_{i=1}^{n}q_i(x)U_0\frac{\partial U_0}{\partial x_i}+p(x)(U_0+u)^2-p(x)\,U_0^2,w\right).(6.13)$$

Let us study the properties of the operator $F(u,\varepsilon)$. Note first of all that $F(0,\varepsilon)\equiv 0$. The following statements hold.[182]

**Lemma 6.1.** *For any* $u\in D(F)$ *the operator* $F(u,\varepsilon)$ *has the Gâteaux derivative defined by the relation:*

$$(F'(u,\varepsilon)\,h,w)\;=\;(\mu\nabla h,\nabla w)+\left(\sum_{i=1}^{n}b_i(x)\frac{\partial h}{\partial x_i},w\right)+(a(x)\,h,w)$$

$$+\varepsilon\left(\sum_{i=1}^{n}c_i(x)\frac{\partial h}{\partial x_i}+d(x)\,h,w\right)$$

$$+\varepsilon^{M+1}\left(\sum_{i=1}^{n}q_i(x)(U_0+u)\frac{\partial h}{\partial x_i}\right.$$

$$\left.+\sum_{i=1}^{n}q_i(x)h\frac{\partial(U_0+u)}{\partial x_i}+2p(x)(U_0+u)\,h,w\right),\quad w\in H_1.$$

**Lemma 6.2.** *The operator* $F'(u,\varepsilon)\colon H_1\to H_1^*$ *is bounded.*

**Lemma 6.3.** *The operator* $F'(u,\varepsilon)\colon H_1\to H_1^*$ *is continuous in* $u$ *on* $D(F)=H_1$ *and satisfies the Lipschitz condition*

$$\|F'(u_1,\varepsilon)-F'(u_2,\varepsilon)\|_{H_1\to H_1^*}\le\alpha\|u_1-u_2\|_{H_1},\qquad(6.14)$$

*where*

$$\alpha=2\varepsilon^{M+1}c^2(\Omega)\left(\left\|\left(\sum_{i=1}^{n}q_i^2(x)\right)^{1/2}\right\|_{L_\infty(\Omega)}+\|p\|_{L_\infty(\Omega)}c_1(\Omega)\right),$$

$$c(\Omega)=\begin{cases}2(\operatorname{mes}\Omega)^{1/4},&n=2,\\[2mm]4(\operatorname{mes}\Omega)^{1/2},&n=3,\end{cases}$$

$$c_1(\Omega)=\begin{cases}\frac{3}{4}(\operatorname{mes}\Omega)^{1/2},&n=2,\\[2mm]4(\operatorname{mes}\Omega)^{1/3},&n=3.\end{cases}$$

From Lemmas 6.1–6.3 it follows that $F'(u, \varepsilon)$ is the Frechet derivative of the operator $F(u, \varepsilon)$. Hence, using the well-known results of non-linear analysis (see Chapter 1), we arrive at the following corollaries.

**Corollary 6.1.** *For $\varepsilon \in [0, 1]$ the following representation holds:*

$$F(u, \varepsilon) = F(v, \varepsilon) + \int_0^1 F'(v + \zeta(u - v), \varepsilon) \, d\zeta(u - v) \quad \forall \, u, v \in D(F).$$

**Corollary 6.2.** *The operator $F(u, \varepsilon)$ is represented in the form*

$$F(u, \varepsilon) = \int_0^1 F'(\zeta u, \varepsilon) \, d\zeta u \equiv A(u, \varepsilon) \, u, \qquad (6.15)$$

*where the linear operator*

$$A(u, \varepsilon) = \int_0^1 F'(\zeta u, \varepsilon) \, d\zeta$$

*maps $H_1$ into $H_1^*$ with the domain $D(A) = H_1$ and is defined by the equality*

$$(A(u, \varepsilon) h, w) = (\mu \nabla h, \nabla w) + \left( \sum_{i=1}^n b_i(x) \frac{\partial h}{\partial x_i}, w \right) + (a(x) h, w)$$

$$+ \varepsilon \left( \sum_{i=1}^n c_i(x) \frac{\partial h}{\partial x_i} + d(x) h, w \right)$$

$$+ \varepsilon^{M+1} \left( \sum_{i=1}^n q_i(x) \left( U_0 + \frac{u}{2} \right) \frac{\partial h}{\partial x_i} \right.$$

$$\left. + \sum_{i=1}^n q_i(x) h \frac{\partial (U_0 + u/2)}{\partial x_i} + 2p \left( U_0 + \frac{u}{2} \right) h, w \right) \quad (6.16)$$

*with $h \in D(A), w \in H_1$.*

Using Corollary 6.2, we write the perturbed problem (6.10) in an equivalent form

$$A(u, \varepsilon) u = f(\varepsilon). \qquad (6.17)$$

## 6.4. Properties of the operator $A(u, \varepsilon)$

Note first that, by Lemma 6.2, for a fixed $u \in H_1$ the operator $A(u, \varepsilon)$ is a linear bounded operator mapping $H_1$ into $H_1^*$ with the domain $D(A) = H_1$.

Consider now the operator $A(0, \varepsilon)$. Let

$$\beta = \frac{1}{2}\left\|\sum_{i=1}^{n}\frac{\partial q_i}{\partial x_i}\right\|_{L_\infty(\Omega)}c_1(\Omega) + \frac{1}{2}\left\|\left(\sum_{i=1}^{n}q_i^2\right)^{1/2}\right\|_{L_\infty(\Omega)} + 2\|p\|_{L_\infty(\Omega)}c_1(\Omega).$$

The following lemma holds.

**Lemma 6.4.** *For* $|\varepsilon| < (\mu_0/(\beta c^2(\Omega)\|U_0\|_{H_1}))^{1/(M+1)}$ *the operator* $A(0, \varepsilon)$:
$\overset{\circ}{W_2^1}(\Omega) \to \overset{\circ}{W_2^{-1}}(\Omega)$ *is continuously invertible and*

$$\|A(0, \varepsilon)v\|_{H_1^*} \geq m_\varepsilon\|v\|_{H_1} \quad \forall\, v \in H_1, \tag{6.18}$$

*where*

$$m_\varepsilon = \mu_0 - \beta c^2(\Omega)\|U_0\|_{H_1}\varepsilon^{M+1},$$

*and the constants* $c(\Omega)$ *and* $c_1(\Omega)$ *are defined by (6.14).*

*Proof.* By putting $u = 0$, $h = w = v \in H_1$, we get from (6.16)

$$(A(0, \varepsilon)v, v) = (\mu\nabla v, \nabla v) + \left(\sum_{i=1}^{n}b_i(x)\frac{\partial v}{\partial x_i}, v\right) + (a(x)v, v)$$

$$+\varepsilon\left(\sum_{i=1}^{n}c_i(x)\frac{\partial v}{\partial x_i} + d(x)v, v\right)$$

$$+\varepsilon^{M+1}\left(\sum_{i=1}^{n}q_i(x)\left[U_0\frac{\partial v}{\partial x_i} + v\frac{\partial U_0}{\partial x_i}\right] + 2p(x)U_0 v, v\right).$$

Taking into account the assumption of sufficient regularity of $b_i$, $c_i$, and the conditions

$$\sum_{i=1}^{n}\frac{\partial b_i}{\partial x_i} = \sum_{i=1}^{n}\frac{\partial c_i}{\partial x_i} = 0,$$

and integrating by parts, we get

$$\left(\sum_{i=1}^{n}b_i\frac{\partial v}{\partial x_i}, v\right) = \left(\sum_{i=1}^{n}c_i\frac{\partial v}{\partial x_i}, v\right) = 0,$$

$$\left(\sum_{i=1}^{n}q_i(x)U_0\frac{\partial v}{\partial x_i}, v\right) = -\frac{1}{2}\left(\sum_{i=1}^{n}\frac{\partial q_i U_0}{\partial x_i}v, v\right).$$

Then

$$(A(0, \varepsilon)v, v) = (\mu\nabla v, \nabla v) + ((a + \varepsilon d)v, v) + \varepsilon^{M+1}(\xi v, v),$$

where

$$\xi = \frac{1}{2}\sum_{i=1}^{n} q_i(x)\frac{\partial U_0}{\partial x_i} - \frac{1}{2}\sum_{i=1}^{n} U_0 \frac{\partial q_i}{\partial x_i} + 2p(x)\,U_0.$$

The following inequalities hold

$$|(\xi v, v)| \le \|\xi\|_{L_2(\Omega)}\|v\|_{L_4(\Omega)} \le \beta\|U_0\|_{\overset{\circ}{W_2^1}(\Omega)}\, c^2(\Omega)\|v\|^2_{\overset{\circ}{W_2^1}(\Omega)},$$

where the constants $c(\Omega)$ and $c_1(\Omega)$ are defined by (6.14). Hence, in view of the non-negativity of $a$ and $d$,

$$(A(0, \varepsilon)\,v, v) \ge m_\varepsilon \|v\|^2_{H_1}.$$

This proves the lemma.

### 6.5. The adjoint operator $A^*(u, \varepsilon)$

Introduce now the operator $A^*(u, \varepsilon)$ adjoint to $A(u, \varepsilon)$, according to the classical definition (see Chapter 1). The following lemma holds.[262]

**Lemma 6.5.** *The adjoint operator $A^*(u, \varepsilon)$ is defined by the equality*

$$(h, A^*(u, \varepsilon)\,w) = (\mu\nabla h, \nabla w) - \left(h, \sum_{i=1}^{n} b_i(x)\frac{\partial w}{\partial x_i}\right) + (h, a(x)\,w)$$

$$+\varepsilon\left(h, -\sum_{i=1}^{n} c_i(x)\frac{\partial w}{\partial x_i} + d(x)\,w\right)$$

$$+\varepsilon^{M+1}\left(h, -\sum_{i=1}^{n}\left(U_0 + \frac{u}{2}\right)\frac{\partial q_i w}{\partial x_i} + 2p(x)\left(U_0 + \frac{u}{2}\right)w\right)$$

*for any $h, w \in H_1$; it is a linear bounded operator mapping $H_1$ into $H_1^*$ with the domain $D(A^*) = H_1$.*

It is readily seen that under the hypotheses of Lemma 6.4 the operator $A^*(0, \varepsilon): H_1 \to H_1^*$ is continuously invertible and

$$\|A^*(0, \varepsilon)\,w\|_{H_1^*} \ge m_\varepsilon \|w\|_{H_1} \quad \forall\, w \in H_1. \tag{6.19}$$

Hence, in view of Lemma 6.4, we conclude that the equations

$$A(0, \varepsilon)\,v = f, \quad f \in H_1^*,$$

$$A^*(0, \varepsilon)\,w = p, \quad p \in H_1^*,$$

are correctly solvable. The operators $A(0, \varepsilon)$ and $A^*(0, \varepsilon)$ are closed as they are bounded. Therefore[105], the last-listed equations are solvable everywhere in $H_1^*$. Hence, the following lemma is true.

**Lemma 6.6.** *For* $|\varepsilon| < (\mu_0/(\beta c^2(\Omega)\|U_0\|_{H_1}))^{1/(M+1)}$ *the operators* $A(0,\varepsilon)$
*and* $A^*(0,\varepsilon)$ *map* $H_1$ *into* $H_1^*$ *with the domains* $D(A) = D(A^*) = H_1$, *and*
*the ranges* $R(A) = R(A^*) = H_1^*$. *They have continuous inverse operators,*
*and*

$$\|A^{-1}(0,\varepsilon)\|_{H_1^* \to H_1} \leq 1/m_\varepsilon, \quad \|(A^*(0,\varepsilon))^{-1}\|_{H_1^* \to H_1} \leq 1/m_\varepsilon, \qquad (6.20)$$

*where*

$$m_\varepsilon = \mu_0 - \beta c^2(\Omega)\|U_0\|_{H_1}\varepsilon^{M+1}.$$

## 6.6. The solution existence for the perturbed problem

Under the hypotheses of Lemma 6.6 the perturbed problem $A(u,\varepsilon)\,u = f(\varepsilon)$
is equivalent to the following one

$$u = A^{-1}(0,\varepsilon)([A(0,\varepsilon) - A(u,\varepsilon)]\,u + f(\varepsilon)),$$

or

$$u = T(u,\varepsilon), \qquad (6.21)$$

where

$$T(u,\varepsilon) = A^{-1}(0,\varepsilon)([A(0,\varepsilon) - A(u,\varepsilon)]\,u + f(\varepsilon)).$$

Thus, the perturbed problem is reduced to an equation of the form (6.21)
which is convenient to study using the contraction principle. The following
theorem is valid.

**Theorem 6.1.** *Let* $S(0,R)$ *be a ball in* $H_1$ *of the radius* $R$. *Under the*
*restrictions*

$$\frac{2\left(\left\|\left(\sum_{i=1}^n q_i^2\right)^{1/2}\right\|_{L_\infty(\Omega)} + \|p\|_{L_\infty(\Omega)}c_1(\Omega)\right)c^2(\Omega)\,\varepsilon^{M+1}}{\mu_0 - \beta c^2(\Omega)\|U_0\|_{H_1}\varepsilon^{M+1}}R < 1,$$

$$\|f(\varepsilon)\|_{H_1^*} \leq \frac{1}{2}\left(\mu_0 - \beta c^2(\Omega)\|U_0\|_{H_1}\varepsilon^{M+1}\right)R$$

*the equation* $A(u,\varepsilon)\,u = f(\varepsilon)$ *has a unique solution* $u \in S(0,R)$ *being repre-*
*sented as the series* $u = \sum_{i=1}^n \varepsilon^i u_i$ *convergent for* $0 \leq \varepsilon \leq \varepsilon_0$, *where*

$$\varepsilon_0 = \min\left(\mu_0 \middle/ \left(\mu_0 + c(\Omega)\left\|\sum_{i=1}^n c_i^2\right\|_{L_2(\Omega)}^{1/2} + c^2(\Omega)\|d\|_{L_2(\Omega)}\right),\right.$$

$$\mu_0 \middle/ \left(\mu_0 + 2c^2(\Omega)\left(\left\|\left(\sum_{i=1}^n q_i^2\right)^{1/2}\right\|_{L_\infty(\Omega)}\right.\right.$$

$$\left.\left.\left. + c_1(\Omega)\|p\|_{L_\infty(\Omega)}\right)\|U_0\|_{H_1}\right)\right). \qquad (6.22)$$

*Proof.* Let us show that under the hypotheses of the theorem the non-linear operator $T(u, \varepsilon)$ from (6.21) is a contraction and maps $S(0, R)$ into $S(0, R)$. Let $u_1, u_2 \in S(0, R)$. Using Lemma 6.6, we get

$$\|T(u_1, \varepsilon) - T(u_2, \varepsilon)\|_{H_1} \leq \frac{1}{m_\varepsilon} \|[A(0, \varepsilon) - A(u_1, \varepsilon)] u_1$$

$$- [A(0, \varepsilon) - A(u_2, \varepsilon)] u_2\|_{H_1}$$

$$\leq \frac{1}{m_\varepsilon} \|[A(u_1, \varepsilon) - A(u_2, \varepsilon)] u_1\|_{H_1}$$

$$+ \frac{1}{m_\varepsilon} \|[A(0, \varepsilon) - A(u_2, \varepsilon)](u_1 - u_2)\|_{H_1} .$$

From Lemma 6.3 and Corollary 6.2, we find

$$\|[A(u_1, \varepsilon) - A(u_2, \varepsilon)] u_1\|_{H_1} = \left\| \int_0^1 [F'(\xi u_1, \varepsilon) - F'(\xi u_2, \varepsilon)] \, d\xi u_1 \right\|_{H_1}$$

$$\leq \alpha \int_0^1 \xi \|u_1 - u_2\|_{H_1} \|u_1\|_{H_1} \, d\xi$$

$$= \frac{\alpha}{2} \|u_1 - u_2\|_{H_1} \|u_1\|_{H_1},$$

$$\|[A(0, \varepsilon) - A(u_2, \varepsilon)](u_1 - u_2)\|_{H_1} \leq \frac{\alpha}{2} \|u_1 - u_2\|_{H_1} \|u_2\|_{H_1}.$$

Since $\|u_i\|_{H_1} \leq R$, $i = 1, 2$, the above inequalities give

$$\|T(u_1, \varepsilon) - T(u_2, \varepsilon)\|_{H_1} \leq \frac{1}{m_\varepsilon} \alpha R \|u_1 - u_2\|_{H_1} .$$

Hence, the operator $T(u, \varepsilon)$ is a contraction on $S(0, R)$ if $\alpha R / m_\varepsilon < 1$. Furthermore, since

$$\|T(u, \varepsilon)\|_{H_1} \leq \frac{1}{m_\varepsilon} \|[A(0, \varepsilon) - A(u, \varepsilon)] u\|_{H_1} + \frac{1}{m_\varepsilon} \|f(\varepsilon)\|_{H_1^*}$$

$$\leq \frac{1}{2} \frac{\alpha R}{m_\varepsilon} \|u\|_{H_1} + \frac{1}{m_\varepsilon} \|f(\varepsilon)\|_{H_1^*}$$

if $u \in S(0, R)$, then under the hypotheses of the theorem we obtain that $\|T(u, \varepsilon)\|_{H_1} \leq R$, i.e. the operator $T(u, \varepsilon)$ maps $S(0, R)$ into $S(0, R)$.

Hence, according to the contraction principle[290], equation (6.52), as well as the equation $A(u, \varepsilon) u = f(\varepsilon)$, has a unique solution $u \in S(0, R)$. This solution may be constructed as a limit of a sequence $u^{(n)}$ defined by the following iterative process:

$$u^{(n+1)} = T(u^{(n)}, \varepsilon), \quad n = 0, 1, 2, \ldots$$

Write this process in the form

$$A(0,\varepsilon)\,u^{(n+1)} = [A(0,\varepsilon) - A(u^{(n)},\varepsilon)]\,u^{n)} + f(\varepsilon), \quad n = 0,1,2,\ldots \quad (6.23)$$

The operator $A(0,\varepsilon)$ may be represented as

$$A(0,\varepsilon) = A_0 + \varepsilon A_1 + \varepsilon^{M+1} A_{M+1},$$

where

$$(A_0 h, w) = (\mu \nabla h, \nabla w) + \left(\sum_{i=1}^{n} b_i(x)\frac{\partial h}{\partial x_i}, w\right) + (a(x)\,h, w),$$

$$(A_1 h, w) = \left(\sum_{i=1}^{n} c_i(x)\frac{\partial h}{\partial x_i} + d(x)\,h, w\right),$$

$$(A_{M+1} h, w) = \left(\sum_{i=1}^{n} q_i(x)\left[U_0\frac{\partial h}{\partial x_i} + h\frac{\partial U_0}{\partial x_i}\right] + 2p(x)U_0 h, w\right), \quad h, w \in H_1.$$

The operators $A_0$, $A_1$, and $A_{M+1}$ map $H_1$ into $H_1^*$ with the domain $D(A) = H_1$ and are linear and bounded. Since $A_0 = A(0,0)$, the operator $A_0^{-1}$ exists and, by Lemma 6.6, $\|A_0^{-1}\|_{H_1^* \to H_1} \leq 1/\mu_0$. The following inequalities hold[262]

$$\|A_1 A_0^{-1}\|_{H_1^* \to H_1^*} \leq M_A, \qquad \|A_{M+1} A_0^{-1}\|_{H_1^* \to H_1^*} \leq M_A,$$

where

$$M_A = \frac{1}{\mu_0}\max\left(c(\Omega)\left\|\sum_{i=1}^{n} c_i^2\right\|_{L_2(\Omega)}^{1/2} + c^2(\Omega)\|d\|_{L_2(\Omega)},\right.$$

$$\left.2c^2(\Omega)\left(\left\|\left(\sum_{i=1}^{n} q_i^2\right)^{1/2}\right\|_{L_\infty(\Omega)} + c_1(\Omega)\|p\|_{L_\infty(\Omega)}\right)\|U_0\|_{H_1}\right).$$

For the function $f(\varepsilon)$ defined by (6.12), the estimate

$$\|f(\varepsilon)\|_{H_1^*} \leq c_1\varepsilon + c_2\varepsilon^{M+1}$$

is valid, where the constants $c_1$ and $c_2$ are independent of $U_0$. The above inequalities show that for any $n$ the solution $u^{(n+1)}$ of equation (6.23) may be represented as a series in powers $\varepsilon$ convergent as $|\varepsilon| < \varepsilon_0$, where $\varepsilon_0 = 1/(M_A + 1)$. Taking into account the explicit form of $M_A$, we find that $\varepsilon_0$ is equal to (6.22). Hence, the solution $u$ of the problem $A(u,\varepsilon)\,u = f(\varepsilon)$ is represented as a    series    $u$    $=$    $\sum_{i=1}^{\infty} \varepsilon^i u_i$    convergent    for $0 \le \varepsilon < \varepsilon_0$.

## 6.7. Perturbation algorithm

Theorem 6.1 makes it possible to use the regular perturbation algorithm for finding the solution $u$ of the problem $A(u,\varepsilon)\,u = f(\varepsilon)$. The algorithm consists of the following. Substituting the expansion $u = \sum_{i=1}^{\infty} \varepsilon^i u_i$ into the equation $A(u,\varepsilon)\,u = f(\varepsilon)$ and equating the terms involving the same powers of $\varepsilon$, we come to the set of equations for finding the corrections $u_i$:

$$A_0 u_0 = 0,$$

$$A_0 u_1 = f_1 - A_1 u_0,$$

$$A_0 u_k = -A_1 u_{k-1}, \quad k \le M,$$

$$A_0 u_k = f_{M+1}\delta_{k,M+1} - A_1 u_{k-1} - A_{M+1} u_{k-M-1}$$
$$- \sum_{i+j+M+1=k} F(u_i, u_j), \quad k = 1, 2, \ldots,$$

where

$$(f_1, w) = -\left( \sum_{i=1}^{n} c_i(x)\frac{\partial U_0}{\partial x_i} + d(x)\, U_0, w \right),$$

$$(f_{M+1}, w) = -\left( \sum_{i=1}^{n} q_i(x) U_0 \frac{\partial U_0}{\partial x_i} + p(x)\, U_0^2, w \right), \quad w \in H_1,$$

$\delta_{k,M+1}$ is the Kronecker delta, $M > 0$, and $F(h, g) \in H_1^*$ is defined by the equality

$$(F(h,g), w) = \left( \sum_{i=1}^{n} \frac{1}{2}q_i(x)\left( h\frac{\partial g}{\partial x_i} + g\frac{\partial h}{\partial x_i} \right) + p(x)\, gh, w \right), \quad w \in H_1,$$

$$2F(h, U_0) = A_{M+1}h.$$

If $M = 0$, we get

$$A_0 u_0 = 0,$$

$$A_0 u_k = (f_1 + f_{M+1})\delta_{1k} - A_1 u_{k-1} - A_{M+1} u_{k-1} - \sum_{i+j+1=k} F(u_i, u_j), \quad k \le 1,$$

or, taking into account the form of the operators $A_0$, $A_1$, and $A_{M+1}$,

$$u_0 = 0,$$

$$
(\mu\nabla u_1, \nabla w) + \left(\sum_{i=1}^{n} b_i(x)\frac{\partial u_1}{\partial x_i}, w\right) + (a(x)\, u_1, w)
$$
$$
= -\left(\sum_{i=1}^{n} c_i(x)\frac{\partial U_0}{\partial x_i} + d(x)\, U_0, w\right)
$$
$$
-\left(\sum_{i=1}^{n} q_i(x) U_0 \frac{\partial U_0}{\partial x_i} + p(x)\, U_0^2, w\right) \quad \forall\, w \in H_1,
$$

$$
(\mu\nabla u_k, \nabla w) + \left(\sum_{i=1}^{n} b_i(x)\frac{\partial u_k}{\partial x_i}, w\right) + (a(x)\, u_k, w)
$$
$$
= -\left(\sum_{i=1}^{n} c_i(x)\frac{\partial u_{k-1}}{\partial x_i} + d(x)\, u_{k-1}, w\right)
$$
$$
-\left(\sum_{i=1}^{n} q_i(x)\left[U_0\frac{\partial u_{k-1}}{\partial x_i} + u_{k-1}\frac{\partial U_0}{\partial x_i}\right] + 2p(x)U_0 u_{k-1}, w\right)
$$
$$
-\sum_{i+j+1=k}\left(\sum_{l=2}^{n} q_i(x)\frac{1}{2}\left(u_i\frac{\partial u_j}{\partial x_l} + u_j\frac{\partial u_i}{\partial x_l}\right) + p(x)u_i u_j, w\right),
$$

(6.24)

where $w \in H_1$, $k = 2, 3, \ldots$ The following statement holds.

**Theorem 6.2** $(M = 0)$. *Let $S(0, R)$ be a ball in $H_1$ of the radius $R$. Under the restriction $0 \le \varepsilon < \varepsilon_1$, where*

$$
\varepsilon_1 = \min\left(\mu_0 \bigg/ \left(c^2(\Omega)\left[2\left(\left\|\left(\sum_{i=1}^{n} q_i^2\right)^{1/2}\right\|_{L_\infty(\Omega)}\right.\right.\right.\right.
$$
$$
\left.\left.\left.\left. +\|p\|_{L_\infty(\Omega)}c_1(\Omega)\right)R + \beta\|U_0\|_{H_1}\right]\right),\right.
$$
$$
\left.\mu_0 R\bigg/\left(2\left(f_0 + \frac{R}{2}c^2(\Omega)\,\beta\|U_0\|_{H_1}\right)\right)\right),
$$

(6.25)

$$
f_0 = \left(c(\Omega)\left\|\sum_{i=1}^{n} c_i^2\right\|_{L_2(\Omega)}^{1/2} + c^2(\Omega)\|d\|_{L_2(\Omega)} + c^2(\Omega)\left(\left\|\left(\sum_{i=1}^{n} q_i^2\right)^{1/2}\right.\right.\right.
$$
$$
\left.\left.\left. +c_1(\Omega)\|p\|_{L_\infty(\Omega)}\right)\|U_0\|_{H_1}\right)\|U_0\|_{H_1},
$$

*the perturbed problem $A(u, \varepsilon)\, u = f(\varepsilon)$ has a unique solution $u \in S(0, R)$ being represented as the series $u = \sum_{i=1}^{\infty} \varepsilon^i u_i$ convergent as $0 \le \varepsilon \le \varepsilon_0$, where*

$$\varepsilon_0 = \min\left(\varepsilon_1, \mu_0 \middle/ \left(2c(\Omega)\left\|\sum_{i=1}^{n} c_i^2\right\|_{L_2(\Omega)}^{1/2} + 2c^2(\Omega)\|d\|_{L_2(\Omega)}\right.\right.$$

$$+4\left(\left\|\left(\sum_{i=1}^{n} q_i^2\right)^{1/2}\right\|_{L_\infty(\Omega)} + c_1(\Omega)\|p\|_{L_\infty(\Omega)}\right)$$

$$\left.\left.\times c^2(\Omega)\|U_0\|_{H_1}\right)\right). \tag{6.26}$$

*The perturbation algorithm (6.24) for finding $u = U - U_0$ is convergent, and the following estimate holds:*

$$\|u - u^{(N)}\|_{H_1} \le (\varepsilon/\varepsilon_0)^{N+1}\|U_0\|_{H_1}, \tag{6.27}$$

*where*

$$u^{(N)} = \sum_{i=1}^{N} \varepsilon^i u_i, \quad 0 \le \varepsilon < \varepsilon_0.$$

*Proof.* Let $M = 0$. For $f(\varepsilon)$ the following inequality is true:

$$\|f(\varepsilon)\|_{H_1^*} \le \varepsilon f_0,$$

where $f_0$ is defined by (6.25). If $0 \le \varepsilon < \varepsilon_1$, where $\varepsilon_1$ is given by (6.25), then

$$\frac{2\left(\left\|\left(\sum_{i=1}^{n} q_i^2\right)^{1/2}\right\|_{L_\infty(\Omega)} + \|p\|_{L_\infty(\Omega)}c_1(\Omega)\right)c^2(\Omega)\,\varepsilon}{\mu_0 - \beta c^2(\Omega)\|u_0\|_{H_1}\varepsilon}\,R < 1,$$

$$\|f(\varepsilon)\|_{H_1^*} \le \varepsilon f_0 \le \frac{1}{2}\left(\mu_0 - \beta c^2(\Omega)\|u_0\|_{H_1}\varepsilon\right)R.$$

Thus, the hypotheses of Theorem 6.1 for $M = 0$ are satisfied. Hence, equation $A(u, \varepsilon)\, u = f(\varepsilon)$ has a unique solution. To find $u$ let us use the perturbation algorithm. Computing the corrections $u_k$, $k = 1, 2, \ldots$, from (6.24) successively, we construct the function $u^{(N)} = \sum_{i=1}^{N} \varepsilon^i u_i$ and show that $u^{(N)} \to u$ as $N \to \infty$. To do this, we make use of the technique presented in Section 5. Similarly to the proof of Theorem 5.1, we come to the conclusion that under the restriction $0 \le \varepsilon < \varepsilon_0$, where $\varepsilon_0$ satisfies the inequality

$$f_0 \le \|U_0\|_{H_1}\left(\frac{\mu_0}{\varepsilon_0} - c(\Omega)\left\|\sum_{i=1}^{n} c_i^2\right\|_{L_2(\Omega)}^{1/2} - c^2(\Omega)\|d\|_{L_2(\Omega)}\right)$$

$$-3c^2(\Omega)\left(\left\|\left(\sum_{i=1}^{n} q_i^2\right)^{1/2}\right\|_{L_\infty(\Omega)}\right.$$

$$\left.+c_1(\Omega)\|p\|_{L_\infty(\Omega)}\right)\|U_0\|_{H_1}\right), \tag{6.28}$$

$u^{(N)} \to u$ as $N \to \infty$ in the norm of $H_1$, and the estimate (6.27) is valid.

Solving the inequality (6.28) with respect to $\varepsilon_0$ and taking into account the form of $f_0$, we find

$$\varepsilon_0 \leq \mu_0 \left/ \left( 2c(\Omega)\left\|\sum_{i=1}^{n} c_i^2\right\|_{L_2(\Omega)}^{1/2} + 2c^2(\Omega)\|d\|_{L_2(\Omega)} + 5c^2(\Omega)\frac{\sqrt{n}}{2}\|U_0\|_{H_1}\right.\right.$$

$$\left.\left.+4c^2(\Omega)\left(\left\|\left(\sum_{i=1}^{n} q_i^2\right)^{1/2}\right\|_{L_\infty(\Omega)} + \|p\|_{L_\infty(\Omega)}c_1(\Omega)\right)\|U_0\|_{H_1}\right).\right.$$

Taking $\varepsilon_0$ equal to (6.26) proves the theorem.

Note in conclusion that the above-stated results remain valid (only specific constants vary) for a more general perturbing non-linear operator. An example is the case when the perturbed problem involves the term $\varepsilon^{M+1}F(U,U)$, where $F(U,U)$: $H_1 \times H_1 \to H_1^*$ is an arbitrary 2-power operator satisfying the inequality

$$\|F(u_1, u_2)\|_{H_1^*} \leq c\|u_1\|_{H_1}\|u_2\|_{H_1}, \quad c = \text{constant} > 0.$$

# Chapter 6

# Adjoint equations and the $N$-th order perturbation algorithms in non-linear problems of transport theory

In deciding on a principle of construction of adjoint equations, one is to take proper account of the objectives to be pursued in the problems under consideration (finding the corrections to solutions, computing the functionals, choosing the control parameters, etc.). In addition, one should ensure the solvability of adjoint equations If the problem on development of perturbation algorithms is attacked, the adjoint equations are wanted to be correctly solvable as it is a relatively simple matter in this case to justify perturbation algorithms both of small and of the $N$-th order. It may be also suggested that, in general, one may be hard pressed to find the 'optimal' principle of construction of adjoint equations, and it would be appropriate to solve this problem for each specific case. This chapter considers similar cases that have arisen in transport theory problems. We discuss, as a rule, the possibility of application of perturbation algorithms for solving the problem under consideration, based on the solution of main and adjoint equations.

For simplicity we consider only one-velocity problems of transport theory. As will readily be observed, the formulated statements may be easily extended to cover the case of multi-group transport problems.

## 1. SOME PROBLEMS OF TRANSPORT THEORY

**1.1.** Let $D$ be a domain of $\mathbf{R}^3$ (for simplicity $D$ is assumed to be convex) with the boundary $\partial D$, $\overline{D} = D \cup \partial D$;

$$\Omega = \left\{ s\colon s = (s_1, s_2, s_3);\ |s| = \left(\sum_{i=1}^{3} s_i^2\right)^{1/2} = 1 \right\}$$

is the unit sphere; $(s, \nabla)v = \sum_{i=1}^{3} s_i\, \partial v/\partial x_i$ is the derivative of $v$ with respect to the direction $s$; $x = (x_1, x_2, x_3) \in \overline{D}$; $n = (n_1, n_2, n_3)$ is the unit vector of outward normal at the point $x \in \partial D$; and $Su \equiv u_0 \equiv \int_0^{2\pi} d\psi \int_0^{\pi} \sin \vartheta u(s, x)\, d\vartheta$ is the integral of $u$ over the sphere $\Omega$.

Consider the following problem for the one-velocity transport equation

$$(s, \nabla)\varphi + \sigma(x)\,\varphi = \frac{1}{4\pi} \int_\Omega \sigma_{is,f}(x)\,\varphi(s', x)\,ds' + F(s, x),$$

$$\varphi(s, x) = 0, \quad x \in \partial D, \quad (s, n) < 0, \quad s \in \Omega, \tag{1.1}$$

where $F(s, x)$ is the external source function, $0 < \sigma_0 \leq \sigma(x) \leq \sigma_1 < \infty$, $\sigma_0, \sigma_1 = $ constant; $\sigma_{is,f}(x) = \sigma_{is}(x) + \sigma_f \nu H_f(x)$; $\sigma(x) = \sigma_a(x) + \sigma_a^{(f)} N_f(x) + \sigma_a^{(y)} N_y(x)$; $\nu, \sigma_a^{(f)}, \sigma_f, \sigma_a^{(y)} = $ constant $> 0$; and $N_f(x)$, $N_y(x)$, $\sigma_a(x)$ are non-negative measurable functions bounded almost everywhere.

In the event that external sources are absent $(F(s, x) \equiv 0)$, it does not always happen that problem (1.1) has a non-trivial solution. Such a solution exists only for a specific choice of the coefficients $\sigma(x), \ldots, N_y(x)$. To find these coefficients one considers an eigenvalue problem of the form

$$(s, \nabla)\varphi + \sigma(x)\,\varphi - \frac{1}{4\pi} \int_\Omega \sigma_{is,f}(x)\,\varphi(s', x)\,ds' = \lambda\varphi(s, x),$$

$$\varphi(s, x) = 0, \quad x \in \partial D, \quad (s, n) < 0, \quad s \in \Omega \tag{1.2}$$

(this is one of the eigenvalue problems of transport theory[261]).

If it happens that $\lambda = 0$, in this case we denote the solution of problem (1.2) by $\varphi_{(0)}$ and assign the corresponding coefficients the extra upper index (0). Thus, $\varphi_{(0)}$ is a non-trivial solution of the problem

$$(s, \nabla)\varphi_{(0)} + \sigma^{(0)}(x)\,\varphi_{(0)} - \frac{1}{4\pi} \int_\Omega \sigma_{is,f}^{(0)}(x)\,\varphi_{(0)}(s', x)\,ds' = 0,$$

$$\varphi_{(0)}(s, x) = 0, \quad x \in \partial D, \quad (s, n) < 0, \quad s \in \Omega, \tag{1.3}$$

where $\sigma^{(0)}(x) = \sigma_a(x) + \sigma_a^{(f)} N_f^{(0)}(x) + \sigma_a^{(y)} N_y^{(0)}(x)$; $\sigma_{is,f}^{(0)}(x) = \sigma_{is}(x) + \nu\sigma_f N_f^{(0)}(x)$. It is known[262] that the eigenvalue $\lambda = 0$ is simple, and the eigenfunction $\varphi_{(0)}(s, x)$ may be chosen to be non-negative almost everywhere and such that the 'integral stream function' $S\varphi_{(0)} = \int_\Omega \varphi_{(0)}\,ds$ is positive in $D$. As a normalizing condition for $\varphi_{(0)}$ one can consider the relation

$$J_p(\varphi_{(0)}) \equiv \int_{\Omega \times D} \varphi_{(0)}(s, x)\,p(x)\,ds\,dx = W_{(0)}, \tag{1.4}$$

where $p(x)$ is a bounded non-negative function, and $W_{(0)}$ is a given positive constant. This is the condition of normalizing the 'critical state' of the system governed by problem (1.3) to a prescribed level $W_{(0)}$ of energy derived from the physical process.

Notice that perturbation algorithms are widely used for solving the problems of the form (1.2), (1.3). Some of these algorithms are considered below.

**1.2.** Let a physical process in a medium taking up a volume $D$ be governed by (1.3) under condition (1.4) at the time $t = 0$ and by the time-dependent problem at $t > 0$:

$$\frac{\partial \varphi}{\partial t} + (s, \nabla)\varphi + \sigma(t, x, \varphi)\,\varphi - \frac{1}{4\pi}\int_\Omega \sigma_{is,f}(t, x, \varphi)\,\varphi(s', x)\,ds' = 0,$$

$$\varphi|_{t=0} = \varphi_{(0)}, \tag{1.5}$$

$$\varphi = 0, \quad x \in \partial D, \quad (s, n(x)) < 0,$$

where

$$\sigma(t, x, \varphi) = \sigma_a(x) + \sigma_a^{(f)} N_f^{(0)}(x)$$

$$\times \exp\left\{-\sigma_a^{(f)} \int_0^t \int_\Omega \varphi(t', s, x)\,ds\,dt'\right\} + \sigma_a^{(y)} N_y(t, x),$$

$$N_y(t, x) = N_y^{(0)}(x) + \sum_{i=1}^\infty \frac{t^i N_y^{(i)}(x)}{i!},$$

$$\sigma_{is,f}(t, x, \varphi) = \sigma_{is}(x) + \nu\sigma_f N_f^{(0)}(x)$$

$$\times \exp\left\{-\sigma_a^{(f)} \int_0^t \int_\Omega \varphi(t', s, x)\,ds\,dt'\right\},$$

$N_y^{(i)}(x)$ are prescribed functions.

Suppose that our interest is not only the solution $\varphi(t, s, x)$ of problem (1.5), but also the value of the functional

$$J_{p,\chi}(\varphi) = \int_{\Omega \times D} \int_0^T \varphi(t, s, x)\,p(x)\,\chi(t)\,dt\,dx\,ds, \tag{1.6}$$

where

$$\chi(t) = \begin{cases} 1/(t_2 - t_1), & t \in (t, t_2), \\ 0, & t \notin (t_1, t_2) \end{cases} \tag{1.7}$$

($t_1$, $t_2$ are real values from the interval $(0, T)$, $T < \infty$). Note that the functional $J_{p,\chi}$ is the mean value of the energy released in $D$ over the period $(t_1, t_2)$.

Along with problem (1.5), one can consider the eigenvalue problem

$$(s, \nabla)\Phi_t + \sigma(t, x, \varphi)\Phi_t - \frac{1}{4\pi}\int_\Omega \sigma_{is,f}(t, x, \varphi)\Phi_t(s, x)\,ds' = \lambda_t\Phi_t, \tag{1.8}$$

$$\Phi_t = 0, \quad x \in \partial D, \quad (s, n) < 0,$$

where $\Phi_t$ and $\lambda_t$ depend parametrically on $t$, and the $\varphi$ is assumed to be known, with the normalizing condition

$$J_p(\Phi_t) = \int_{\Omega \times D} \Phi_t(s, x)\,p(x)\,ds\,dx = W_{(p)}. \tag{1.9}$$

If it happens that $\lambda_{t_0} = 0$ at a time $t_0$, then through the simplicity of the eigenvalue $\lambda_{t_0} = 0$ we get

$$\Phi_{t_0}(s, x) = \varphi_{(0)}(s, x). \tag{1.10}$$

**1.3.** Suppose that the solution of problem (1.5) may be represented as the series

$$\varphi(t, s, x) = \varphi_{(0)}(s, x) + \sum_{i=1}^{\infty} \frac{t^{(i)} \varphi_{(i)}(s, x)}{i!}, \tag{1.11}$$

where the functions $\varphi_{(i)}(s, x)$, $i = 1, 2, \ldots$, satisfy the boundary condition in (1.5). To determine $\varphi_{(s)}(s, x)$ let us substitute (1.11) into (1.5) and put $t = 0$. As a result we get

$$\left. \frac{\partial \varphi}{\partial t} \right|_{t=0} = \varphi_{(1)} = 0. \tag{1.12}$$

To find $\varphi_{(2)}$ differentiate the first equation in (1.5) with respect to $t$. By setting $t = 0$ we get

$$\begin{aligned}
\left. \frac{\partial^2 \varphi}{\partial t^2} \right|_{t=0} &= \varphi_{(2)}(s, x) \\
&= \left[ (\sigma_a^{(f)})^2 N_f^{(0)}(x)(S\varphi_{(0)}) - \sigma_a^{(y)} N_y^{(1)}(x) \right] \varphi_{(0)}(s, x) \\
&\quad - \frac{1}{4\pi} \nu \sigma_f \sigma_a^{(f)} N_f^{(0)}(x)(S\varphi_{(0)})^2.
\end{aligned} \tag{1.13}$$

Hereafter we assume that

$$N_f^{(0)}(x) = 0 \quad \text{on } \partial D. \tag{1.14}$$

Then, it follows from (1.13) that $\varphi_{(2)}(s, x)$ satisfies the boundary condition in (1.5).

Thus, if $N_y^{(0)}(x)$ and $N_y^{(1)}(x)$ are prescribed functions and $\varphi_{(0)}(s, x)$ is predetermined, then we have

$$\varphi(t, s, x) = \varphi_{(0)}(s, x) + \frac{\varphi_{(2)}(s, x)\, t^2}{2} + O(t^3), \tag{1.15}$$

where the function $\varphi_{(2)}(s, x)$ is defined by formula (1.13).

Using (1.15), we find

$$\exp\left\{ -\sigma_a^{(f)} \int_0^t (S\varphi)\, dt' \right\} = \left[ 1 - \sigma_a^{(f)} t (S\varphi_{(0)}) \right] + O(t^2). \tag{1.16}$$

Then problem (1.8) may be written as

$$(s, \nabla)\, \Phi_t + \sigma(t, x)\, \Phi_t - \frac{1}{4\pi} \int_\Omega \sigma_{is,f}(t, x)\, \Phi_t(s', x)\, ds' = \lambda_t \Phi_t,$$
$$\Phi_t = 0, \quad x \in \partial D, \quad (s, n) < 0, \quad s \in \Omega, \tag{1.17}$$

where

$$\sigma(t, x) = \sigma_a(x) + \sigma_a^{(f)} N_f^{(0)}(x) \left[ 1 - \sigma_a^{(f)} t(S\varphi_{(0)}) + \frac{(\sigma_a^{(f)})^2 (S\varphi_{(0)})^2 t^2}{2} \right]$$

$$+ \sigma_a^{(y)} \left[ N_y^{(0)} + t N_y^{(1)} + \frac{t^2 N_y^{(2)}(x)}{2} \right] + O(t^3), \tag{1.18}$$

$$\sigma_{is,f}(t, x) = \sigma_{is}(x) + \nu \sigma_f N_f^{(0)}(x) \left[ 1 - \sigma_a^{(f)} t(S\varphi_{(0)}) \right.$$

$$+ \left. \frac{(\sigma_a^{(f)})^2 (S\varphi_{(0)})^2 t^2}{2} \right] + O(t^3). \tag{1.19}$$

and the functional $J_{p,\chi}(\varphi)$ may be represented in the form

$$J_{p,\chi}(\varphi) = J_p(\varphi_{(0)}) + J_p(\varphi_{(2)}) \left( \frac{t_2^2 + t_1 t_2 + t_1^2}{6} \right) + O(t_2^3). \tag{1.20}$$

Therefore, for $t \in (0, T)$, $T \ll 1$, one can compute the energy functional by formula (1.20) to sufficient accuracy, where

$$J_p(\varphi_{(0)}) = W_{(0)},$$

$$J_p(\varphi_{(2)}) = \int_D p(x) \left\{ (\sigma_a^{(f)})^2 N_f^{(0)}(x)(S\varphi_{(0)}) \right. \tag{1.21}$$

$$- \sigma_a^{(y)} N_y^{(1)}(x) - \nu \sigma_f \sigma_a^{(f)} N_f^{(0)}(x)(S\varphi_{(0)}) \Big\} (S\varphi_{(0)}) \, dx.$$

*Remark* 1.1. If $\sup(p(x)) \cap \sup(N_y^{(1)}(x)) = \emptyset$, then

$$J_p(\varphi_{(2)}) = \int_D p(x) \{ \sigma_a^{(f)} - \nu \sigma_f \} \sigma_a^{(f)} N_f^{(0)}(x)(S\varphi_{(0)}) \, dx. \tag{1.22}$$

The above-formulated problems are mathematical models which are of our main concern in this chapter. We shall use perturbation algorithms for studying and solving these problems. The next chapter considers perturbation algorithms, as applied to eigenvalue problems.

## 2. THE $N$-TH ORDER PERTURBATION ALGORITHM FOR AN EIGENVALUE PROBLEM

**2.1.** Consider the eigenvalue problem

$$A\varphi = \lambda \varphi, \tag{2.1}$$

where $A$ is a linear operator mapping the Hilbert space $X$ into $X$ with the domain $D(A)$ dense in $X$. Let $\lambda_{(0)}$ be a simple eigenvalue of problem (2.1), $\varphi_{(0)}$ be the corresponding eigenfunction, and $\lambda_{(0)}^* = \bar{\lambda}_{(0)}$, $\varphi_{(0)}^*$ the solutions of the adjoint problem

$$A^* \varphi_{(0)}^* = \lambda_{(0)}^* \varphi_{(0)}^*. \tag{2.2}$$

Let the normalizing condition

$$(\varphi_{(0)}, \varphi_{(0)}^*) = 1 \tag{2.3}$$

be satisfied. Consider the problems perturbed with respect to (2.1), (2.2):

$$(A + \varepsilon \delta A)\tilde{\varphi} = \tilde{\lambda}\tilde{\varphi}, \tag{2.4}$$

$$(A^* + \varepsilon \delta A^*)\tilde{\varphi}^* = \tilde{\lambda}^* \tilde{\varphi}^*, \tag{2.5}$$

where $\delta A$ is a linear operator mapping $X$ into $X$ with the domain $D(A)$, $\delta A^* = (\delta A)^*$, $\varepsilon$ a small parameter, $0 < \varepsilon \leq 1$.

We will seek the perturbed eigenvalues and eigenfunctions in the form

$$\tilde{\varphi} = \varphi_{(0)} + \varepsilon\varphi_1 + \varepsilon^2\varphi_2 + \ldots, \quad \tilde{\varphi}^* = \varphi_{(0)}^* + \varepsilon\varphi_1^* + \varepsilon^2\varphi_2^* + \ldots,$$

$$\tilde{\lambda} = \lambda_{(0)} + \varepsilon\lambda_1 + \varepsilon^2\lambda_2 + \ldots, \quad \tilde{\lambda}^* = \lambda_{(0)}^* + \varepsilon\lambda_1^* + \varepsilon^2\lambda_2^* + \ldots, \tag{2.6}$$

where $\{\varphi_i\}_{i=1}^\infty$, $\{\varphi_i^*\}_{i=1}^\infty$ are functions, and $\{\lambda_i\}_{i=1}^\infty$, $\{\lambda_i^*\}_{i=1}^\infty$ numbers to be determined.

Consider the following normalizing conditions for $\tilde{\varphi}$ and $\tilde{\varphi}^*$:

$$(\tilde{\varphi}, \varphi_{(0)}^*) = 1, \quad (\varphi_{(0)}, \tilde{\varphi}^*) = 1. \tag{2.7}$$

Using the well-known technique, we obtain the formulae for $\lambda_i$, $\lambda_i^*$ and the equations

$$A\varphi_{(0)} = \lambda_{(0)}\varphi_{(0)},$$

$$A\varphi_1 - \lambda_{(0)}\varphi_1 = -\delta A\varphi_{(0)} + \lambda_1\varphi_{(0)},$$

$$\ldots$$

$$A^*\varphi_{(0)}^* = \lambda_{(0)}^*\varphi_{(0)}^*,$$

$$A^*\varphi_1^* - \lambda_{(0)}^*\varphi_1^* = -\delta A^*\varphi_{(0)}^* + \lambda_1^*\varphi_{(0)}^*, \tag{2.8}$$

$$\ldots$$

$$(\varphi_i, \varphi_{(0)}^*) = 0, \quad (\varphi_{(0)}, \varphi_i^*) = 0, \quad i = 1, 2, \ldots,$$

$$\lambda_1 = (\delta A\varphi_{(0)}, \varphi_{(0)}^*), \quad \lambda_1^* = (\delta A^*\varphi_{(0)}^*, \varphi_{(0)}) = \bar{\lambda}_1,$$

$$\lambda_i^* = \bar{\lambda}_i, \quad i = 1, 2, \ldots, \tag{2.9}$$

and for $i \geq N + 1$ the following relation holds[103]

$$\lambda_i = \bar{\lambda}_i^* = (\delta A \varphi_{i-N-1}, \varphi_N^*) - \sum_{l=1}^{N} \sum_{j=1}^{i-N-1} \lambda_{i-l-j}(\varphi_j, \varphi_l^*). \tag{2.10}$$

Thus, one can compute $\{\lambda_i\}_{i=N+1}^{2N+1}$, $\{\lambda_i^*\}_{i=N+1}^{2N+1}$ using the functions $\{\varphi_i\}_{i=0}^{N}$, $\{\varphi_i^*\}_{i=0}^{N}$ and the values $\{\lambda_i\}_{i=0}^{N}$, $\{\lambda_i^*\}_{i=0}^{N}$.

In particular, for the first three corrections we find

$$\lambda_1 = \bar{\lambda}_1^* = (\delta A \varphi_{(0)}, \varphi_{(0)}^*),$$

$$A\varphi_1 - \lambda_{(0)}\varphi_1 = -\delta A \varphi_{(0)} + \lambda_1 \varphi_{(0)}, \quad (\varphi_1, \varphi_{(0)}^*) = 0,$$

$$A^* \varphi_1^* - \lambda_{(0)}^* \varphi_1^* = -\delta A^* \varphi_{(0)}^* + \lambda_1^* \varphi_{(0)}^*, \quad (\varphi_{(0)}, \varphi_1^*) = 0, \tag{2.11}$$

$$\lambda_2 = (\delta A \varphi_1, \varphi_{(0)}^*) = \bar{\lambda}_2^*,$$

$$\lambda_3 = (\delta A \varphi_1, \varphi_1^*) - \lambda_1(\varphi_1, \varphi_1^*) = \bar{\lambda}_3^*.$$

We shall use these formulae to determine corrections in the transport problem.

**2.2.** Consider the eigenvalue problem for the transport equation

$$(s, \nabla)\varphi + \sigma(x)\,\varphi(s, x) - \frac{\sigma_{is,f}(x)}{4\pi} \int_\Omega \varphi(s', x)\, ds' = \lambda \varphi(s, x), \tag{2.12}$$

$$\varphi(s, x) = 0, \quad x \in \partial D, \quad (s, n) < 0.$$

Let $\lambda_{(0)}$ be an eigenvalue with the smallest modulus. As stated above, this eigenvalue is real and simple, and the corresponding eigenfunction $\varphi_{(0)}(s, x)$ is non-negative. The adjoint to the (2.12) problem reads

$$-(s, \nabla)\psi + \sigma\psi - \frac{\sigma_{is,f}}{4\pi} \int_\Omega \psi\, ds' = \lambda^* \psi, \tag{2.13}$$

$$\psi(s, x) = 0, \quad x \in \partial D, \quad (s, n) > 0.$$

Consider the normalizing condition for $\varphi_{(0)}$, $\psi_{(0)}$ in the form

$$\int_{\Omega \times D} \varphi_{(0)}(s, x)\, \psi_{(0)}(s, x)\, ds\, dx = 1. \tag{2.14}$$

Here

$$\sigma = \sigma_a(x) + \sigma_a^{(f)} N_f^{(0)}(x) + \sigma_a^{(y)} N_y^{(0)}(x),$$

$$\sigma_{is,f} = \sigma_{is}(x) + \nu \sigma_f N_f^{(0)}(x). \tag{2.15}$$

Consider the perturbed problems

$$(s, \nabla)\tilde{\varphi} + (\sigma + \delta\sigma)\tilde{\varphi} = \frac{\sigma_{is,f} + \delta\sigma_{is,f}}{4\pi} \int_{\Omega} \tilde{\varphi}(s', x) \, ds' = \tilde{\lambda}\tilde{\varphi},$$

$$\tilde{\varphi} = 0, \quad x \in \partial D, \quad (s, n) < 0, \qquad (2.16)$$

$$-(s, \nabla)\tilde{\psi} + (\sigma + \delta\sigma)\tilde{\psi} = \frac{\sigma_{is,f} + \delta\sigma_{is,f}}{4\pi} \int_{\Omega} \tilde{\psi} \, ds' = \tilde{\lambda}^*\tilde{\psi},$$

$$\tilde{\psi} = 0, \quad x \in \partial D, \quad (s, n) > 0, \qquad (2.17)$$

where

$$\delta\sigma = t \left[ \sigma_a^{(y)} N_y^{(1)}(x) - (\sigma_a^{(f)})^2 N_f^{(0)}(x) \left( \int_{\Omega} \varphi_{(0)} \, ds \right) \right]$$

$$+ t^2 \left[ \sigma_a^{(y)} N_y^{(2)}(x) + \frac{1}{2}(\sigma_a^{(f)})^3 N_f^{(0)}(x) \left( \int_{\Omega} \varphi_{(0)} \, ds \right)^2 \right], \quad (2.18)$$

$$\delta\sigma_{is,f} = t \left[ -\nu\sigma_f \sigma_a^{(f)} N_f^{(0)}(x) \int_{\Omega} \varphi_{(0)} \, ds \right]$$

$$+ \frac{t^2}{2} \left[ \nu\sigma_f (\sigma_a^{(f)})^2 \left( \int_{\Omega} \varphi_{(0)} \, ds \right)^2 N_f^{(0)}(x) \right], \qquad (2.19)$$

and $t$ is assumed to be a small parameter.

Let us use the algorithm presented in Subsection 2.1 for the approximate solution of problems (2.16), (2.17). Then

$$\tilde{\varphi} = \varphi_{(0)} + t\varphi_{(1)} + t^2\varphi_{(2)} + \ldots, \quad \tilde{\psi} = \psi_{(0)} + t\psi_{(1)} + t^2\psi_{(2)} + \ldots,$$

$$\tilde{\lambda} = \lambda_{(0)} + t\lambda_1 + t^2\lambda_2 + \ldots, \quad \tilde{\lambda}^* = \lambda_{(0)}^* + t\lambda_1^* + t^2\lambda_2^* + \ldots \qquad (2.20)$$

By $\alpha(x)$ we denote

$$\alpha(x) = \sigma_a^{(y)} N_y^{(1)}(x) + [\nu\sigma_f \sigma_a^{(f)} N_f^{(0)}(x) - (\sigma_a^{(f)})^2 N_f^{(0)}(x)] \int_{\Omega} \varphi_{(0)} \, ds,$$

and consider the formulae for the first three corrections to the unperturbed eigenvalue:

$$\lambda_1 = \lambda_1^* = \int_D \alpha(x) \int_{\Omega} \varphi_{(0)}(s, x) \psi_{(0)}(s, x) \, ds \, dx,$$

$$\lambda_2 = \lambda_2^* = \int_D \alpha(x) \int_{\Omega} \varphi_{(1)}(s, x) \psi_{(0)}(s, x) \, ds \, dx, \qquad (2.21)$$

$$\lambda_3 = \lambda_3^* = \int_D \alpha(x) \int_\Omega \varphi_{(1)}(s, x)\, \psi_{(1)}(s, x)\, ds\, dx$$

$$-\lambda_1 \int_{\Omega \times D} \varphi_{(1)}(s, x)\, \psi_{(1)}(s, x)\, ds\, dx,$$

where the functions $\varphi_{(1)}(s, x)$, $\psi_{(1)}(s, x)$ are the solutions to the problems:

$$(s, \nabla)\, \varphi_{(1)} + \sigma\varphi_{(1)} - \frac{\sigma_{is,f}}{4\pi} \int_\Omega \varphi_{(1)}\, ds$$

$$= \frac{\alpha(x)}{4\pi} \int_\Omega \varphi_{(0)}\, ds + \lambda_{(1)}\varphi_{(0)} + \lambda_{(0)}\varphi_{(1)},$$

$$\varphi_{(1)} = 0, \quad x \in \partial D, \quad (s, n) < 0, \quad \int_{\Omega \times D} \varphi_{(1)}\varphi_{(0)}\, ds\, dx = 0; (2.22)$$

$$-(s, \nabla)\, \psi_{(1)} + \sigma\psi_{(1)} - \frac{\sigma_{is,f}}{4\pi} \int_\Omega \psi_{(1)}\, ds$$

$$= \frac{\alpha(x)}{4\pi} \int_\Omega \psi_{(0)}\, ds + \lambda_{(1)}\psi_{(0)} + \lambda_{(0)}\psi_{(1)},$$

$$\psi_{(1)} = 0, \quad x \in \partial D, \quad (s, n) > 0, \quad \int_{\Omega \times D} \varphi_{(0)}\psi_{(1)}\, ds\, dx = 0. (2.23)$$

Notice that in this case

$$\begin{aligned}\varphi_{(0)}(s, x) &= \psi_{(0)}(-s, x), \\ \varphi_{(1)}(s, x) &= \psi_{(1)}(-s, x).\end{aligned} \quad (2.24)$$

By this is meant that if the function $\varphi_{(0)}(s, x)$ is known, it will suffice to solve problem (2.22) and find the corrections $\lambda_1$, $\lambda_2$, and $\lambda_3$.

**2.3** In the event that $\lambda_{(0)} = 0$, we obtain the following expression for $\tilde{\lambda}(t)$:

$$\tilde{\lambda}(t) = \lambda_1 t + \lambda_2 t^2 + \lambda_3 t^3 + O(t^4), \quad (2.25)$$

where

$$\lambda_1 = \int_D \alpha(x) \int_\Omega \varphi_{(0)}(s, x)\, \varphi_{(0)}(-s, x)\, ds\, dx,$$

$$\lambda_2 = \int_D \alpha(x) \int_\Omega \varphi_{(1)}(s, x)\, \varphi_{(0)}(-s, x)\, ds\, dx,$$

$$\lambda_3 = \int_D \alpha(x) \int_\Omega \varphi_{(1)}(s, x)\, \varphi_{(1)}(-s, x)\, ds\, dx$$

$$-\lambda_1 \int_{\Omega \times D} \varphi_{(1)}(s, x)\, \varphi_{(1)}(-s, x)\, ds\, dx.$$

These formulae for corrections may be advantageous for approximate solution of the control problem considered in the next section.

## 3. A PROBLEM OF CONTROL AND ITS APPROXIMATE SOLUTION WITH THE USE OF PERTURBATION ALGORITHMS

**3.1.** Let a physical process in $D$ be described by the eigenfunction $\varphi_{(0)}(s, x)$ corresponding to $\lambda_{(0)} = 0$ and satisfying (2.12). Let thereafter, for $t > 0$, the fuel begin to burn away and the process is governed by the models presented in Section 1. Suppose that we need to find a function $N_y^{(1)}(x)$ such that the physical system remains 'critical', i.e. $\tilde{\lambda} \cong 0$ is satisfied as before. The problem of finding $N_y^{(1)}(x)$ is said to be a problem of control.

Consider problem (2.16) and assume that the first eigenvalue of (2.12) equals zero. To ensure that the system remains critical, the equality $\delta\lambda_{(0)} = \tilde{\lambda} - \lambda_{(0)} = 0$ would suffice. In this case, owing to the simplicity of the eigenvalue $\lambda_{(0)}$, the eigenfunctions coincide (within a normalizing factor), i.e. the system remains in its original state. To get $\delta\lambda_{(0)} \cong 0$, put

$$\sigma_a^{(y)} N_y^{(1)}(x) + [\nu\sigma_f\sigma_a^{(f)} N_f^{(0)}(x) - (\sigma_a^{(f)})^2 N_f^{(0)}(x)] \int_\Omega \varphi_{(0)} \, ds = 0. \qquad (3.1)$$

From (3.1) we find

$$N_y^{(1)}(x) = (\sigma_a^{(f)} - \nu\sigma_f)\sigma_a^{(f)} / \left( \sigma_a^{(y)} \left( \int_\Omega \varphi_{(0)}(s, x) \, ds \right) N_f^{(0)}(x) \right), \qquad (3.2)$$

which results in $\lambda_1 = \lambda_2 = \lambda_3 = 0$. Hence,

$$\delta\lambda_{(0)} = O(t^4), \quad 0 < t \ll 1. \qquad (3.3)$$

Note also that (3.2) gives

$$J(\varphi_{(2)}) = 0, \quad J_{p,\chi}(\tilde{\varphi}) \cong J_p(\varphi_{(0)}) + O(t^3). \qquad (3.4)$$

**3.2.** Thus, a solution of the control problem is given by (3.2). However, in cases of practical importance, the supports of the functions $N_y^{(1)}(x)$ and $N_f^{(0)}(x)$ do not intersect. Then condition (3.2) should be treated as a suggestion for determining the 'averaged' form of $N_y^{(1)}(x)$ or, otherwise, one should resolve the problem on finding $N_y^{(1)}(x)$ with supplementary conditions on $N_y^{(1)}(x)$. Let us suppose, for example, that $N_y^{(1)}(x)$ is of the form

$$N_y^{(1)}(x) = a^{(1)} \sum_{i=1}^{I} \chi_i(x), \qquad (3.5)$$

where $a^{(1)} = $ constant , and $\chi_i(x)$ is the characteristic function of the sub-domain $D_i \subset D$. Assume also that the subdomains $D_i$ do not intersect each other. We take $a^{(1)}$ such that $\lambda_1 = 0$. To satisfy $\lambda_1 = 0$, it will suffice to put

$$
a^{(1)} = \frac{\int_D [\sigma_a^{(f)} - \nu\sigma_f]\sigma_a^{(f)} \left(\int_\Omega \varphi_{(0)} \, ds\right) N_f^{(0)}(x) \left(\int_\Omega \varphi_{(0)}\psi_{(0)} \, ds\right) \, dx}{\sigma_a^{(y)} \sum_{i=1}^I \int_{D_i} dx \int_\Omega \varphi_{(0)}\psi_{(0)} \, ds}, \tag{3.6}
$$

where $\psi_{(0)} = \varphi_{(0)}(-s, x)$.

**3.3.** If we define $a^{(1)}$ and $N_y^{(1)}$ by formulae (3.6), (3.5), we get $\lambda_1 = 0$, but $J_p(\varphi_{(2)}) \not\equiv 0$, $\lambda_2 \neq 0$, $\lambda_3 \neq 0$. To satisfy the conditions $\lambda_1 = 0$, $J_p(\varphi_2) = 0$, as an example, let us suppose that $N_y^{(1)}(x)$ is of the form

$$
N_y^{(1)}(x) = a^{(1)} \sum_{i=1}^{I_0} \chi_i(x) + a^{(2)} \sum_{i=I_0+1}^I \chi_i(x), \tag{3.7}
$$

where $I_0 < I$; $a^{(1)}, a^{(2)} = $ constant. Substituting (3.7) into (1.21) and into the expression for $\lambda_1$, we obtain the set of two equations for determining $a^{(1)}$ and $a^{(2)}$:

$$
a^{(1)} \left( \sum_{i=1}^{I_0} \int_{D_i} dx \int_\Omega \varphi_{(0)}(s, x)\, \varphi_{(0)}(-s, x)\, ds \right)
$$
$$
+ a^{(2)} \left( \sum_{i=I_0+1}^I \int_{D_i} dx \int_\Omega \varphi_{(0)}(s, x)\, \varphi_{(0)}(-s, x)\, ds \right)
$$
$$
= \frac{1}{\sigma_a^{(y)}} \int_D [\sigma_a^{(f)} - \nu\sigma_f]\sigma_a^f \left(\int_\Omega \varphi_{(0)} \, ds\right)
$$
$$
\times N_f^{(0)}(x) \left(\int_\Omega \varphi_{(0)}(s, x)\, \varphi_{(0)}(-s, x)\, ds\right) \, dx, \tag{3.8}
$$
$$
a^{(1)} \left( \sum_{i=1}^{I_0} \int_{D_i} p(x) \left(\int_\Omega \varphi_{(0)} \, ds\right) \, dx \right)
$$
$$
+ a^{(2)} \left( \sum_{i=I_0+1}^I \int_{D_i} p(x) \left(\int_\Omega \varphi_{(0)} \, ds\right) \, dx \right)
$$
$$
= \frac{1}{\sigma_a^{(y)}} \int_D p(x)[\sigma_a^{(f)} - \nu\sigma_f]\sigma_a^{(f)} N_f^{(0)}(x) \left(\int_\Omega \varphi_{(0)} \, ds\right)^2 \, dx.
$$

Solving this system, we find $a^{(1)}$, $a^{(2)}$ which result in $\lambda_1 = 0$, $J_p(\varphi_{(2)}) = 0$. Note that there is no difficulty in formulating the solvability condition for (3.8). It is easily seen also that if the support of the function $p(x)$ does not intersect $\cup_{i=1}^I D_i$, the system (3.8) is degenerate and $J_p(\varphi_{(2)}) \not\equiv 0$, i.e. in general, the energy functional will vary. However, the critical conditions will be satisfied if we require $\lambda_1 = 0$, $\lambda_2 = 0$ (hence, $\lambda_3 = 0$). From these conditions we obtain equations for $a^{(1)}$ and $a^{(2)}$. The first equation coincides with the first one in (3.8), and the second equation takes the form

$$
a^{(1)} \left( \sum_{i=1}^{I_0} \int_{D_i} dx \int_{\Omega} \varphi_{(1)}(s, x)\, \varphi_{(1)}(-s, x)\, ds \right)
$$
$$
+ a^{(2)} \left( \sum_{i=I_0+1}^{I} \int_{D_i} dx \int_{\Omega} \varphi_{(1)}(s, x)\, \varphi_{(1)}(-s, x)\, ds \right)
$$
$$
= \frac{1}{\sigma_a^{(y)}} \int_D [\sigma_a^{(f)} - \nu \sigma_f]\, \sigma_a^f \left( \int_{\Omega} \varphi_{(0)}\, ds \right)
$$
$$
\times N_f^{(0)}(x) \left( \int_{\Omega} \varphi_{(1)}(-s, x)\, \varphi_{(1)}(s, x)\, ds \right)\, dx.
$$

The solvability condition for this system is as follows:

$$
\left( \sum_{i=1}^{I_0} \int_{D_i} dx \int_{\Omega} \varphi_{(0)}(s, x)\, \varphi_{(0)}(-s, x)\, ds \right)
$$
$$
\times \left( \sum_{i=I_0+1}^{I} \int_{D_i} dx \int_{\Omega} \varphi_{(1)}(s, x)\, \varphi_{(1)}(-s, x)\, ds \right)
$$
$$
\neq \left( \sum_{i=I_0+1}^{I} \int_{D_i} dx \int_{\Omega} \varphi_{(0)}(s, x)\, \varphi_{(0)}(-s, x)\, ds \right)
$$
$$
\times \left( \sum_{i=1}^{I_0} \int_{D_i} dx \int_{\Omega} \varphi_{(1)}(s, x)\, \varphi_{(1)}(-s, x)\, ds \right).
$$

Now, one can readily formulate in greater detail the algorithm on finding the control parameters ($a^{(1)}$, $a^{(2)}$, the disposition and sizes of $D_i$, $i = 1, \ldots, I$, etc.).

## 4. INVESTIGATION AND APPROXIMATE SOLUTION OF A NON-LINEAR PROBLEM FOR THE TRANSPORT EQUATION

**4.1.** Assume that $N_y(t, x) \equiv 0$ in (1.5), $t \equiv \varepsilon \ll 1$, and terms of the order $O(t)$ are taken into account. Consider this problem in the plane-parallel geometry. Then we come to the following mathematical model:

$$
\mu \frac{\partial \varphi}{\partial x} + \sigma(x)\, \varphi = \frac{1}{2} \int_{-1}^{1} \sigma_s(x, \varphi, \varepsilon)\, \varphi(\mu', x)\, d\mu' + F(\mu, x), \qquad (4.1)
$$

$$
\varphi(\mu, x)|_{\Gamma_-} = 0 \qquad (4.2)
$$

(i.e. $\varphi(\mu, 0) = 0$, $0 < \mu \leq 1$, $\varphi(\mu, H) = 0$, $-1 \leq \mu < 0$), where $x \in (0, H) \equiv D \subset \mathbf{R}^1$, $\mu \in [-1, 1] \equiv \Omega$, $F(\mu, x) \in L_2(\Omega \times D)$, $\varepsilon \in [0, \varepsilon_0]$, $\varepsilon_0 < \infty$, $0 < \sigma_0 \leq \sigma(x) \leq \sigma_1 < \infty$, $\sigma_0, \sigma_1 = $ constant, and the function $\sigma_s(x, \varphi, \varepsilon)$ is defined by

$$\sigma_s(x, \varphi, \varepsilon) = \sigma_{s,0}(x) + \sigma_f(x) \, N_f(x) \left( 1 - \varepsilon \sigma_f \frac{1}{2} \int_\Omega \varphi(\mu, x) \, d\mu \right),$$

where $\sigma_{s,0}(x)$, $\sigma_f(x)$, $N_f(x)$ are non-negative functions bounded almost everywhere. We assume that

$$(\sigma_{s,0}(x) + \sigma_f(x) \, N_f(x))/\sigma(x) \leq k_0 = \text{constant} < 1.$$

Let us introduce the Hilbert space $H_2^1(\Omega \times D)$ with the norm

$$\|\varphi\|_{H_2^1(\Omega \times D)} = \left( \|\varphi\|_{L_2(\Omega \times D)}^2 + \left\| \mu \frac{\partial \varphi}{\partial x} \right\|_{L_2(\Omega \times D)}^2 \right)^{1/2}$$

defined through the scalar product, and the operator $\Phi$ given by

$$\Phi(\varphi, \varepsilon) = \mu \frac{\partial}{\partial x} \varphi + \sigma \varphi - \frac{1}{2} \int_\Omega \sigma_s(x, \varphi, \varepsilon) \, \varphi(\mu', x) \, d\mu'$$

and mapping $H_2^1(\Omega \times D)$ into $L_2(\Omega \times D)$ with the domain $D(\Phi) = \{\varphi(\mu, x): \varphi \in H_2^1(\Omega \times D), \ \varphi|_{\Gamma_-} = 0\}$. Then problem (4.1), (4.2) may written as an operator equation

$$\Phi(\varphi, \varepsilon) = F. \tag{4.3}$$

Suppose that we need to solve this equation and to complete the value of the functional

$$J(\varphi) = (\varphi, p) \equiv \int_{\Omega \times D} \varphi(\mu, x) \, p(x) \, d\mu \, dx, \tag{4.4}$$

where $p(x)$ is a prescribed function bounded almost everywhere. One can use the perturbation algorithm to find an approximate solution of equation (4.3) and to compute the functional (4.4). To formulate and justify this algorithm, we study first the properties of the Gâteaux derivative of the operator $\Phi$.

**4.2.** The Gâteaux derivative $\Phi'$ of the operator $\Phi$ has the form

$$\Phi'(\varphi, \varepsilon) h = \mu \frac{\partial}{\partial x} h + \sigma h - \frac{\sigma_{s,0} + \sigma_f N_f}{2} \int_\Omega h(\mu', x) \, d\mu'$$

$$+ \frac{\varepsilon \sigma_f^2 N_f}{2} \int_\Omega \varphi(\mu', x) \, d\mu' \int_\Omega h(\mu', x) \, d\mu'. \tag{4.5}$$

To prove the continuity of $\Phi'$, we invoke the following inequality:

$$\left\| \frac{1}{2} \int_\Omega \varphi(\mu, x) \, d\mu \right\|_{W_2^\beta(D)} \leq c_1 \|\varphi\|_{H_2^1(\Omega \times D)}, \tag{4.6}$$

satisfied for $\varphi(\mu, x) \in D(\Phi)$, where $\beta$ is any number of the interval $(0, 1/2]$, $W_2^\beta(D)$ the Sobolev space with the fractional index $\beta$, and $c_1$ a constant independent of $\varphi$.

**Lemma 4.1.** *The operator* $\Phi': H_2^1(\Omega \times D) \to L_2(\Omega \times D)$ *is continuous.*

*Proof.* For arbitrary $U, V \in D(\Phi)$, $h \in D(\Phi')$ we get

$$\|(\Phi'(U, \varepsilon) - \Phi'(V, \varepsilon)) h\|_{L_2(\Omega \times D)}$$

$$\leq c\varepsilon \left\| \frac{1}{2} \int_\Omega (U - V) \, d\mu' \frac{1}{2} \int_\Omega h(\mu', x) \, d\mu' \right\|_{L_2(D)}$$

$$\leq c\varepsilon \left\| \frac{1}{2} \int_\Omega (U - V) \, d\mu \right\|_{L_4(D)} \left\| \frac{1}{2} \int_\Omega h \, d\mu \right\|_{L_4(D)}.$$

We take $\beta > 1/4$ in (4.6). Then

$$\left\| \frac{1}{2} \int_\Omega h \, d\mu \right\|_{L_4(D)} \leq c \left\| \frac{1}{2} \int_\Omega h \, d\mu \right\|_{W_2^\beta(D)} \leq cc_1 \|h\|_{H_2^1(\Omega \times D)},$$

$$\left\| \frac{1}{2} \int_\Omega (U - V) \, d\mu \right\|_{L_4(D)} \leq cc_1 \|U - V\|_{H_2^1(\Omega \times D)}.$$

Hence,

$$\|\Phi'(U, \varepsilon) - \Phi'(V, \varepsilon)\|_{H_2^1 \to L_2} \leq cc_1 \varepsilon \|U - V\|_{H_2^1} \to 0$$

as $\|U - V\|_{H_2^1} \to 0$, i.e. $\Phi'$ is continuous.

**Corollary.** *The following formula holds:*

$$\Phi(\varphi_0 + u, \varepsilon) = \Phi(\varphi_0, \varepsilon) + \int_0^1 \Phi'(\varphi_0 + tu, \varepsilon) \, dtu, \qquad (4.7)$$

*where*

$$\int_0^1 \Phi'(\varphi_0 + tu, \varepsilon) \, dtu = \mu \frac{\partial}{\partial x} h + \sigma h - \frac{\sigma_{s,0} + \sigma_f N_f}{2} \int_\Omega h \, d\mu$$

$$+ \frac{\varepsilon \sigma_f^2 N_f}{2} \int_\Omega \left( \varphi_0 + \frac{u}{2} \right) d\mu' \int_\Omega h(\mu, x) \, d\mu,$$

$$\varphi, \varphi_0 \in D(\varphi), \quad h, u \in D(\Phi').$$

**Lemma 4.2.** *The equation*

$$\Phi'(\varphi_0, \varepsilon)\omega = f, \quad \varphi_0 \in D(\Phi), \quad \varepsilon \in [0, \varepsilon_0], \qquad (4.8)$$

*is solvable correctly and everywhere in* $L_2(\Omega \times D)$ *if* $\varepsilon_0$ *is sufficiently small.*

*Proof.* For arbitrary $\varphi_0 \in D(\Phi)$, $\omega \in H^1_2(\Omega \times D)$ we get

$$\left\| \frac{\varepsilon \sigma_f^2 N_f}{2} \int_\Omega \varphi_0 \, d\mu' \int_\Omega \omega \, d\mu' \right\|_{L_2} \leq \varepsilon c \|\varphi_0\|_{H^1_2} \|\omega\|_{H^1_2},$$

$$\|\Phi'(\varphi_0, 0)\omega\|_{L_2} - \varepsilon c \|\varphi_0\|_{H^1_2} \|\omega\|_{H^1_2} \leq \|\Phi'(\varphi_0, \varepsilon)\omega\|_{L_2}.$$

The following inequalities hold:

$$c_2 \|\omega\|_{H^1_2(\Omega \times D)} \leq \|\Phi'(\varphi_0, 0)\omega\|_{L_2(\Omega \times D)} \leq c_3 \|\omega\|_{H^1_2(\Omega \times D)},$$

where $\omega \in D(\Phi)$; $c_2$ and $c_3$ are positive constants independent of $\omega$. Hence,

$$\left( c_2 - \varepsilon c \|\varphi_0\|_{H^1_2} \right) \|\omega\|_{H^1_2} \leq \|\Phi'(\varphi_0, \varepsilon)\omega\|_{L_2(\Omega \times D)}.$$

If $\varepsilon_0$ is sufficiently small, we find that $m_\varepsilon = c_2 - \varepsilon c \|\varphi_0\|_{H^1_2} \geq c_2 - \varepsilon_0 c \|\varphi_0\|_{H^1_2} =$ constant $> 0$. Therefore, equation (4.8) is solvable correctly and everywhere in $L_2(\Omega \times D)$.

It is easily shown that the operator $\Phi'(\varphi_0, 0)$ is closed. Using Lemma 4.2 and taking into account the continuity of $\Phi'$, we come to the conclusion that the operator $\Phi'(\varphi_0, \varepsilon)$ is also closed for $\varepsilon \in [0, \varepsilon_0]$ if $\varepsilon_0$ is sufficiently small.

Lemmas 4.1 and 4.2 make it possible to prove that the solution $\varphi(\mu, x)$ of equation (4.3) exists and may be represented as the series:

$$\varphi = \varphi_0 + \sum_{i=1}^{\infty} \varepsilon^i u_i(\mu, x), \tag{4.9}$$

where $\varphi_0$ is the solution of the problem

$$\mu \frac{\partial}{\partial x} \varphi_0 + \sigma \varphi_0 = \frac{\sigma_{s,0} + \sigma_f N_f}{2} \int_\Omega \varphi_0 \, d\mu' + F(\mu, x),$$

$$\varphi_0(\mu, x)|_\Gamma = 0. \tag{4.10}$$

Fix then an element $\varphi_0 \in D(\Phi)$ which is the solution of problem (4.10), set $f(\varepsilon) = F - \Phi(\varphi_0, \varepsilon)$, and from (4.3) come to the equation

$$A(u, \varepsilon) u = f(\varepsilon), \tag{4.11}$$

where

$$A(u, \varepsilon) = \int_0^1 F'(tu, \varepsilon) \, dt = \int_0^1 \Phi'(\varphi_0 + tu, \varepsilon) \, dt,$$

$$F(u, \varepsilon) \equiv A(u, \varepsilon) u = \Phi(\varphi_0 + u, \varepsilon) - \Phi(\varphi_0, \varepsilon),$$

$$u = \varphi - \varphi_0.$$

Here the operator $A(u, \varepsilon)$ maps $H^1_2(\Omega \times D)$ into $L_2(\Omega \times D)$ with the domain $D(F) = \{u: (\varphi_0 + u) \in D(\Phi)\}$. Note that $A(0, \varepsilon) = \Phi'(\varphi_0, \varepsilon)$. Hence, instead of $\Phi'(\varphi, \varepsilon)$, Lemmas 4.1 and 4.2 may be reformulated for the operator $A(0, \varepsilon)$. From the proofs of Lemmas 4.1 and 4.2 we obtain the following inequalities:

$$\|A(0, \varepsilon)\,\omega\|_{L_2(\Omega \times D)} \geq m_\varepsilon \|\omega\|_{L_2(\Omega \times D)}, \tag{4.12}$$

$$\sup_{h \in D(\Phi')} \frac{\|(\Phi'(\varphi_0 + u, \varepsilon) - \Phi'(\varphi_0 + v, \varepsilon))\,h\|_{L_2(\Omega \times D)}}{\|h\|_{H_2^1(\Omega \times D)}} \leq \varepsilon k \|u - v\|_{H_2^1(\Omega \times D)}, \tag{4.13}$$

where $m_\varepsilon = c_2 - \varepsilon c \geq c_2 - \varepsilon_0 c = m$; $k$ is a constant independent of $\varepsilon$, $u$ and $v$;

$$w \in S(0, R) = \left\{ u\colon \ \varphi_0 + u \in D(\Phi), \|u\|_{H_2^1(\Omega \times D)} \leq R \right\}.$$

If $\varepsilon_0$ is sufficiently small, $m$ is positive, and, therefore, we assume hereinafter that $m_\varepsilon \geq m = \ $ constant $> 0$.

We prove now the existence theorem for equation (4.11).

**Theorem 4.1.** *Let the following conditions be satisfied:*

$$\varepsilon k R / m_\varepsilon \leq q = \ \text{constant} \ < 1;$$
$$\tag{4.14}$$
$$\|f(\varepsilon)\|_{L_2(\Omega \times D)} \leq R(m_\varepsilon - \varepsilon k/2).$$

*Then equation (4.11) has a unique solution $u(\mu, x)$ in $S(0, R)$ which may be represented as the series*

$$u = \sum_{i=1}^{\infty} \varepsilon^i u_i(\mu, x), \tag{4.15}$$

*where $u_i(\mu, x)$, $i = 1, 2, \ldots$, are functions of $D(F)$ independent of $\varepsilon$.*

*Proof.* Equation (4.14) is equivalent to

$$u = T(u, \varepsilon), \tag{4.16}$$

where

$$T(u, \varepsilon) = B(u, \varepsilon)\,u + \bar{f}, \quad \bar{f} = A^{-1}(0, \varepsilon)\,f(\varepsilon),$$
$$B(u, \varepsilon) = A^{-1}(0, \varepsilon)(A(0, \varepsilon) - A(u, \varepsilon)).$$

However, $\|A^{-1}(0, \varepsilon)\|_{L_2 \to H_2^1} \leq 1/m_\varepsilon$. Hence,

$$\|T(u, \varepsilon) - T(v, \varepsilon)\|_{H_2^1} \ \leq \ \|(B(u, \varepsilon) - B(v, \varepsilon))\,u\|_{H_2^1} + \|B(u, \varepsilon)(u - v)\|_{H_2^1}$$

$$\leq \ \frac{\varepsilon k}{m_\varepsilon} \int_0^1 \|t(u - v)\|_{H_2^1} \|u\|_{H_2^1} \ dt$$

$$+ \frac{\varepsilon k}{m_\varepsilon} \int_0^1 \|tu\|_{H_2^1} \|u - v\|_{H_2^1} \ dt$$

$$\leq \ \frac{\varepsilon k R}{m_\varepsilon} \|u - v\|_{H_2^1} \leq \|u - v\|_{H_2^1},$$

i.e. $T(u, \varepsilon)$ is a contraction on $\bar{S}(0, R)$. Since

$$\|T(u, \varepsilon)\|_{H_2^1} \leq \frac{\varepsilon k R}{2m_\varepsilon} + \frac{\|f(\varepsilon)\|_{L_2}}{m_\varepsilon} \leq \frac{\varepsilon k R}{2m_\varepsilon} + \frac{R(m_\varepsilon - \varepsilon k/2)}{m_\varepsilon} = R,$$

for $u \in \bar{S}(0, R)$, $T(u, \varepsilon)$ maps $\bar{S}(0, R)$ into $\bar{S}(0, R)$. Then according to the contraction principle, equation (4.16), as well as (4.11), has a unique solution $u(\mu, x)$. The representation of $u(\mu, x)$ in the form (4.15) follows from the equations

$$u^{(n+1)} = B(u^{(n)}, \varepsilon) \, u^{(n)} + \bar{f}, \quad n = 0, 1, \ldots,$$

the successive approximation method, and the analyticity in $\varepsilon$ of the operators $A^{-1}(0, \varepsilon)$, $B(v, \varepsilon)$, given the analyticity of the element $v$ in $\varepsilon$.

**Corollary.** *Under the hypotheses of Theorem 4.1 equation (4.3) has a unique solution $\varphi(\mu, x)$ represented as*

$$\varphi(\mu, x) = \varphi_0(\mu, x) + \sum_{i=1}^{\infty} \varepsilon^i u_i(\mu, x). \tag{4.17}$$

The above-stated theorem makes it possible to use the perturbation algorithm for finding the elements $\varphi_0, u_1, u_2, \ldots$ in the expansion (4.17). By this is meant that the theorem ensures the justification of the perturbation algorithm, as applied to the problem under consideration. According to this algorithm, the function $\varphi_0(\mu, x)$ is defined as a solution to problem (4.10). The function $u_1(\mu, x)$ satisfies the equations

$$\mu \frac{\partial}{\partial x} u_1 + \sigma u_1 = \frac{\sigma_{s,0} + \sigma_f N_f}{2} \int_\Omega u_1 \, d\mu' + f_1(x),$$

$$u_1(\mu, x)|_{\Gamma_-} = 0, \tag{4.18}$$

where

$$f_1(x) = -\frac{\sigma_f^2 N_f}{2} \left( \int_\Omega \varphi_0(\mu, x) \, d\mu \right)^2.$$

Having solved (4.10) and (4.18), one can compute $J(\varphi)$ approximately by the formula

$$J(\varphi) \cong J^{(1)} = (\varphi_0, p) + \varepsilon(u_1, p). \tag{4.19}$$

It is easy to write down also the equations for $u_2, u_3, \ldots, u_N$. Having solved them, we find more accurate the $N$th order approximations to $\varphi$ and $J(\varphi)$:

$$\varphi_N = \varphi_0 + \sum_{i=1}^{N} \varepsilon^i u_i,$$

$$J_N = J(\varphi_N) = (\varphi_0, p) + \sum_{i=1}^{N} \varepsilon^i (u_i, p). \tag{4.20}$$

**4.3.** Let us formulate the perturbation algorithm for finding an approximate value of the functional $J(\varphi)$, using the solution of the adjoint equation

$$A^*(u, \varepsilon)\, u^* = p, \quad u \in D(\Phi'), \tag{4.21}$$

where

$$A^*(u, \varepsilon) = \int_0^1 (\Phi'(\varphi_0 + tu, \varepsilon))^*\, dt.$$

The operator $(\Phi'(\varphi, \varepsilon))^*$ maps $L_2(\Omega \times D)$ into $H_2^{-1}(\Omega \times D) \equiv (H_2^1(\Omega \times D))^*$ and is defined by the equality

$$((\Phi'(\varphi, \varepsilon))^*\, u^*, \omega)_{L_2} = (u^*, \Phi'(\varphi, \varepsilon)\, \omega)_{L_2}, \tag{4.22}$$

where $\omega$ is an arbitrary function of $D(\Phi')$. In view of (4.22), we conclude that (4.21) is equivalent to the equality

$$(A^*(u, \varepsilon)\, u^*, \omega)_{L_2} = (p, \omega)_{L_2}, \quad \omega \in D(\Phi'), \tag{4.23}$$

where

$$(A^*(u, \varepsilon)\, u^*, \omega)_{L_2} = \left( u^*, \int_0^1 \Phi'(\varphi_0 + tu, \varepsilon)\, dt\omega \right)_{L_2}$$

$$= \left( u^*, \mu \frac{\partial}{\partial x} \omega + \sigma\omega - \frac{\sigma_{s,0} + \sigma_f N_f}{2} \int_\Omega \omega\, d\mu' \right.$$

$$\left. + \varepsilon \frac{\sigma_f^2 N_f}{2} \int_\Omega \left( \varphi_0 + \frac{u}{2} \right) d\mu' \int_\Omega \omega(\mu', x)\, d\mu' \right)_{L_2}.$$

According to the above-listed statements, the equation $A(u, \varepsilon)\omega = f$ is solvable everywhere in $L_2(\Omega \times D)$ if $\varepsilon_0$ is sufficiently small, and the operator $A(u, \varepsilon)$ is closed. Then equation (4.21) is correctly solvable and

$$\|u^*\|_{L_2} \le c\|p\|_{H_2^{-1}(\Omega \times D)}. \tag{4.24}$$

Now, by analogy with the proof of Theorem 4.1, it is not difficult to show that the solution of equation (4.21) is represented as the series $u^* = \sum_{i=0}^\infty \varepsilon^i u_i^*$. The elements $u_i^* \in L_2(\Omega \times D)$, $i = 0, 1, \ldots$, may be determined using the perturbation algorithm which is convergent for this problem. In particular, equations for $u_0^*$, $u_1^*$ are

$$\left( u_0^*, \mu \frac{\partial}{\partial x} \omega + \sigma\omega - \frac{\sigma_{s,0} + \sigma_f N_f}{2} \int_\Omega \omega\, d\mu' \right)_{L_2} = (p, \omega)_{L_2}, \tag{4.25}$$

$$\left( u_1^*, \mu \frac{\partial}{\partial x} v + \sigma v + \frac{\sigma_{s,0} + \sigma_f N_f}{2} \int_\Omega v\, d\mu' \right)_{L_2} = (G_1, v)_{L_2}, \tag{4.26}$$

where

$$G_1 = -\frac{\sigma_f^2 N_f}{2} \int_\Omega \varphi_0 \, d\mu' \int_\Omega u_0^* \, d\mu', \tag{4.27}$$

and $\omega$, $v$ are arbitrary functions of $D(\Phi')$. Solving (4.10), (4.25), and (4.26), one can compute the value

$$J^{(3)}(\varphi) = (\varphi_0, p)_{L_2} + \varepsilon(f_1, u_0^*)_{L_2} + \varepsilon^2(f_1, u_1^*)_{L_2}$$

$$- \varepsilon^3 \left( \frac{\sigma_f^2 N_f}{2} \int_\Omega \varphi_0 \, d\mu' \int_\Omega u_1 \, d\mu', u_1^* \right)_{L_2}, \tag{4.28}$$

approximating $J(\varphi)$ with an accuracy of $O(\varepsilon^4)$.

Thus, it has been possible to prove the continuity of the operator $\Phi'$ mapping $H_2^1(\Omega \times D)$ into $L_2(\Omega \times D)$, make the above-applied transformations of non-linear problems under consideration (representations (4.7), (4.11) and others), prove the corresponding assertions on solvability of equations (4.11) and (4.21), show the validity of the representations of their solutions in the form of a power series in $\varepsilon$, and justify thereby the application of the regular perturbation algorithm for the original problem.

In conclusion note that the formulation and justification of the above-stated algorithm may be easily extended to cover the case when $D \subset \mathbf{R}^3$.

# Chapter 7

# Adjoint equations and perturbation algorithms for a quasilinear equation of motion

This chapter considers mixed initial-boundary value problems for a specific class of non-linear first order hyperbolic equations, and the functionals of their solutions. Such equations arise in many problems of wave propagation. These are so-called kinematic waves, since the governing equations are obtained from the conservation laws. Examples are flood waves in rivers, waves in glaciers, waves in traffic flow, and certain wave phenomena in chemical reactions. Different functionals of solutions (and especially their deviations from the standard values), which play an important role in these problems, need to be computed. The aim of this chapter is to develop and justify the algorithms for computing the functionals on the basis of adjoint equation technique and perturbation theory.

## 1. STATEMENT OF THE PROBLEM. BASIC ASSUMPTIONS. OPERATOR FORMULATION

Consider the initial-boundary value problem for the quasilinear first-order differential equation of the form

$$
\begin{cases}
\dfrac{\partial \varphi}{\partial t} + \dfrac{\partial G(\varphi, \varepsilon)}{\partial x} = f(t, x, \varepsilon), & t \in (0, T], \quad x \in (a, b], \\[2mm]
\varphi(0, x) = \varphi_0(x, \varepsilon), \\[2mm]
\varphi(t, a) = \varphi_1(t, \varepsilon),
\end{cases}
\tag{1.1}
$$

where $\varphi(t, x)$ is an unknown distribution function, and $G(\varphi, \varepsilon)$, $f(t, x, \varepsilon)$, $\varphi_0(x, \varepsilon)$ and $\varphi_1(t, \varepsilon)$ are prescribed functions, $\varepsilon \in [0, 1]$.

Equations of the form (1.1) arise when describing kinematic waves (flood waves, waves in glaciers, waves in traffic flow, and wave phenomena in chemical reactions).[269,287,314] One of the important practical problems here is to compute the deviation of some functional $J(\varphi)$ from the standard value. We consider the functional of the form $J(\varphi)$

$$J(\varphi) = (\varphi, p) = \int_0^T \int_a^b \varphi(t, x) \, p(t, x) \; dt \; dx, \qquad (1.2)$$

where $p$ is a given function, and $\varphi$ is the solution to problem (1.1) which is said to be a perturbed problem.

Let us know the solution $\varphi^{(0)}$ of the unperturbed problem which is given by (1.1) for $\varepsilon = 0$, and it has the form

$$\begin{cases} \dfrac{\partial \varphi^{(0)}}{\partial t} + \dfrac{\partial G(\varphi^{(0)}, 0)}{\partial x} = f(t, x, 0), \quad t \in (0, T], \quad x \in (a, b], \\[2mm] \varphi^{(0)}(0, x) = \varphi_0(x, 0), \\[2mm] \varphi^{(0)}(t, a) = \varphi_1(t, 0). \end{cases} \qquad (1.3)$$

We need to compute an approximate value of functional (1.2) without solving the non-linear problem (1.1).

We find for $J(\varphi)$:

$$J(\varphi) = J(\varphi^{(0)}) + \delta J, \qquad (1.4)$$

where $\delta J$ is the sought-for deviation of the functional from the standard value $J(\varphi^{(0)})$, $\delta J = (u, p)$, $u = \varphi - \varphi^{(0)}$.

Assume that (for a fixed $\varepsilon \in [0, 1]$) $f \in L_2(\Omega)$, $\varphi_0 \in L_2(\Gamma_1)$, $\varphi_1 \in L_\infty(\Gamma_2)$, where $\Omega = (0, T) \times (a, b)$, $\Gamma_1 = \{(t, x) \in \partial\Omega: t = 0, \; a \le x \le b\}$, $\Gamma_2 = \{(t, x) \in \partial\Omega: x = a, \; 0 \le t \le T\}$, $p \in L_2(\Omega)$. Let the function $g(u, \varepsilon) = \partial G(u, \varepsilon)/\partial u$ have the derivative $\partial g/\partial u$ and

$$|g(u, \varepsilon)| \le g_0 + g_1 |u|^{\alpha_1} + g_2 |u|^{\alpha_2}, \qquad (1.5)$$

$$\left| \frac{\partial g}{\partial u}(u, \varepsilon) \right| \le g_0' + g_1' |u|^{\alpha_1'} + g_2' |u|^{\alpha_2'}, \qquad (1.6)$$

where $g_i, g_i' = \text{constant} \ge 0$ $(i = 0, 1, 2)$, $\sum_{i=0}^2 g_i > 0$, $\sum_{i=0}^2 g_i' > 0$, $\alpha_i > 0$, $\alpha_1' > 0$ $(i = 1, 2)$.

*Remark* 1.1. Constraints of forms (1.5) and (1.6) on $g$ arise in the above-mentioned applications.[269,314] For example, in the problems of chemical exchange processes, the function $g$ has the form

$$g(\varphi, \varepsilon) = \frac{v(k_2 B + \varepsilon(k_1 - k_2)\varphi)^2}{k_1 k_2 AB + (k_2 B + \varepsilon(k_1 - k_2)\varphi)^2}, \qquad v, k_1, k_2, A, B = \text{constant} > 0.$$

It is readily seen that (1.5) and (1.6) are satisfied for $g_1 = g_2 = g_1' = g_2' = 0$, $g_0 = v$, $g_0' = 3\sqrt{3}|k_1 - k_2| \left(\sqrt{k_1 k_2 AB}\right)^{-1} \varepsilon v/8$.

We shall consider weak solutions of problems (1.1) and (1.3) (see Oleinik[228], Rozhdestvenskii-Yanenko[253]). The function $\varphi \in L_\infty(\Omega)$ is said to be a weak solution of problem (1.1) if the equality

$$[\varphi, \psi]_\varepsilon = (f, \psi) \tag{1.7}$$

is satisfied for any $\psi \in W^1_{2,b,T}(\Omega)$, where

$$[\varphi, \psi]_\varepsilon = -\left(\varphi, \frac{\partial \psi}{\partial t}\right) - \left(G(\varphi, \varepsilon), \frac{\partial \psi}{\partial x}\right)$$

$$-\int_0^T G(\varphi_1(t, \varepsilon), \varepsilon)\,\psi(t, a)\ dt$$

$$-\int_a^b \varphi_0(x, \varepsilon)\,\psi(0, x)\ dx,$$

$$W^1_{2,b,T}(\Omega) = \left\{u \in W^1_2(\Omega): u|_{t=T} = 0,\ u|_{x=b} = 0\right\}.$$

The same weak formulation of problem (1.1) was considered by Shutyaev and Seleznev[269,270] where the properties of the weak solution were discussed. Without going into detail we shall assume that the generalized solution $\varphi \in L_\infty(\Omega)$ of problem (1.1) exists, is unique and sufficiently regular.

By $W^{-1}_{2,b,T}(\Omega)$ we denote the space identified with the space of linear continuous functionals on $W^1_{2,b,T}(\Omega)$ with the norm

$$\|u\|_{W^{-1}_{2,b,T}(\Omega)} = \sup_{w \in W^1_{2,b,T}(\Omega),\ w \neq 0} \frac{|(u, w)|}{\|w\|_{W^1_2(\Omega)}}. \tag{1.8}$$

**Lemma 1.1.** *To the form $[\varphi, \psi]_\varepsilon$ corresponds a non-linear operator $\Phi$ mapping $L_2(\Omega)$ into $W^{-1}_{2,b,T}(\Omega)$ with the domain $D(\Phi) = L_\infty(\Omega)$ and defined by the equality*

$$(\Phi(\varphi, \varepsilon), \psi) = [\varphi, \psi]_\varepsilon, \quad \psi \in W^1_{2,b,T}(\Omega). \tag{1.9}$$

*Proof.* For a fixed element $\varphi \in D(\Phi) = L_\infty(\Omega)$ the form $[\varphi, \psi]_\varepsilon$ is a linear bounded functional with respect to $\psi \in W^1_{2,b,T}(\Omega)$. In fact, we have

$$|[\varphi, \psi]_\varepsilon| \leq \|\varphi\|_{L_2(\Omega)}\left\|\frac{\partial \psi}{\partial t}\right\|_{L_2(\Omega)} + \|G(\varphi, \varepsilon)\|_{L_2(\Omega)}\left\|\frac{\partial \psi}{\partial x}\right\|_{L_2(\Omega)}$$

$$+ \left|\int_0^T G(\varphi_1(t, \varepsilon), \varepsilon)\,\psi(t, a)\ dt\right|$$

$$+ \left|\int_a^b \psi(a, x)\,\varphi_0(x, \varepsilon)\ dx\right|. \tag{1.10}$$

In view of (1.5),

$$|G(\varphi, \varepsilon)| \leq g_0|\varphi| + \frac{g_1}{\alpha_1 + 1} |\varphi|^{\alpha_1+1} + \frac{g_2}{\alpha_2 + 1} |\varphi|^{\alpha_2+1}. \tag{1.11}$$

Then, taking into account that $\varphi \in L_\infty(\Omega)$, we find

$$\|G(\varphi, \varepsilon)\|_{L_2(\Omega)} \leq \sqrt{T(b-a)} \bigg( g_0\|\varphi\|_{L_\infty(\Omega)} \tag{1.12}$$

$$+ \frac{g_1}{\alpha_1 + 1} \|\varphi\|_{L_\infty(\Omega)}^{\alpha_1+1} + \frac{g_2}{\alpha_2 + 1} \|\varphi\|_{L_\infty(\Omega)}^{\alpha_2+1} \bigg) \equiv c_0.$$

For $\psi \in W_2^1(\Omega)$ we have[210] $\psi|_{\partial\Omega} \in L_2(\partial\Omega)$ and $\|\psi|_{\partial\Omega}\|_{L_2(\partial\Omega)} \leq c_1\|\psi\|_{W_2^1(\Omega)}$ ($c_1 = $ constant $> 0$). From (1.11) we thus obtain the estimates

$$\left| \int_0^T G(\varphi_1(t, \varepsilon), \varepsilon) \, \psi(t, a) \, dt \right|$$

$$\leq \left( \int_0^T G^2(\varphi_1(t, \varepsilon), \varepsilon) \, dt \right)^{1/2} \left( \int_0^T \varphi^2(t, a) \, dt \right)^{1/2}$$

$$\leq c_1 T^{1/2} \bigg( g_0\|\varphi_1\|_{L_\infty(\Gamma_2)} + \frac{g_1}{\alpha_1 + 1} \|\varphi_1\|_{L_\infty(\Gamma_2)}^{\alpha_1+1}$$

$$+ \frac{g_2}{\alpha_2 + 1} \|\varphi_1\|_{L_\infty(\Gamma_2)}^{\alpha_2+1} \bigg) \|\psi\|_{W_2^1(\Omega)} \equiv c_2\|\psi\|_{W_2^1(\Omega)}. \tag{1.13}$$

Since

$$\left| \int_a^b \psi(0, x) \varphi_0(x, \varepsilon) \, dx \right| \leq \left( \int_a^b \varphi_0^2(x, \varepsilon) \, dx \right)^{1/2} \left( \int_a^b \psi^2(0, x) \, dx \right)^{1/2}$$

$$\leq c_3\|\psi\|_{W_2^1(\Omega)}, \tag{1.14}$$

from (1.10)–(1.14), we arrive at the inequality

$$|[\varphi, \psi]_\varepsilon| \leq \left( \|\varphi\|_{L_2(\Omega)} + c_0 + c_2 + c_3 \right) \|\psi\|_{W_2^1(\Omega)}. \tag{1.15}$$

Hence, $[\varphi, \psi]_\varepsilon$ is a linear bounded functional (with respect to $\psi \in W_2^1(\Omega)$). By the Riesz theorem, there exists $\tilde{\varphi} \in W_2^1(\Omega)$ such that

$$[\varphi, \psi]_\varepsilon = (\tilde{\varphi}, \psi)_{W_2^1(\Omega)}. \tag{1.16}$$

However, it has been known[14] that there exists a linear operator $\overline{S}$ mapping $W_2^1(\Omega)$ into $W_2^{-1}(\Omega)$ with the domain $D(\overline{S}) \subset W_2^1(\Omega)$ such that

$$(u, v)_{W_2^1(\Omega)} = (\overline{S}u, v) \tag{1.17}$$

for any $u, v \in W_2^1(\Omega)$. From (1.16) and (1.17) we get

$$[\varphi, \psi]_\varepsilon = (\overline{S}\tilde{\varphi}, \psi). \tag{1.18}$$

Therefore, to each function $\varphi \in L_\infty(\Omega)$ corresponds the function $\Theta = \overline{S}\tilde{\varphi} \in W_2^{-1}(\Omega)$. This correspondence defines a non-linear operator $\Phi$, i.e.

$$\Phi(\varphi, \varepsilon) = \overline{S}\tilde{\varphi} \in W_2^{-1}(\Omega) \subset W_{2,b,T}^{-1}(\Omega).$$

The operator $\Phi$ maps $L_2(\Omega)$ into $W_{2,b,T}^{-1}$ and has the domain $D(\Phi) = L_\infty(\Omega)$. In view of (1.18), the equality (1.9) is valid.

*Remark* 1.2. For the problems of chemical exchange processes we have $g_1 = g_2 = g_1' = g_2' = 0$ in (1.5),(1.6). Therefore, Lemma 1.1 and all the reasonings hold true also if $\varphi_1 \in L_2(\Gamma_2)$, $\varphi \in L_2(\Omega)$.

Equality (1.7) and Lemma 1.1 yield

**Lemma 1.2.** *Problem* (1.1) *has a generalized solution* $\varphi \in L_\infty(\Omega)$ *if and only if the equation*

$$(\Phi(\varphi, \varepsilon), \psi) = (f, \psi) \quad \forall\, \psi \in W_{2,b,T}^1(\Omega) \tag{1.19}$$

*has a solution* $\varphi \in D(\Phi)$. *These solutions coincide.*

Henceforward, we shall write equation (1.19) in the brief form:

$$\Phi(\varphi, \varepsilon) = f. \tag{1.20}$$

This is an operator form of the weak formulation of problem (1.1). We shall consider (1.20) as an operator equation in $W_{2,b,T}^{-1}(\Omega)$ with the operator $\Phi$. Here we have $\varepsilon \in [0, 1]$, and if we put $\varepsilon = 0$, then we come to the unperturbed problem:

$$\Phi(\varphi, 0) = f|_{\varepsilon=0}. \tag{1.21}$$

By $\varphi^{(0)}$ we denote the solution of this problem. Equation (1.21) is equivalent to the equality

$$-\left(\varphi, \frac{\partial\psi}{\partial t}\right) - \left(G(\varphi, 0), \frac{\partial\psi}{\partial x}\right)$$

$$-\int_0^T G(\varphi_1(t, 0), 0)\, \psi(t, a)\, dt - \int_a^b \varphi_0(x, 0)\, \psi(0, x)\, dx$$

$$= (f|_{\varepsilon=0}, \psi) \quad \forall\, \psi \in W_{2,b,T}^1(\Omega). \tag{1.22}$$

Hereinafter, we assume that problems (1.20) and (1.21) have the solutions $\varphi, \varphi^{(0)} \in L_\infty(\Omega)$.

## 2. TRANSFORMATION OF THE PROBLEM.
## PROPERTIES OF THE NON-LINEAR OPERATOR

By subtracting $\Phi(\varphi^{(0)}, \varepsilon)$ from the left- and right-hand sides of equation (1.20) we come to the equality

$$F(u, \varepsilon) = f(\varepsilon), \tag{2.1}$$

where

$$F(u, \varepsilon) = \Phi(\varphi, \varepsilon) - \Phi(\varphi^{(0)}, \varepsilon) = \Phi(\varphi^{(0)} + u, \varepsilon) - \Phi(\varphi^{(0)}, \varepsilon),$$

$$u = \varphi - \varphi^{(0)}, \quad f(\varepsilon) = f - \Phi(\varphi^{(0)}, \varepsilon).$$

The operator $F(u, \varepsilon)$ is defined by the expression

$$(F(u, \varepsilon), w) = \Big(\Phi(\varphi^{(0)} + u, \varepsilon), w\Big) - \Big(\Phi(\varphi^{(0)}, \varepsilon), w\Big) \quad \forall\, w \in W^1_{2,b,T}(\Omega). \tag{2.2}$$

It maps $L_2(\Omega)$ into $W^{-1}_{2,b,T}(\Omega)$ and has the domain $D(F) = L_\infty(\Omega)$. We find for $f(\varepsilon)$:

$$(f(\varepsilon), w) = (f, w) + \left(G(\varphi^{(0)}, \varepsilon), \frac{\partial w}{\partial x}\right) + \left(\varphi^{(0)}, \frac{\partial w}{\partial t}\right)$$

$$+ \int_0^T G(\varphi_1(t, \varepsilon), \varepsilon)\, w(t, a)\, dt$$

$$+ \int_a^b \varphi_0(x, \varepsilon)\, w(0, x)\, dx, \quad w \in W^1_{2,b,T}(\Omega). \tag{2.3}$$

Using the definition of the operator $\Phi$ (see Lemma 1.1), we can rewrite equality (2.2) in the explicit form:

$$(F(u, \varepsilon), w) = -\left(G(\varphi^{(0)} + u, \varepsilon) - G(\varphi^{(0)}, \varepsilon), \frac{\partial w}{\partial x}\right) - \left(u, \frac{\partial w}{\partial t}\right),$$

$$w \in W^1_{2,b,T}(\Omega). \tag{2.4}$$

Let us study the properties of the operator $F(u, \varepsilon)$. Note first that

$$F(0, \varepsilon) = 0. \tag{2.5}$$

**Lemma 2.1.** *The operator $F(u, \varepsilon)$ has the Gâteaux derivative $F'(u, \varepsilon)$ at any point $u \in D(F)$ defined by the equality*

$$(F'(u, \varepsilon)\, h, w) = -\left(g(\varphi^{(0)} + u, \varepsilon)\, h, \frac{\partial w}{\partial x}\right) - \left(h, \frac{\partial w}{\partial t}\right),$$

$$w \in W^1_{2,b,T}(\Omega), \quad h \in L_\infty(\Omega). \tag{2.6}$$

*Proof.* We find for the first addend in the right-hand side of (2.4):

$$\lim_{\xi \to 0} -\frac{1}{\xi}\left[\left(G(\varphi^{(0)} + u + \xi h, \varepsilon) - G(\varphi^{(0)}, \varepsilon), \frac{\partial w}{\partial x}\right)\right.$$

$$\left. - \left(G(\varphi^{(0)} + u, \varepsilon) - G(\varphi^{(0)}, \varepsilon), \frac{\partial w}{\partial x}\right)\right]$$

$$= \lim_{\xi \to 0}\left\{-\frac{1}{\xi}\int_0^\xi \left(\frac{\partial G}{\partial \varphi}(\varphi^{(0)} + u + \xi' h, \varepsilon) h, \frac{\partial w}{\partial x}\right) d\xi'\right\}$$

$$= -\left(\frac{\partial G}{\partial \varphi}(\varphi^{(0)} + u, \varepsilon) h, \frac{\partial w}{\partial x}\right)$$

$$= -\left(g(\varphi^{(0)} + u, \varepsilon) h, \frac{\partial w}{\partial x}\right). \tag{2.7}$$

In view of (1.5) and the fact that $\varphi^{(0)}, u \in L_\infty(\Omega)$, we get

$$\left|\left(g(\varphi^{(0)} + u, \varepsilon) h, \frac{\partial w}{\partial x}\right)\right| \le c\|h\|_{L_2(\Omega)}\|w\|_{W_2^1(\Omega)} < \infty,$$

where

$$c = g_0 + g_1\|\varphi^{(0)} + u\|_{L_\infty(\Omega)}^{\alpha_1} + g_2\|\varphi^{(0)} + u\|_{L_\infty(\Omega)}^{\alpha_2}.$$

Therefore, the addend considered is differentiable. The operator $A_2$ defined by the second addend in the right-hand side of (2.4) as

$$(A_2 u, w) = -\left(u, \frac{\partial w}{\partial t}\right), \quad w \in W_{2,b,T}^1(\Omega)$$

is a linear bounded operator mapping $L_2(\Omega)$ into $W_{2,b,T}^{-1}(\Omega)$. Hence, the Gâteaux derivative $A_2'$ of $A_2$ coincides with $A_2$:

$$(A_2' h, w) = -\left(u, \frac{\partial w}{\partial t}\right), \quad w \in W_{2,b,T}^1(\Omega). \tag{2.8}$$

From (2.7) and (2.8) we arrive at (2.6).

**Lemma 2.2.** *The operator $F'(u, \varepsilon)$ mapping $L_2(\Omega)$ into $W_{2,b,T}^{-1}$ is bounded.*

*Proof.* From (2.6) we have for $u \in D(F) = L_\infty(\Omega)$

$$|(F'(u, \varepsilon) h, w)| \le \text{constant } \|h\|_{L_2(\Omega)}\|w\|_{W_2^1(\Omega)}. \tag{2.9}$$

Then

$$\|F'(u, \varepsilon) h\|_{W_{2,b,T}^{-1}(\Omega)} = \sup_{w \in W_{2,b,T}^{-1}(\Omega),\ w \neq 0} \frac{|(F'(u, \varepsilon) h, w)|}{\|w\|_{W_2^1(\Omega)}} \le \text{constant } \|h\|_{L_2(\Omega)}.$$

**Lemma 2.3.** *The operator $F'(u, \varepsilon)$: $L_2(\Omega) \to W_{2,b,T}^{-1}(\Omega)$ is continuous on $D(F)$ with respect to $u$, and the estimate*

$$\|F'(u_1, \varepsilon) - F'(u_2, \varepsilon)\|_{L_2(\Omega) \to W_{2,b,T}^{-1}(\Omega)} \leq c\|u_1 - u_2\|_{L_\infty(\Omega)} \qquad (2.10)$$

*holds for any $u_1, u_2 \in D(F)$.*

*Proof.* Let $u_1, u_2 \in D(F) = L_\infty(\Omega)$. From (2.6) we have for the quantity $I = ((F'(u_1, \varepsilon) - F'(u_2, \varepsilon)) h, w)$ that

$$I = - \left( (g(\varphi^{(0)} + u_1, \varepsilon) - g(\varphi^{(0)} + u_2, \varepsilon)) h, \frac{\partial w}{\partial x} \right).$$

Since

$$g(\varphi^{(0)} + u_1, \varepsilon) - g(\varphi^{(0)} + u_2, \varepsilon) = \int_0^1 g_u'(\varphi^{(0)} + u_2 + \tau(u_1 - u_2), \varepsilon) \, d\tau(u_1 - u_2),$$

then, in view of (1.6),

$$|I| \leq c\|(u_1 - u_2) h\|_{L_2(\Omega)} \left\| \frac{\partial w}{\partial x} \right\|_{L_2(\Omega)}$$

$$\leq c\|u_1 - u_2\|_{L_\infty(\Omega)} \|h\|_{L_2(\Omega)} \|w\|_{W_2^1(\Omega)},$$

where

$$c = g_0' + g_1' \left( \|\varphi^{(0)} + u_2\|_{L_\infty(\Omega)} + \|u_1 - u_2\|_{L_\infty(\Omega)} \right)^{\alpha_1'}$$

$$+ g_2' \left( \|\varphi^{(0)} + u_2\|_{L_\infty(\Omega)} + \|u_1 - u_2\|_{L_\infty(\Omega)} \right)^{\alpha_2'}.$$

Hence,

$$\|(F'(u_1, \varepsilon) - F'(u_2, \varepsilon)) h\|_{W_{2,b,T}^{-1}(\Omega)} \leq c\|u_1 - u_2\|_{L_\infty(\Omega)} \|h\|_{L_2(\Omega)},$$

and the estimate (2.10) is valid.

*Remark* 2.1. The proof of Lemma 2.3 depends upon the fact that $u_1, u_2 \in L_\infty(\Omega)$. The constant $c$ in (2.10) depends in general on $\varphi^{(0)}$, $u_1$, $u_2$. For the problems of chemical exchange processes we have $g_1 = g_2 = g_1' = g_2' = 0$ in (1.5), (1.6) and, hence, $c = g_0'$ does not depend on $\varphi^{(0)}$, $u_1$, $u_2$.

From Lemmas 2.1–2.3 and well-known propositions of non-linear analysis given in Chapter 1, we come to the following corollaries.

**Corollary 2.1.** *The formula*

$$F(u, \varepsilon) = F(v, \varepsilon) + \int_0^1 F'(v + \xi(u - v), \varepsilon) \, d\xi(u - v)$$

*holds for any $u, v \in D(F)$, $\varepsilon \in [0, 1]$.*

**Corollary 2.2.** *The operator $F(u, \varepsilon)$ can be represented in the form*

$$F(u, \varepsilon) = \int_0^1 F'(\xi u, \varepsilon) \, d\xi u = A(u, \varepsilon) \, u, \qquad (2.11)$$

*where the linear operator $A(u, \varepsilon) = \int_0^1 F'(\xi u, \varepsilon) \, d\xi$ maps $L_2(\Omega)$ into $W_{2,b,T}^{-1}(\Omega)$ and is defined by the equality*

$$(A(u, \varepsilon) \, h, w) = -\left( h, \frac{\partial w}{\partial t} \right) - \left( \left( \int_0^1 g(\varphi^{(0)} + \xi u, \varepsilon) \, d\xi \right) h, \frac{\partial w}{\partial x} \right) \qquad (2.12)$$

*for any $h \in D(A) = L_2(\Omega)$, $w \in W_{2,b,T}^1(\Omega)$.*

*Remark 2.2.* Corollary 2.2 results from (2.6) and Corollary 2.1 for $v \equiv 0$. The function $\int_0^1 g(\varphi^{(0)} + \xi u, \varepsilon) \, d\xi$ can be represented in the form:

$$\int_0^1 g(\varphi^{(0)} + \xi u, \varepsilon) \, d\xi = \frac{1}{u} \int_{\varphi^{(0)}}^{u + u^{(0)}} g(\xi', \varepsilon) \, d\xi'$$

$$= \frac{1}{u} \left[ G(u + \varphi^{(0)}, \varepsilon) - G(\varphi^{(0)}, \varepsilon) \right].$$

By Corollary 2.2, the perturbed problem (2.1) can be written in the form

$$A(u, \varepsilon) \, u = f(\varepsilon), \qquad (2.13)$$

where the operator $A(u, \varepsilon)$: $L_2(\Omega) \to W_{2,b,T}^{-1}(\Omega)$ with the domain $D(A) = L_2(\Omega)$ is defined by (2.12).

## 3. ADJOINT EQUATION

Consider the equation

$$\Phi(\varphi, \varepsilon) = f \qquad (3.1)$$

and the functional

$$J(\varphi) = (\varphi, p), \quad p \in L_2(\Omega). \qquad (3.2)$$

Let $\varphi^{(0)}$ be a solution of the unperturbed problem

$$\Phi(\varphi^{(0)}, 0) = f|_{\varepsilon=0} \qquad (3.3)$$

which is supposed to be known. Then

$$J(\varphi) = J(\varphi^{(0)}) + \delta J = (\varphi^{(0)}, p) + \delta J = (\varphi^{(0)}, p) + (u, p), \qquad (3.4)$$

where $u = \varphi - \varphi^{(0)}$.

The equation for $u$ is given by (2.13). Therefore, one can find $u$, construct $\delta J = (u, p)$ and, then, compute $J(\varphi)$ using (3.4). But equation (2.13) is non-linear in $u$. In view of this, we make use of the equation adjoint to (2.13) and compute $J(\varphi)$ using the perturbation algorithm.

Consider the operator $A(u, \varepsilon)$ defined by (2.12) for a fixed $u \in L_\infty(\Omega)$. This operator maps $L_2(\Omega)$ into $W_{2,b,T}^{-1}(\Omega)$ and it is linear and bounded:

$$\|A(u, \varepsilon) h\|_{W_{2,b,T}^{-1}(\Omega)} = \sup_{w \in W_{2,b,T}^1(\Omega),\, w \neq 0} \frac{|(A(u, \varepsilon) h, w)|}{\|w\|_{W_2^1(\Omega)}} \leq c\|h\|_{L_2(\Omega)}, \quad (3.5)$$

where the constant $c$ is determined from (1.5):

$$c = 1 + g_0 + g_1 \left( \|\varphi^{(0)}\|_{L_\infty(\Omega)} + \|u\|_{L_\infty(\Omega)} \right)^{\alpha_1}$$

$$+ g_2 \left( \|\varphi^{(0)}\|_{L_\infty(\Omega)} + \|u\|_{L_\infty(\Omega)} \right)^{\alpha_2}.$$

Introduce now the adjoint operator $A^*(u, \varepsilon)$ in a manner standard in linear operator theory (see Chapter 1, Section 1). The operator $A^*(u, \varepsilon)$ maps $\left( W_{2,b,T}^{-1}(\Omega) \right)^* \equiv W_{2,b,T}^1(\Omega)$ into $(L_2(\Omega))^* \equiv L_2(\Omega)$, and the following equality holds:

$$(A(u, \varepsilon) h, w) = (h, A^*(u, \varepsilon) w), \quad h \in D(A), \quad w \in D(A^*). \quad (3.6)$$

**Lemma 3.1.** *The adjoint operator $A^*(u, \varepsilon)$ mapping $W_{2,b,T}^1(\Omega)$ into $L_2(\Omega)$ with the domain $D(A^*) = W_{2,b,T}^1(\Omega)$ is defined by*

$$A^*(u, \varepsilon) w = -\frac{\partial w}{\partial t} - \left( \int_0^1 g(\varphi^{(0)} + \xi u, \varepsilon) \, d\xi \right) \frac{\partial w}{\partial x} \quad (3.7)$$

*and is linear and bounded.*

*Proof.* Let $h \in D(A)$, $w \in W_{2,b,T}^1(\Omega)$. Consider the functional $g(h) = (A(u, \varepsilon) h, w)$. This functional is linear and bounded with respect to $h \in L_2(\Omega)$, since, in view of (3.5),

$$|g(h)| \leq \|A(u, \varepsilon) h\|_{W_{2,b,T}^{-1}(\Omega)} \|w\|_{W_{2,b,T}^1(\Omega)} \leq c\|h\|_{L_2(\Omega)} \|w\|_{W_{2,b,T}^1(\Omega)}.$$

By the Riesz theorem, there exists $\tilde{h} \in L_2(\Omega)$ such that $g(h) = (h, \tilde{h})$; hence,

$$(A(u, \varepsilon) h, w) = (h, \tilde{h}), \quad (3.8)$$

the element $\tilde{h}$ being unique. Therefore, $w \in D(A^*)$ and $A^*(u, \varepsilon) w = \tilde{h}$. From (3.8) we get (3.6). Since $w$ is an arbitrary element of $W_{2,b,T}^1(\Omega)$, then $D(A^*) = W_{2,b,T}^1(\Omega)$. From (3.8) and (2.12) we find

$$-\left( h, \frac{\partial w}{\partial t} \right) - \left( \left( \int_0^1 g(\varphi^{(0)} + \xi u, \varepsilon) \, d\xi \right) h, \frac{\partial w}{\partial x} \right) = (h, A^*(u, \varepsilon) w)$$

for any $h \in D(A)$. Hence,

$$A^*(u, \varepsilon) w = -\frac{\partial w}{\partial t} - \left( \int_0^1 g(\varphi^{(0)} + \xi u, \varepsilon) \, d\xi \right) \frac{\partial w}{\partial x}.$$

This operator is linear and bounded, since

$$\|A^*(u, \varepsilon) w\|_{L_2(\Omega)} = \left\| \frac{\partial w}{\partial t} + \left( \int_0^1 g(\varphi^{(0)} + \xi u, \varepsilon) \, d\xi \right) \frac{\partial w}{\partial x} \right\|_{L_2(\Omega)} \le c\|w\|_{W_2^1(\Omega)}.$$

*Remark* 3.1. The notations $\partial w/\partial t$ and $\partial w/\partial x$ in (3.7) are considered to mean the weak derivatives of $w$ with respect to $t$ and $x$, respectively. If $w \in W_{2,b,T}^1(\Omega)$, then $\partial w/\partial t$, $\partial w/\partial x \in L_2(\Omega)$.

Consider now the adjoint equation

$$A^*(u, \varepsilon) u^* = p, \tag{3.9}$$

where $p \in L_2(\Omega)$ is the function determining the functional (3.2). Equation (3.9) is an operator equation in $L_2(\Omega)$. In view of (3.7), this equation can be reformulated as the following problem: find $u^* \in W_2^1(\Omega)$ such that

$$\begin{cases} -\dfrac{\partial u^*}{\partial t} - \left( \displaystyle\int_0^1 g(\varphi^{(0)} + \xi u, \varepsilon) \, d\xi \right) \dfrac{\partial u^*}{\partial x} = p, \\ u^*(T, x) = 0, \\ u^*(t, b) = 0. \end{cases} \tag{3.10}$$

**Lemma 3.2.** *Let* $\partial \varphi^{(0)}/\partial x$, $\partial u/\partial x \in L_\infty(\Omega)$ *and* $p$ *belong to the range* $R(A^*(u, \varepsilon))$ *of the operator* $A^*(u, \varepsilon)$. *Then the adjoint equation (3.9) has a unique solution* $u^*$ *and the following estimate holds:*

$$\|u^*\|_{L_2(\Omega)} \le c\|p\|_{L_2(\Omega)}, \quad c = \text{constant} > 0. \tag{3.11}$$

*Proof.* If $p \in R(A^*(u, \varepsilon))$, there exists $u^* \in D(A^*) = W_{2,b,T}^1(\Omega)$ such that (3.9) is satisfied. We write (3.9) in the form (3.10). By multiplying equation (3.10) by $u^*$ and integrating it with respect to $x$ over $[a, b]$, we obtain the equality

$$\int_a^b \frac{\partial u^*}{\partial t} u^* \, dx + \int_a^b \tilde{a}(t, x) \frac{\partial u^*}{\partial x} u^* \, dx = -\int_a^b p u^* \, dx \tag{3.12}$$

which holds almost everywhere on $[0, T]$ with $\tilde{a}(t, x) = \int_0^1 g(\varphi^{(0)} + \xi u, \varepsilon) \, d\xi$. Since the function $g(u, \varepsilon)$ is differentiable with respect to $u$, condition (1.6) is valid and $\partial\varphi^{(0)}/\partial x$, $\partial\varphi/\partial x \in L_\infty(\Omega)$, then there exists the weak derivative $\partial\tilde{a}/\partial x \in L_\infty(\Omega)$. Hence, we may integrate the second addend in the left-hand side of (3.12) by parts:

$$\int_a^b \frac{\partial u^*}{\partial t} u^* \, dx + \frac{1}{2}\tilde{a}u^{*2}\Big|_{x=a}^{x=b} - \frac{1}{2}\int_a^b \frac{\partial \tilde{a}}{\partial x} u^{*2} \, dx = -\int_a^b pu^* \, dx. \qquad (3.13)$$

Integrate (3.13) with respect to $t' \in [t, T]$. Since $u^*(T, x) = 0$ and

$$\int_t^T dt' \int_a^b \frac{\partial u^*}{\partial t} u^* \, dx = \int_a^b dx \int_t^T \frac{\partial u^*}{\partial t} u^2 t \, dt' = \int_a^b u^{*2}\Big|_{t'=t}^{t'=T} dx$$

$$- \int_a^b \left( \int_t^T u^* \frac{\partial u^*}{\partial t} \, dt' \right) dx = -\frac{1}{2}E(t),$$

where

$$E(t) = \int_a^b u^{*2}(t, x) \, dx,$$

then, in view of $u^*(t, b) = 0$, we obtain from (3.13)

$$\frac{1}{2}E(t) + \int_t^T \frac{1}{2}\tilde{a}u^{*2}(t, a) \, dt = -\frac{1}{2}\int_t^T dt \int_a^b u^{*2}\frac{\partial \tilde{a}}{\partial x} \, dx$$

$$+ \int_t^T \int_a^b pu^* \, dx \, dt.$$

Hence, the following estimate holds:

$$E(t) \leq C_1 + c_2 \int_t^T E(\tau) \, d\tau, \quad t \in [0, T],$$

where $c_1 = 2\|p\|_{L_2(\Omega)}\|u^*\|_{L_2(\Omega)}$,

$$c_2 = \left\|\frac{\partial \tilde{a}}{\partial x}\right\|_{L_2(\Omega)} \leq \left[ g_0' + g_1' \left( \|\varphi^{(0)}\|_{L_\infty(\Omega)} + \|u\|_{L_\infty(\Omega)} \right)^{\alpha_1'} \right.$$

$$\left. + g_2' \left( \|\varphi^{(0)}\|_{L_\infty(\Omega)} + \|u\|_{L_\infty(\Omega)} \right)^{\alpha_2'} \right]$$

$$\times \left( \left\|\frac{\partial \varphi^{(0)}}{\partial x}\right\|_{L_\infty(\Omega)} + \left\|\frac{\partial u}{\partial x}\right\|_{L_\infty(\Omega)} \right).$$

Then, by the Gronwall lemma[269], we get $E(t) \leq c_1 e^{(T-t)c_2}$. Integrating this inequality in $t$ over $[0, T]$, we find

$$\int_0^T \int_a^b u^{*2}(t, x) \, dx \leq c_1 \frac{e^{Tc_2} - 1}{c_2}.$$

Therefore, $\|u^*\|^2_{L_2(\Omega)} \leq c_3\|u^*\|_{L_2(\Omega)}\|p\|_{L_2(\Omega)}$, where $c_3 = 2\left(e^{Tc_2} - 1\right)/c_2$. Hence,

$$\|u^*\|_{L_2(\Omega)} \leq c_3\|p\|_{L_2(\Omega)}, \tag{3.14}$$

where $c_3 = 2\left(e^{Tc_2} - 1\right)/c_2$.

*Remark 3.2.* If we treat the operator $A^*(u, \varepsilon)$ as an operator mapping $L_2(\Omega)$ into $L_2(\Omega)$ with the domain $W^1_{2,b,T}(\Omega)$, then estimate (3.14) gives us the correct solvability of the adjoint equation (3.9).

We write the adjoint problem (3.10) in the form

$$\begin{cases} -\dfrac{\partial u^*}{\partial t} - \tilde{a}(t, x)\dfrac{\partial u^*}{\partial x} = p, \quad t \in [0, T), \quad x \in [a, b), \\[2mm] u^*|_{t=T} = 0, \\[2mm] u^*|_{x=b} = 0, \end{cases} \tag{3.15}$$

where

$$\tilde{a}(t, x) = \int_0^1 g\left(\varphi^{(0)}(t, x) + \xi u(t, x), \varepsilon\right) \, d\xi. \tag{3.16}$$

The problems of the form (3.15) were studied by Shutyaev[265]. We make use of the following

**Theorem 3.1.** *Let* (1) $p(t, x)$, $\partial p/\partial x \in L_\infty(\Omega)$; (2) $a_1 \leq \tilde{a}(t, x) \leq a_2$, $a_1, a_2 = $ *constant* $> 0$; (3) *the Lipschitz condition in $x$ be satisfied for almost each $t \in [0, T]$:*

$$|\tilde{a}(t, x_1) - \tilde{a}(t, x_2)| \leq k_a|x_1 - x_2|, \quad k_a = \text{constant} > 0, \quad \forall \, x_1, x_2 \in [a, b].$$

*Then problem (3.15) has a unique solution $u^* \in W^1_{2,b,T}(\Omega)$, it is continuous on $\overline{\Omega}$, and the following estimate holds:*

$$\|u^*\|_{W^1_2(\Omega)} \leq c_1\|p\|_{L_\infty(\Omega)} + c_2 \left\|\frac{\partial p}{\partial x}\right\|_{L_\infty(\Omega)}, \tag{3.17}$$

*where* $c_1 = \sqrt{(b-a)T}(1+T+(1+a_2)\,e^{k_a T}/a_1)$, $c_2 = \sqrt{(b-a)T}(1+a_2)\,Te^{k_a T}$.

In this case the function $\tilde{a}(t, x)$ is defined by (3.16). As stated above, in view of (1.6) and $\partial\varphi^{(0)}/\partial x$, $\partial u/\partial x \in L_\infty(\Omega)$, there exists the weak derivative $\partial\tilde{a}/\partial x \in L_\infty(\Omega)$. Then

$$|\tilde{a}(t, x_1) - \tilde{a}(t, x_2)| = \left|\int_{x_1}^{x_2} \frac{\partial\tilde{a}}{\partial x}(t, \xi) \, d\xi\right|$$

$$\leq \left\|\frac{\partial\tilde{a}}{\partial x}\right\|_{L_\infty(\Omega)} |x_1 - x_2| \equiv k_a|x_1 - x_2|,$$

i.e hypothesis (3) of Theorem 3.1 is satisfied. By virtue of (1.5), we get

$$|\tilde{a}(t,x)| \le g_0 + g_1 \left(\|\varphi^{(0)}\|_{L_\infty(\Omega)} + \|u\|_{L_\infty(\Omega)}\right)^{\alpha_1}$$

$$+g_2 \left(\|\varphi^{(0)}\|_{L_\infty(\Omega)} + \|u\|_{L_\infty(\Omega)}\right)^{\alpha_2} \equiv a_2,$$

i.e. one of the inequalities in hypothesis (2) of Theorem 3.1 is also satisfied. As far as the inequality $\tilde{a}(t,x) \ge a_1$ is concerned, it does not hold in all cases; we should require in addition that it be valid.

We arrive thus at the following

**Corollary 3.1.** *Let* $p \in L_\infty(\Omega)$, $\partial p/\partial x \in L_\infty(\Omega)$, $\partial \varphi/\partial x \in L_\infty(\Omega)$, $\partial \varphi^{(0)}/\partial x \in L_\infty(\Omega)$ *and the function* $g$ *satisfy conditions (1.5), (1.6) and*

$$\int_0^1 g\left(\varphi^{(0)}(t,x) + \xi u(t,x), \varepsilon\right) d\xi \ge a_1, \quad a_1 = \text{constant} > 0. \quad (3.18)$$

*Then the adjoint problem (3.10) has a unique solution* $u^* \in W^1_{2,b,T}(\Omega)$, *it is continuous on* $\overline{\Omega}$, *and estimate (3.17) holds.*

*Remark* 3.3. Condition (3.18) is satisfied in many perturbed problems[269], if the solutions $\varphi^{(0)}$ and $\varphi$ are positive. In this case

$$a_1 = \inf_{\xi,t,x} g\left(\varphi^{(0)}(t,x) + \xi u(t,x), \varepsilon\right) > 0.$$

For example, in problems of chemical exchange processes[269,314] we get for positive $\varphi^{(0)}$, $\varphi$, and $k_2 > k_1$

$$\tilde{a}(t,x) = \int_0^1 g\left(\varphi^{(0)} + \xi u, \varepsilon\right) d\xi \ge \frac{v\,(k_2 B)^2}{k_1 k_2 AB + (k_2 B)^2} \equiv a_1 > 0$$

with $\tilde{a}(t,x) \le v \equiv a_2$.

## 4. AN ALGORITHM FOR COMPUTING THE FUNCTIONAL

Let us turn back to the functional (3.4) and derive a formula to compute it using the adjoint problem. Assume that we know the solution $u^*$ of the adjoint problem (3.9). From (2.13) we have

$$(A(u,\varepsilon)\,u, w) = (f(\varepsilon), w) \quad \forall\, w \in W^1_{2,b,T}(\Omega). \quad (4.1)$$

Hence, in view of (3.6),

$$(u, A^*(u,\varepsilon)\,w) = (f(\varepsilon), w) \quad \forall\, w \in W^1_{2,b,T}(\Omega). \quad (4.2)$$

We take $w = u^* \in W^1_{2,b,T}(\Omega)$. Then $A^*(u,\varepsilon)\,u^* = p$ and from (4.2) we find

$$(u, p) = (f(\varepsilon), u^*). \quad (4.3)$$

Therefore, we obtain for the functional $\delta J = (u, p)$ an alternative representation of the form

$$\delta J = (f(\varepsilon), u^*). \tag{4.4}$$

Hence,

$$J(\varphi) = (\varphi^{(0)}, p) + (f(\varepsilon), u^*). \tag{4.5}$$

But formulae (4.3)–(4.5) involve the solution $u^*$ of the adjoint problem (3.9) which depends, in general, on the unknown solution $u$ of the original non-linear problem (2.13). Instead of $u^*$, we consider another function $u_0^*$ approximating $u^*$ in some sense.

Let $u_0^*$ be a solution to the equation

$$A^*(0, \varepsilon)\, u_0^* = p, \qquad\qquad , \ (4.6)$$

where the operator $A^*(0, \varepsilon)$ is defined by (3.7) for $u = 0$ and has the form

$$A^*(0, \varepsilon)\, w = -\frac{\partial w}{\partial t} - g(\varphi^{(0)}, \varepsilon)\frac{\partial w}{\partial x}, \quad w \in D(A^*) = W^1_{2,b,T}(\Omega). \tag{4.7}$$

If $\partial\varphi^{(0)}/\partial x \in L_\infty(\Omega)$, $p \in R(A^*(0, \varepsilon))$, then, by Lemma 3.2, equation (4.6) has a unique solution $u_0^*$ and the following estimate holds:

$$\|u^*\|_{L_2(\Omega)} \le c\|p\|_{L_2(\Omega)}. \tag{4.8}$$

Moreover, Corollary 3.1 gives the following

**Lemma 4.1.** *Let* $p$, $\partial p/\partial x$, $\varphi^{(0)}$, $\partial\varphi^{(0)}/\partial x \in L_\infty(\Omega)$ *and the function* $g$ *satisfy conditions (1.2), (1.3), and*

$$g\left(\varphi^{(0)}(t, x), \varepsilon\right) \ge a_1, \quad a_1 = \text{constant} > 0. \tag{4.9}$$

*Then the adjoint problem (4.6) has a unique solution* $u^* \in W^1_{2,b,T}(\Omega)$, *it is continuous on* $\overline{\Omega}$, *and the following estimate*

$$\|u_0^*\|_{W^1_2(\Omega)} \le c_1\|p\|_{L_\infty(\Omega)} + c_2\left\|\frac{\partial p}{\partial x}\right\|_{L_\infty(\Omega)}, \quad c_1, c_2 = \text{constant} > 0 \ (4.10)$$

*is valid.*

We prove now that $J(\varphi)$ and $\tilde{J} = J(\varphi^{(0)}) + (f(\varepsilon), u_0^*)$ are close in some sense.

**Theorem 4.1.** *Under the hypotheses of Lemma 4.1 the difference* $|J(\varphi) - \tilde{J}|$ *is small if the value* $\|u\|_{L_\infty(\Omega)}$ *is small. The following estimate holds:*

$$|J(\varphi) - \tilde{J}| \le c\|u\|^2_{L_\infty(\Omega)}\left(\|p\|_{L_\infty(\Omega)} + \left\|\frac{\partial p}{\partial x}\right\|_{L_\infty(\Omega)}\right), \quad c = \text{constant} > 0. \tag{4.11}$$

*Proof.* By $X$ we denote the space of functions $v(t, x) \in L_\infty(\Omega)$ such that there exists the weak derivative $\partial v/\partial x \in L_\infty(\Omega)$, with the norm

$$\|v\|_X = \|v\|_{L_\infty(\Omega)} + \left\|\frac{\partial v}{\partial x}\right\|_{L_\infty(\Omega)}$$

Let $X^*$ be the space dual to $X$. Consider the linear functional

$$I(v) = (f(\varepsilon), w), \tag{4.12}$$

where $v \in X$, $f(\varepsilon) \in W_{2,b,T}^{-1}(\Omega)$ and $w$ is the solution to the problem

$$-\frac{\partial w}{\partial t} - g(\varphi^{(0)}, \varepsilon)\frac{\partial w}{\partial x} = v, \quad t \in [0, T), \quad x \in [a, b), \tag{4.13}$$

$$w(T, x) = w(t, b) = 0.$$

Since $v \in X$, under the hypotheses of Lemma 4.1 there exists a unique solution $w \in W_{2,b,T}^1(\Omega)$ of problem (4.13) and the following estimate holds:

$$\|w\|_{W_{2,b,T}^1(\Omega)} \leq c\|v\|_X, \quad c = \text{ constant } > 0.$$

By this is meant that for $f(\varepsilon) \in W_{2,b,T}^{-1}(\Omega)$ the functional $I(v)$ is bounded on $X$, since

$$|I(v)| \leq \|f(\varepsilon)\|_{W_{2,b,T}^{-1}(\Omega)}\|w\|_{W_{2,b,T}^1(\Omega)} \leq c\|f(\varepsilon)\|_{W_{2,b,T}^{-1}(\Omega)}\|v\|_X.$$

Hence, there exists a unique element $u_1 \in X^*$ such that $I(v) = (u_1, v)$, that is,

$$(u_1, v) = (f(\varepsilon), w)$$

and

$$\|u_1\|_{X^*} = \sup_{v \in X, \ v \neq 0} \frac{|I(v)|}{\|v\|_X} \leq c\|f(\varepsilon)\|_{W_{2,b,T}^{-1}(\Omega)}. \tag{4.14}$$

From (4.13) we get

$$\left(u_1, -\frac{\partial w}{\partial t} - g(\varphi^{(0)}, \varepsilon)\frac{\partial w}{\partial x}\right) = (f(\varepsilon), w) \quad \forall \, w \in D, \tag{4.15}$$

where

$$D = \left\{w \in W_{2,b,T}^1(\Omega): \frac{\partial w}{\partial t} + g(\varphi^{(0)}, \varepsilon)\frac{\partial w}{\partial x} \in X\right\}.$$

We have thus proved that there exists a unique solution $u_1 \in X^*$ of the problem (4.15). Let $u_0^*$ be the solution of problem (4.6) for $p \in X$ and $u_1$ be the solution of problem (4.15) with $f(\varepsilon)$ given by (2.1). By setting $w = u_0^*$ in (4.15), we get $(u_1, p) = (f(\varepsilon), u_0^*)$. Hence, in view of (4.3)–(4.5), we find

$$J(\varphi) = \tilde{J} = (f(\varepsilon), u^* - u_0^*) = (u - u_1, p). \tag{4.16}$$

Let us estimate now the norm $\|u - u_1\|_{X^*}$. The function $u$ satisfies equation (2.13) which, in accord with (2.12), may be written in the form

$$\left(u, -\frac{\partial w}{\partial t} - \left(\int_0^1 g(\varphi^{(0)} + \xi u, \varepsilon) \, d\xi\right) \frac{\partial w}{\partial x}\right) = (f(\varepsilon), w) \quad \forall \, w \in W_{2,b,T}^1(\Omega).$$
(4.17)

From (4.15) and (4.17) we obtain for any $w \in D$

$$\left(u - u_1, -\frac{\partial w}{\partial t} - g(\varphi^{(0)}, \varepsilon)\frac{\partial w}{\partial x}\right)$$

$$= \left(u, \left(\int_0^1 \left[g(\varphi^{(0)} + \xi u, \varepsilon) - g(\varphi^{(0)}, \varepsilon)\right] d\xi\right) \frac{\partial w}{\partial x}\right). \qquad (4.18)$$

Consider the linear functional

$$I_1(w) = \left(u, \left(\int_0^1 \left[g(\varphi^{(0)} + \xi u, \varepsilon) - g(\varphi^{(0)}, \varepsilon)\right] d\xi\right) \frac{\partial w}{\partial x}\right), \quad w \in W_{2,b,T}^1(\Omega).$$

Since

$$g(\varphi^{(0)} + \xi u, \varepsilon) - g(\varphi^{(0)}, \varepsilon) = \int_0^1 \frac{\partial g}{\partial \varphi}(\varphi^{(0)} + \xi \tau u, \varepsilon) \, d\tau(\xi u),$$

then under the hypotheses of the theorem the following estimate holds:

$$|I_1(w)| \leq c\|u\|_{L_2(\Omega)}\|u\|_{L_\infty(\Omega)}\|w\|_{W_{2,b,T}^1(\Omega)},$$

that is, $I_1(w)$ is a linear bounded functional on $W_{2,b,T}^1(\Omega)$. Hence, there exists a unique element $w^* \in W_{2,b,T}^{-1}(\Omega)$ such that $I_1(w) = (w^*, w)$ for any $w \in W_{2,b,T}^1(\Omega)$ and

$$\|w^*\|_{W_{2,b,T}^{-1}(\Omega)} = \sup_{w \in W_{2,b,T}^1(\Omega), \, w \neq 0} \frac{|I_1(w)|}{\|w\|_{W_{2,b,T}^1(\Omega)}} \leq c\|u\|_{L_2(\Omega)}\|u\|_{L_\infty(\Omega)}. \quad (4.19)$$

Consider the restriction of the functional $I_1(w)$ on $D$. Then (4.18) may be written in the form:

$$\left(u - u_1, -\frac{\partial w}{\partial t} - g(\varphi^{(0)}, \varepsilon)\frac{\partial w}{\partial x}\right) = (w^*, w) \quad \forall \, w \in D. \qquad (4.20)$$

Since $u - u_1$ is the solution of (4.20), the estimate of the form (4.14) holds:

$$\|u - u_1\|_{X^*} \leq c\|w^*\|_{W_{2,b,T}^{-1}(\Omega)}$$

and, in view of (4.19),

$$\|u - u_1\|_{X^*} \leq c\|u\|_{L_2(\Omega)}\|u\|_{L_\infty(\Omega)} \leq c\|u\|_{L_\infty(\Omega)}^2.$$

From (4.16) we thus obtain

$$|J(\varphi) - \tilde{J}| \leq \|u - u_1\|_X \cdot \|p\|_X \leq c\|u\|_{L_\infty(\Omega)}^2 \left( \|p\|_{L_\infty(\Omega)} + \left\| \frac{\partial p}{\partial x} \right\|_{L_\infty(\Omega)} \right).$$

**Corollary 4.1.** *If the hypotheses of Theorem 4.1 are satisfied, one can put approximately*

$$J(u) \cong \tilde{J} \equiv J(\varphi^{(0)}) + J_1, \tag{4.21}$$

*where* $J_1 = (f(\varepsilon), u_0^*)$.

*Remark* 4.1. In problems of chemical exchange processes, the function $g(\varphi, \varepsilon)$ satisfies, in addition, the inequality (see Remark 1.1)

$$\left| \frac{\partial g}{\partial \varphi} \right| \leq \Theta\varepsilon, \quad \Theta = \text{ constant } > 0.$$

This makes it possible to derive a sharper estimate for $J(\varphi) - \tilde{J}$.

Corollary 4.1 leads to the following algorithm for computing the functional $J(\varphi)$:

(1) Find the solution $u_0^*$ of the adjoint problem

$$\begin{cases} -\dfrac{\partial u_0^*}{\partial t} - g(\varphi^{(0)}, \varepsilon) \dfrac{\partial u_0^*}{\partial x} = p, & t \in [0, T], \quad x \in [a, b], \\ u_0^*(T, x) = u_0^*(t, b) = 0. \end{cases} \tag{4.22}$$

(2) Compute the correction $J_1$ using (4.21) and (2.3) for $w = u_0^*$:

$$J_1 = (f(\varepsilon), u_0^*) = (f, u_0^*) + \left( G(\varphi^{(0)}, \varepsilon), \frac{\partial u_0^*}{\partial x} \right) + \left( \varphi^{(0)}, \frac{\partial u_0^*}{\partial t} \right)$$

$$+ \int_0^T G(\varphi_1(t, \varepsilon), \varepsilon) \, u_0^*(t, a) \, dt$$

$$+ \int_a^b \varphi_0(x, \varepsilon) \, u_0^*(0, x) \, dx.$$

(3) Find an approximate value of $J(\varphi)$ by the formula

$$J(\varphi) \equiv J(\varphi^{(0)}) + J_1 = (\varphi^{(0)}, p) + J_1.$$

# 5. THE PROBLEM ON CHEMICAL EXCHANGE PROCESSES

Consider a perturbed problem on chemical exchange processes of the form[269]:

$$
\begin{cases}
-\dfrac{\partial \varphi}{\partial t} + g(\varphi, \varepsilon)\dfrac{\partial \varphi}{\partial x} = f(t, x, \varepsilon), & t \in [0, T), \quad x \in [a, b), \\[2mm]
\varphi(0, x) = \varphi_0(x, \varepsilon), \\[2mm]
\varphi(t, a) = \varphi_1(t, \varepsilon),
\end{cases}
\tag{5.1}
$$

where

$$
g(\varphi, \varepsilon) = \frac{v(k_2 B + (k_1 - k_2)\varphi\varepsilon)^2}{k_1 k_2 AB + (k_2 B + (k_1 - k_2)\varphi\varepsilon)^2};
$$

$v, k_1, k_2, A, B =$ constant $> 0$, $k_2 > k_1$; $f$, $\varphi_0$, $\varphi_1$ are given functions satisfying the conditions of Section 1.

The non-perturbed problem is derived from (5.1) for $\varepsilon = 0$. In this case, it is linear and has the form

$$
\begin{cases}
-\dfrac{\partial \varphi^{(0)}}{\partial t} + g_0 \dfrac{\partial \varphi^{(0)}}{\partial x} = f(t, x, 0), & t \in [0, T), \quad x \in [a, b), \\[2mm]
\varphi^{(0)}(0, x) = \varphi_0(x, 0), \\[2mm]
\varphi^{(0)}(t, a) = \varphi_1(t, 0),
\end{cases}
\tag{5.2}
$$

where $g_0 = vk_2 B/(k_1 + k_2 B)$.

Assume that the weak solutions $\varphi$ and $\varphi^{(0)}$ of problems (5.1) and (5.2) do exist (in the sense of (1.7)), belong to $L_\infty(\Omega)$ and are non-negative functions. Let, moreover, $\partial \varphi / \partial x \in L_\infty(\Omega)$. The solution $\varphi^{(0)}$ is supposed to be known, and we need to compute the functional

$$
J(\varphi) = (\varphi, p) = \int_0^T dt \int_a^b \varphi(t, x)\, p(t, x)\, dt,
\tag{5.3}
$$

where $p$ is a given function in $L_2(\Omega)$. Here, as before, $\Omega = (0, T) \times (a, b)$.

For the functional $J(\varphi)$ formula (4.5) is valid:

$$
J(\varphi) = (\varphi^{(0)}, p) + (f(\varepsilon), u^*),
\tag{5.4}
$$

where the function $f(\varepsilon)$ is defined by (2.3):

$$
(f(\varepsilon), u^*) = (f, u^*) + \left( G(\varphi^{(0)}, \varepsilon), \frac{\partial u^*}{\partial x} \right) + \left( \varphi^{(0)}, \frac{\partial u^*}{\partial t} \right)
$$

$$
+ \int_0^T G(\varphi_1(t, \varepsilon), \varepsilon)\, u^*(t, a)\, dt + \int_a^b \varphi_0(x, \varepsilon)\, u^*(0, x)\, dx,
$$

$$G(\varphi, \varepsilon) = \int_0^\varphi g(u, \varepsilon) \, du \tag{5.5}$$

$$= v \left[ \varphi + \frac{\sqrt{k_1 k_2 AB}}{(k_1 - k_2)\varepsilon} \left( \arctan \frac{\sqrt{k_1 k_2 AB}}{k_2 B + (k_1 - k_2)\varphi\varepsilon} - \arctan \sqrt{\frac{k_1 A}{k_2 B}} \right) \right],$$

and $u^*$ is the solution to problem (3.10):

$$\begin{cases} -\dfrac{\partial u^*}{\partial t} - \left[ \displaystyle\int_0^1 g(\varphi^{(0)} + \xi u, \varepsilon) \, d\xi \right] \dfrac{\partial u^*}{\partial x} = p, \\[2mm] u^*(T, x) = 0, \\[2mm] u^*(t, b) = 0. \end{cases} \tag{5.6}$$

Here, $u = \varphi - \varphi^{(0)}$.

In Section 4, we have derived formula (4.21) to compute an approximate value of the functional $J(\varphi)$:

$$J(\varphi) \cong (\varphi^{(0)}, p) + (f(\varepsilon), u_0^*), \tag{5.7}$$

where $u_0^*$ is the solution of adjoint problem (4.22):

$$\begin{cases} -\dfrac{\partial u_0^*}{\partial t} - g(\varphi^{(0)}, \varepsilon) \dfrac{\partial u_0^*}{\partial x} = p, \quad t \in [0, T), \quad x \in [a, b), \\[2mm] u_0^*(T, x) = 0, \\[2mm] u_0^*(t, b) = 0. \end{cases} \tag{5.8}$$

Under the above-stated hypotheses for $p \in L_\infty(\Omega)$, $\partial p / \partial x \in L_\infty(\Omega)$ the adjoint problems (5.6) and (5.8) have the unique weak solutions $u^*$, $u_0^* \in W^1_{2,b,T}(\Omega)$ which are continuous functions (see Corollary 3.1 and Lemma 4.1).

For $g(\varphi, \varepsilon)$ the following inequality holds:

$$\left| \frac{\partial g}{\partial \varphi} \right| \leq \frac{3\sqrt{3}}{8} |k_1 - k_2| \left( \sqrt{k_1 k_2 AB} \right)^{-1} \varepsilon. \tag{5.9}$$

Then, by Theorem 4.1, for small values $\varepsilon \|u\|_{L_\infty(\Omega)}$ the difference $J(\varphi) - \tilde{J}$ is small and

$$|J(\varphi) - \tilde{J}| \leq c\varepsilon \|u\|^2_{L_\infty(\Omega)} \left( c_1 \|p\|_{L_\infty(\Omega)} + c_2 \left\| \frac{\partial p}{\partial x} \right\|_{L_\infty(\Omega)} \right), \tag{5.10}$$

$c, c_1, c_2 = $ constant $> 0$. In this case, we may thus use formula (5.7) to compute an approximate value of the functional $J(\varphi)$.

Consider a numerical example. Let $0 \le t \le 1$, $-1 \le x \le 1$ and $\varphi(t, x) = tx + \varepsilon$ be the solution of the perturbed problem (5.1). Then

$$f(t, x, \varepsilon) = x + g(\varphi, \varepsilon) t$$

$$= x + \frac{v(k_2 B + (k_1 - k_2)\varepsilon(tx + \varepsilon))^2 t}{k_1 k_2 AB + (k_2 B + (k_1 - k_2)\varepsilon(tx + \varepsilon))^2}, \qquad (5.11)$$

$$\varphi_0(x, \varepsilon) = \varepsilon, \quad \varphi_1(t, \varepsilon) = -t + \varepsilon, \quad T = 1, \quad a = -1, \quad b = 1.$$

Consider the functional $J(\varphi)$ of the form (5.3) for $p = 1$:

$$J(\varphi) = \int_0^1 dt \int_{-1}^1 \varphi(t, x)\, dx. \qquad (5.12)$$

In this case, it can be easily found in the explicit form:

$$J(\varphi) = 2\varepsilon. \qquad (5.13)$$

Now let us make use of formula (5.7) to compute an approximate value of this functional. The unperturbed problem (5.2) has the form

$$\begin{cases} \dfrac{\partial \varphi^{(0)}}{\partial t} + g_0 \dfrac{\partial \varphi^{(0)}}{\partial x} = x + \dfrac{v k_2 Bt}{k_1 A + k_2 B}, \quad t \in (0, T], \quad x \in (-1, 1], \\[2mm] \varphi^{(0)}(0, x) = 0, \\[2mm] \varphi^{(0)}(t, -1) = -t. \end{cases} \qquad (5.14)$$

It is readily seen that problem (5.14) has the solution $\varphi^{(0)} = tx$ and

$$J(\varphi^{(0)}) = (\varphi^{(0)}, p) = \int_0^1 dt \int_{-1}^1 \varphi^{(0)}(t, x)\, dx \equiv 0. \qquad (5.15)$$

To use formula (5.7) we should find the solution of the adjoint problem (5.8):

$$\begin{cases} -\dfrac{\partial u_0^*}{\partial t} - \dfrac{v(k_2 B + (k_1 - k_2) tx\varepsilon)^2}{k_1 k_2 AB + (k_2 B + (k_1 - k_2) tx\varepsilon)^2} \dfrac{\partial u_0^*}{\partial x} = 1, \\[2mm] u_0^*(1, x) = 0, \\[2mm] u_0^*(t, 1) = 0. \end{cases} \qquad (5.16)$$

Knowing the solution $u_0^*$ of this problem, from (5.5), (5.7), and (5.15) we get

$$J(\varphi) \cong J_1 = (f(\varepsilon), u_0^*)$$

$$= \int_0^1 \int_{-1}^1 f(t, x, \varepsilon)\, u_0^*(t, x)\, dt\, dx$$

$$+ \int_0^1 \int_{-1}^1 G(tx, \varepsilon) \frac{\partial u_0^*}{\partial x} \, dt \, dx + \int_0^1 \int_{-1}^1 tx \, \frac{\partial u_0^*}{\partial t} \, dt \, dx$$

$$+ \int_0^1 G(-t + \varepsilon, \varepsilon) u_0^*(t, 1) \, dt + \int_{-1}^1 \varepsilon u_0^*(0, x) \, dx. \qquad (5.17)$$

The adjoint problem (5.16) has been solved numerically with the use of the symmetrical scheme[95] of the second order of accuracy $O(\tau^2 + h^2)$, where $\tau$ and $h$ are the mesh sizes in $t$ and $x$, respectively. To compute the integrals in (5.17) the generalized trapezoidal rule has been used. The results for the case $\tau = h = 0.02$ are given in Table 1, where $J(\varphi)$ is the exact value of the functional (see (5.13)), and $J_1$ is the approximate value of $J(\varphi)$ computed by formula (5.17). As may be seen from the table, the value $J_1$ obtained from (5.17) fits the exact value of the functional $J(\varphi)$ well, even when $\varepsilon = 0.1$ and $\varepsilon = 1$.

**Table 1.**
The values of the functionals $J(\varphi)$ and $J_1$

| $\varepsilon$ | $J(\varphi)$ | $J_1$ |
|---|---|---|
| 1 | 2 | 2.09925 |
| 0.1 | 0.2 | 0.232007 |
| 0.01 | 0.02 | 0.0202820 |
| 0.001 | 0.002 | 0.00197504 |
| 0 | 0 | $-0.00707 \times 10^{-5}$ |

To conclude, note that the above-stated results on the justification of the algorithms for computing the functionals on the basis of the adjoint equation technique and perturbation theory can be extended to more complicated multi-dimensional problems.

# Chapter 8

# Adjoint equations and perturbation algorithms for a non-linear mathematical model of mass transfer in soil

This chapter considers a non-linear model describing the transport of chemicals through soil. An algorithm for computing the unknown coefficients in the governing non-linear parabolic equation is developed and justified on the basis of adjoint equation technique and perturbation theory.

## 1. MATHEMATICAL MODELS OF MASS TRANSFER IN SOIL

The study of herbicide influence on ecology is impossible without a knowledge of the peculiarities of materials being used in natural objects, and mainly in soil. One of the basic components determining the behaviour of herbicides in soil is migration, that is, the combination of both physico-chemical and hydrodynamical processes which give rise to the movement of chemicals through soil. A sophisticated treatment of the mass transport in the field, and even in the laboratory scale, is a complicated and complex experimental problem. Here, the development and study of mathematical models for the mass transfer in soil play an important role.

Mathematical modelling of the movement of the material in soil columns allows one to study its transport under the controlled conditions, establish the input–output relationships, and verify various theoretical transport models by comparison with the experimental data. Moreover, the use of mathematical models terminates in a significant decrease in a long list of experiments when studying properties of the test materials under varying conditions, and helps in solving some optimization problems on performing the experiments.

There are many models for governing and predicting the dynamics of solutes in the soil water solution, beginning with the simplest models, with a narrow field of uses, which involve only one of the basic mechanisms for the movement of solutes[84] :

(a) simple convection model

$$J = v\Theta C, \tag{1.1}$$

where $J$ is the one-dimensional solute flow, $\Theta$ the volumetric water content, $v$ the average pore-water velocity, $C$ the concentration of solute in the soil water solution;

(b) the dispersion equation without diffusion

$$\Theta \frac{\partial C}{\partial t} = \Theta D \frac{\partial^2 C}{\partial x^2}, \tag{1.2}$$

where $D$ is the dispersion coefficient, $x$ the space variable, and $t$ the time.

Recent advances in the development of deterministic models of solute transport in soil give rise to a model involving the following governing equation[78,313]

$$\frac{\partial(\Theta_m C_m)}{\partial t} + \frac{\partial(\Theta_{im} C_{im})}{\partial t} + \frac{\partial(f \rho S_m)}{\partial t} + \frac{\partial[(1-f) \rho S_{im}]}{\partial t}$$
$$= \frac{\partial}{\partial x}\left(\Theta_m D \frac{\partial C_m}{\partial x}\right) - \frac{\partial(\Theta_m v_m C_m)}{\partial x}, \tag{1.3}$$

where the subscripts $m$ and $im$ identify the mobile and immobile phases, respectively, $S$ is the concentration of chemical adsorbed per unit mass of soil, $\rho$ the soil density, and $f$ the adsorption parameter in the mobile phase. This equation should be joined by the following equation governing the mobile and immobile phase exchange[78]:

$$\frac{\partial \Theta_{im} C_{im}}{\partial t} + (1-f) \rho \frac{\partial S_{im}}{\partial t} = \alpha(C_m - C_{im}), \tag{1.4}$$

where $\alpha$ is the mass transfer coefficient.

Using this model, some authors[78,84,313] have initiated the investigations in order to take account of complex soil structure, and to extend adsorption processes in soil to both mobile and immobile phases. Equations (1.3) and (1.4) satisfactorily govern the movement of adsorbed pesticides in structural soil when $D$, $\Theta_{im}$, $f$, $v$, $\alpha$, $\rho$, $\Theta_m$, $S_m$, and $S_{im}$ are known. The central problem[313] with using the model (1.3), (1.4), and the others involving the convection–dispersion equation, is that some parameters (examples are the hydrodynamic dispersion coefficient, the mass exchange coefficient, the parameters $f$, $\Theta_{im}$, and $\Theta_m$) are impossible for the moment to find independently. One works out these coefficients by the trial-and-error methods, making the curves obtained theoretically coincide with the experimental ones. Moreover, the more parameters are unknown, the less is the assurance that the values matted are adequate. On the other hand, simple models like (1.2), (1.3) must be adapted to given soil conditions and water flow, that is, they are not universal. The most-used equation for governing the movement of chemicals through soil by water is currently the convection–diffusion equation with adsorption[34,49,84,259]

$$\frac{\partial \Theta C}{\partial t} + \frac{\partial \rho S}{\partial t} = \frac{\partial}{\partial x} D \frac{\partial \Theta C}{\partial x} - v \frac{\partial \Theta C}{\partial x}. \tag{1.5}$$

In the general case, $S$ is a rather complicated function of concentration and time:

$$S = \Phi(C, t). \tag{1.6}$$

Depending on the way of describing $S$, there arise different models including features peculiar to the adsorption of solutes under study. The basic species of adsorption inherent in chemicals like herbicides may be cited as the following:

(1) equilibrium adsorption with linear isotherm:

$$S = K_\alpha C, \tag{1.7}$$

   where $K_\alpha$ is the distribution coefficient;
(2) equilibrium adsorption with non-linear isotherm (the Freundlich equation)

$$S = KC^{1/n}, \tag{1.8}$$

   where $K$ and $1/n$ are the empirical constants;
(3) non-equilibrium adsorption with both linear and non-linear isotherms.

Attempts to take into consideration the last-named type of adsorption have led to a wide variety of different models[35,50,126,128,212,259]. The most abundant is the following one:

$$\frac{\mathrm{d}\rho S}{\mathrm{d}t} = k_A \Theta C - k_D \rho S, \tag{1.9}$$

where $k_A$ and $k_D$ are the adsorption and desorption coefficients, respectively. An equation like (1.9) may be derived with the assumption that the rate of movement of chemical adsorbed is proportional to the difference between the quantity of chemical adsorbed $\rho S$ and the amount of chemical adsorbed at equilibrium. With different ways of describing the equilibrium adsorption we obtain versions of model (1.9) in the form:

$$\frac{\mathrm{d}\rho S}{\mathrm{d}t} = k_D(\Theta K_d C - \rho S), \quad S = K_d C, \tag{1.10}$$

and

$$\frac{\partial \rho S}{\partial t} = k_D(\Theta K C^{1/n} - \rho S), \quad S = K C^{1/n}. \tag{1.11}$$

The following version takes into consideration essentially non-homogeneous soil composition and, as a result, different mechanisms of adsorption[126]:

$$\frac{\mathrm{d}\rho S_f}{\mathrm{d}t} = k_{A,f} \Theta C - k_{D,f} \rho S_f, \tag{1.12}$$

$$\frac{\mathrm{d}\rho S_s}{\mathrm{d}t} = k_{A,s} \Theta C - k_{D,s} \rho S_s, \tag{1.13}$$

where the subscripts $f$ and $s$ identify the fast and slow adsorption, respectively. This approach may be extended[319] by the following two equations:

(1) outer-diffusion adsorption kinetic equation

$$\frac{\partial S}{\partial t} = \beta(C - C_p), \quad \beta > 0, \quad S = \Phi(C_p), \tag{1.14}$$

where $C_p$ is the concentration at equilibrium with $S$, $\Phi(C)$ the adsorption isotherm, and $\beta$ the kinetic coefficient;

(2) inner-diffusion adsorption kinetic equation

$$\frac{\partial S}{\partial t} = \gamma(\Phi(C) - S), \tag{1.15}$$

where $\gamma$ is the kinetic coefficient.

Notice that in specific cases kinetic effects of adsorption may be taken into account by introducing the so-called efficient coefficients in equation (1.5), for example, efficient hydrodynamic dispersion coefficient[94,245], instead of using equations (1.14), (1.15).

The problem of finding (or estimating) the unknown coefficients of the model is also discussed in the literature. For the most part these coefficients are evaluated by the trial-and-error method or by the so-called compatibility procedure based on the least-square method[49,126,216,313]. In some instances it is possible to find the required coefficient using results of the experiment, and an explicit form of the solution of a specific model[251].

One can consider a more versatile way of finding the unknown coefficients of the problem[51,291]. It is based on the regularization method due to Tykhonov.[285] With the use of this method the mass transfer coefficients and the isotherm parameters may be determined by the known output dynamic curves.

An algorithm for computing the unknown coefficients in the non-linear mathematical model of mass transfer in soil will be given in this chapter on the basis of adjoint equation technique and perturbation theory.

## 2. FORMULATION OF A NON-LINEAR MATHEMATICAL MODEL

### 2.1. Description for the mathematical model. Posing the problem

It is common knowledge that the following linear problem is well suited to governing the movement of chemical in soil column[14,213,276] :

$$\frac{\partial C}{\partial t} = \frac{\partial}{\partial x} D_0 \frac{\partial C}{\partial x} - v \frac{\partial C}{\partial x} - \frac{\rho}{\Theta} \frac{\partial S}{\partial t} + f, \tag{2.1}$$

$$C(x,0) = 0, \qquad\qquad 0 < x \le L,$$

$$C(0,t) = C_0(t), \qquad\quad 0 \le t \le T,$$

$$C(L,t) = 0, \qquad\qquad 0 \le t \le T, \qquad\qquad (2.2)$$

$$S(x,t) = K^{(0)}C(x,t),$$

where $C(x,t)$ is the concentration of solute in the soil water solution $(\mathrm{g\,cm^{-3}})$, $D$ the efficient dispersion coefficient $(\mathrm{cm^2\,h^{-1}})$, $x$ the distance from the input (cm), $v$ the average pore-water velocity $(\mathrm{cm\,h^{-1}})$, $\rho$ the soil density $(\mathrm{g\ cm^{-3}})$, $\Theta$ the porosity $(\mathrm{cm^3\,cm^{-3}})$, $K^{(0)}$ the distribution coefficient $(\mathrm{cm^3\,g^{-1}})$, $L$ the depth of the soil layer which cannot be reached by the solute in a given time $T$ (cm), $S$ the quantity of solute adsorbed per unit mass of soil $(\mathrm{g\,g^{-1}})$, $C_0(t)$ the concentration at the inlet $(x = 0)$ $(\mathrm{g\,cm^{-3}})$, and $f$ the internal herbicide source function (in actual practice $f \equiv 0$).

However, in the event that the adsorption isotherm is not linear, the discrepancy between the predicted and experimental values comes into particular prominence. One can refine the model (2.1), (2.2) by introducing the non-linear Freundlich equation since a time $t_0$[84]:

$$S = KC^{1/n}, \qquad\qquad (2.3)$$

where $K$ and $1/n$ are the empirical constants. Then both isotherms may be written as one equation

$$S = \varepsilon[KC^{1/n} - K^{(0)}C] + K^{(0)}C,$$

$$\delta S = KC^{1/n} - K^{(0)}C, \qquad\qquad (2.4)$$

where $\varepsilon$ is a real parameter. The value $\varepsilon = 0$ corresponds to the linear case, and $\varepsilon = 1$ to the non-linear Freundlich law. The quantity $\delta S$ exhibits the deviation of the isotherm governed by the Freundlich equation from the linear one.

The necessity of considering for the non-linear equation a time $t = t_0$ distinct from $t = 0$ stems from the fact that in the early stage of the column experiment the chemical introduced is grouped at the upper layer of the column, and its concentration is superior to the maximum capacity of soil to absorb. Because of this, the use of the non-linear Freundlich equation on the interval $(0, t_0)$ may affect the true patterns of adsorption. For $t_0$ it would be reasonable to choose the initial time the chemical takes to get to the output of the column. It can be easily calculated for each experiment as $t_0 = H/v$, where $H$ is the length of the column, and $v$ the average linear velocity. The concentration function $C_1(x) \equiv C(x, t_0)$ at $t = t_0$ may be found, for example, by the computer simulation with the use of the linear model (2.1), (2.2) on the interval $(0, t_0)$.

We shall thus consider the non-linear problem on the interval $(t_0, T)$. Substituting (2.4) into (2.1), we come to the problem

$$\frac{\partial \varphi(C, \varepsilon)}{\partial t} - \frac{\partial}{\partial x} D \frac{\partial C}{\partial x} + v \frac{\partial C}{\partial x} = f(x, t), \tag{2.5}$$

$$C(0, t) = C_0(t), \qquad t_0 \leq t \leq T,$$

$$C(L, t) = 0, \qquad t_0 \leq t \leq T, \tag{2.6}$$

$$C(x, t_0) = C_1(x), \qquad 0 < x \leq L,$$

where $\varphi(C, \varepsilon) = C + \rho(\varepsilon(KC^{1/n} - K^{(0)}C) + K^{(0)}C)/\Theta$, $D, v \in L_\infty((0, L) \times (t_0, T))$, $K^{(0)}, K, \rho, \Theta = $ constant $> 0$, $f \in L_2((0, L) \times (t_0, T))$, $d_0 \leq D \leq d_1$, $v_0 \leq v \leq v_1$, $d_0, d_1, v_0, v_1 = $ constant $> 0$, $n \in [1/2, 1]$, $C_0(t)$ and $C_1(x)$ are prescribed concentration functions.

All the parameters of the problem (2.5)–(2.6) are supposed to be known except the dispersion coefficient $D$. Assume that we are given an approximation of this coefficient, which is equal to the efficient dispersion coefficient $D_0$ in the linear model (2.1), (2.2). Now the problem is as follows: using the results of the soil column experiments, find the dispersion coefficient $D = D_0 + \delta D$ in the non-linear model (2.5)–(2.6).

## 2.2. Operator formulation of the problem

Let $\Omega = (0, L) \times (t_0, T)$. Introduce the real Hilbert spaces

$$L_2(\Omega): (u, v) = \int_\Omega u(x, t) v(x, t) \, dx \, dt, \quad \|u\| = (u, u)^{1/2},$$

$$W_2^{1,0}(\Omega): (u, v)_{W_2^{1,0}} = (u, v) + \left( \frac{\partial u}{\partial x}, \frac{\partial v}{\partial x} \right), \quad \|u\|_{W_2^{1,0}} = (u, u)_{W_2^{1,0}}^{1/2},$$

$$\overset{\circ}{W}_2^{1,0}(\Omega) = \{ u \in W_2^{1,0}(\Omega), \ u(0, t) = u(L, t) = 0 \},$$

$$W_2^1(\Omega): (u, v)_{W_2^1} = (u, v) + \left( \frac{\partial u}{\partial x}, \frac{\partial v}{\partial x} \right) + \left( \frac{\partial u}{\partial t}, \frac{\partial v}{\partial t} \right), \quad \|u\|_{W_2^1} = (u, u)_{W_2^1}^{1/2},$$

$$\overset{\circ}{W}_2^1(\Omega) = \{ u \in W_2^1(\Omega), \ u(0, t) = u(L, t) = 0 \},$$

$$\overset{\circ}{W}_{2, t_0}^1(\Omega) = \{ u \in \overset{\circ}{W}_2^1(\Omega), \ u|_{t=t_0} = 0 \},$$

$$\overset{\circ}{W}_{2, T}^1(\Omega) = \{ u \in \overset{\circ}{W}_2^1(\Omega), \ u|_{t=T} = 0 \}.$$

Henceforward, the space $L_2(\Omega)$ is identified with its adjoint. Moreover, we assume that

$$(W_2^1)^* \equiv (W_2^1)^{-1} = W_2^{-1}, \quad (\overset{\circ}{W}_{2, t_0}^1)^* \equiv \overset{\circ}{W}_{2, t_0}^{-1}, \quad (\overset{\circ}{W}_{2, T}^1)^* \equiv \overset{\circ}{W}_{2, T}^{-1}.$$

By $D(\Phi)$ we denote a set of non-negative functions in $W_2^2(\Omega)$ satisfying conditions (2.6). In the following, when considering problem (2.5), (2.6), the initial data of (2.5), (2.6) are supposed to be such that the solution $C$ exists and belongs to the set $D(\Phi)$. (We thus do not study here the existence problem for (2.5), (2.6).)

Then, assuming the solution $C(x,t)$ of problem (2.5), (2.6) to belong to $D(\Phi)$, multiply equation (2.5) scalarly in $L_2(\Omega)$ by an arbitrary function $w \in \overset{\circ}{W}{}^1_{2,T}(\Omega)$. After integrating by parts, we get

$$[C, w]_\varepsilon = (f, w), \qquad (2.7)$$

where

$$[C, w]_\varepsilon = -\left(u(C, \varepsilon), \frac{\partial w}{\partial t}\right) - \int_0^L \varphi(C_1(x), \varepsilon)\, w(x, t_0)\, \mathrm{d}x$$

$$+ \left(D\frac{\partial C}{\partial x}, \frac{\partial w}{\partial x}\right) + \left(v\frac{\partial C}{\partial x}, w\right), \quad 0 \le \frac{1}{n} - 1 \le 1. \quad (2.8)$$

Our immediate task is to show that to the form $[C, w]_\varepsilon$ corresponds a non-linear operator $\Phi$ with the domain $D(\Phi)$, defined by the equality

$$(\Phi(C, \varepsilon), w) = [C, w]_\varepsilon. \qquad (2.9)$$

With this aim in mind consider the form $[C, w]_\varepsilon$. For a fixed function $C \in D(\Phi)$ this form is a linear functional with respect to $w \in W_2^1(\Omega)$, since

$$|[C, w]_\varepsilon| \le \text{ constant } \|w\|_{W_2^1(\Omega)}.$$

Then, by the Riesz theorem, there exists a unique element $\tilde{C} \in W_2^1(\Omega)$ such that $[C, w]_\varepsilon = (\tilde{C}, w)_{W_2^1(\Omega)}$. However, it is common knowledge that there exists linear self-adjoint operator $A$ with the domain $W_2^2(\Omega)$ and the range $L_2(\Omega)$ such that $(u, w)_{W_2^1(\Omega)} = (Au, Aw)_{L_2(\Omega)}$ for any $u, w \in W_2^1(\Omega)$. The domain $D(S)$ of the operator $S \equiv A^2$ therewith is dense in $W_2^1(\Omega)$.[135] Hence, for $u \in D(S)$ we have $(u, w)_{W_2^1(\Omega)} = (Su, w)_{L_2(\Omega)}$.

Let now $u$ be an arbitrary element of $W_2^1(\Omega)$ and a sequence of elements $u_n \in D(S)$, $n = 1, 2, \ldots$, converges to $u$: $\|u_n - u\|_{W_2^1(\Omega)} \to 0$ as $n \to \infty$. Since

$$\|S(u_n - u_m)\|_{W_2^{-1}} = \sup_{w \in W_2^1(\Omega)} \frac{|(Su_n - Su_m, w)_{L_2(\Omega)}|}{\|w\|_{W_2^1(\Omega)}}$$

$$= \sup_{w \in W_2^1(\Omega)} \frac{|(u_n - u_m, w)_{W_2^1(\Omega)}|}{\|w\|_{W_2^1(\Omega)}}$$

$$\le \|u_n - u_m\|_{W_2^1(\Omega)} \to 0,$$

then the sequence of the elements $p_n = Su_n$ is fundamental in $W_2^{-1}(\Omega)$. Denote the limit of this sequence by $p$: $\|p-p_n\|_{W_2^{-1}} \to 0$ as $n \to \infty$. Therefore to each element $u \in W_2^1(\Omega)$ corresponds an element $p \in W_2^{-1}(\Omega)$. This correspondence is given by the operator $\tilde{S}$ ($\tilde{S}u = p$) being an extension of $S$ and mapping $W_2^1(\Omega)$ into $W_2^{-1}(\Omega)$. The domain of the operator $\tilde{S}$ is the entire space $W_2^1(\Omega)$. For any $u, w \in W_2^1(\Omega)$ we have thus the equalities

$$(u, w)_{W_2^1(\Omega)} = (p, w) = (\tilde{S}u, w)_{L_2(\Omega)}.$$

Then we get the relationships

$$[C, w]_\varepsilon = (\tilde{C}, w)_{W_2^1(\Omega)} = (\tilde{S}\tilde{C}, w)_{L_2(\Omega)}, \quad w \in \overset{\circ}{W}{}_{2,T}^1(\Omega)$$

which give the one-to-one correspondence $C \to \tilde{S}\tilde{C}$. This correspondence defines the required non-linear operator $\Phi$:

$$\Phi(C, \varepsilon) = \tilde{S}\tilde{C},$$

mapping $W_2^1(\Omega)$ into $W_2^{-1}(\Omega)$ with the domain $D(\Phi)$, and (2.9) is satisfied. Using (2.9), rewrite the equality (2.7) as

$$(\Phi(C, \varepsilon), w) = (f, w), \quad w \in \overset{\circ}{W}{}_{2,T}^1(\Omega), \tag{2.10}$$

which in turn may be treated as an operator equation

$$\Phi(C, \varepsilon) = f \tag{2.11}$$

with the element $f \in L_2(\Omega) \subset W_2^{-1}(\Omega)$ and the operator $\Phi$ mapping $W_2^1(\Omega)$ into $W_2^{-1}(\Omega)$. This equation is an operator formulation of problem (2.5),(2.6). By the hypothesis, the solution of problem (2.5), (2.6) exists and it is a solution of (2.11).

In the event that $\varepsilon = 0$, we get the equation

$$\Phi(C, 0) = f \tag{2.12}$$

linear in $C$ and equivalent to the equality

$$-\left(\left(1 + \frac{\rho K^{(0)}}{\Theta}\right)C, \frac{\partial w}{\partial t}\right) - \left(1 + \frac{\rho K^{(0)}}{\Theta}\right)\int_0^L C_1(x)\, w(x, t_0)\, \mathrm{d}x$$

$$+ \left(D_0 \frac{\partial C}{\partial x}, \frac{\partial w}{\partial x}\right) + \left(v \frac{\partial C}{\partial x}, w\right) = (f, w), \quad w \in \overset{\circ}{W}{}_{2,T}^1(\Omega). \tag{2.13}$$

Henceforward, by $C^{(0)}(x, t)$ we denote the solution of equation (2.12), associated with $\varepsilon = 0$ and $D \equiv D_0$. (Note that, by the hypothesis, $C^{(0)} \in D(\Phi)$ and, hence, $C^{(0)} \geq 0$.)

*Remark* 2.1. Instead of the operator $\Phi$, one can consider its extension mapping $W_2^{1,0}(\Omega)$ into $W_{2,T}^{-1}(\Omega)$ with a domain larger than $D(\Phi)$ and introduce a weak formulation of problem (2.5), (2.6), and of equation (2.11). It is not our intention to deal with a weak formulation, because we are interested here in algorithmic aspects of the problem, and the solution of (2.5), (2.6) is supposed to exist.

## 3. TRANSFORMATION OF THE PROBLEM. PROPERTIES OF THE NON-LINEAR OPERATOR

### 3.1. Transformation of the problem

Subtracting the element $\Phi(C^{(0)}, \varepsilon)$ from the left- and right-hand sides of equation (2.11), we come to the equation

$$F(u, \varepsilon) = f(\varepsilon), \tag{3.1}$$

where $F(u, \varepsilon) = \Phi(C, \varepsilon) - \Phi(C^{(0)}, \varepsilon) = \Phi(C^{(0)} + u, \varepsilon) - \Phi(C^{(0)}, \varepsilon)$, $u = C - C^{(0)}$, $f(\varepsilon) = f - \Phi(C^{(0)}, \varepsilon)$. The operator $F(u, \varepsilon)$ is defined by the equality

$$(F(u, \varepsilon), w) = (\Phi(C^{(0)} + u, \varepsilon) - \Phi(C^{(0)}, \varepsilon), w), \quad w \in \overset{\circ}{W}_{2,T}^{1}(\Omega).$$

It maps $W_2^1(\Omega)$ into $W_2^{-1}(\Omega)$ with the domain $D(F)$ of functions $u \in \overset{\circ}{W}_2^1(\Omega)$ such that $u + C^{(0)} \in D(\Phi)$. The element $f(\varepsilon)$ in (3.1) is defined by the relationship

$$(f(\varepsilon), w) = (f, w) - (\Phi(C^{(0)}, \varepsilon), w),$$

or

$$(f(\varepsilon), w) = (f, w) + \left(\varphi(C^{(0)}, \varepsilon), \frac{\partial w}{\partial t}\right) + \int_0^L \varphi(C_1(x), \varepsilon) \, w(x, t_0) \, dx$$
$$- \left(D\frac{\partial C^{(0)}}{\partial x}, \frac{\partial w}{\partial x}\right) - \left(v\frac{\partial C^{(0)}}{\partial x}, w\right), \quad w \in \overset{\circ}{W}_{2,T}^{1}(\Omega).$$

### 3.2. Properties of the operator $F(u, \varepsilon)$

Let us investigate some properties of the operator $F(u, \varepsilon)$. Note that

$$F(0, \varepsilon) = 0. \tag{3.2}$$

Let us show that $F(u, \varepsilon)$ is Gâteaux-differentiable for a fixed $\varepsilon \in [0, 1]$.

**Lemma 3.1.** *The operator $F(u, \varepsilon)$ has the Gâteaux derivative at any point $u \in D(F)$, defined by the relationship*

$$(F'(u, \varepsilon) h, w) = -\left(\frac{\partial \varphi}{\partial C}(C^{(0)} + u, \varepsilon) h, \frac{\partial w}{\partial t}\right)$$

$$+ \left( D\frac{\partial h}{\partial x}, \frac{\partial w}{\partial x} \right) + \left( v\frac{\partial h}{\partial x}, w \right), \quad w \in W_2^1(\Omega), \quad (3.3)$$

*where*

$$\left( \frac{\partial \varphi}{\partial c}(C^{(0)} + u, \varepsilon) h, \frac{\partial w}{\partial t} \right)$$

$$= \left( \left( 1 + \frac{\rho}{\Theta} \left( \varepsilon \left( \frac{K(C^{(0)} + u)^{1/n-1}}{n} - K^{(0)} \right) + K^{(0)} \right) \right) h, \frac{\partial w}{\partial t} \right),$$

$1/n - 1 \geq 0$.

*Proof.* Consider the first addend of $(F(u, \varepsilon), w)$. We find

$$\frac{1}{\xi} \left[ \left( \varphi(C^{(0)} + u + \xi h, \varepsilon) - \varphi(C^{(0)}, \varepsilon), \frac{\partial w}{\partial t} \right) \right.$$

$$\left. - \left( \varphi(C^{(0)} + u, \varepsilon) - \varphi(C^{(0)}, \varepsilon), \frac{\partial w}{\partial t} \right) \right]$$

$$= \frac{1}{\xi} \left( \int_0^\xi \frac{d}{d\xi'} \varphi(C^{(0)} + u + \xi'h, \varepsilon) \, d\xi', \frac{\partial w}{\partial t} \right)$$

$$= \frac{1}{\xi} \int_0^\xi \left( \frac{\partial \varphi}{\partial C}(C^{(0)} + u + \xi'h, \varepsilon) h, \frac{\partial w}{\partial t} \right) d\xi',$$

$$\lim_{\xi \to 0} \frac{1}{\xi} \int_0^\xi \left( \frac{\partial \varphi}{\partial C}(C^{(0)} + u + \xi'h, \varepsilon) h, \frac{\partial w}{\partial t} \right) d\xi' = \left( \frac{\partial \varphi}{\partial C}(C^{(0)} + u, \varepsilon) h, \frac{\partial w}{\partial t} \right).$$

Notice that, in view of (2.5) and $C^{(0)} + u \geq 0$, we get

$$\left| \left( \frac{\partial \varphi}{\partial C}(C^{(0)} + u, \varepsilon) h, \frac{\partial w}{\partial t} \right) \right| \leq C\|h\|_{W_2^1(\Omega)}\|w\|_{W_2^1(\Omega)} < \infty.$$

Hence, the addend considered is differentiable. Since the other addends in $F(u, \varepsilon)$ define linear bounded operators mapping $W_2^1(\Omega)$ into $W_2^{-1}(\Omega)$, we come to the conclusion that $F(u, \varepsilon)$ is Gâteaux differentiable and (3.3) is valid.

**Lemma 3.2.** *The operator $F'(u, \varepsilon)$: $W_2^1(\Omega) \to W_2^{-1}(\Omega)$ is bounded.*

*Proof.* Since for $u \in D(\Phi)$ the following inequalities

$$|(F'(u, \varepsilon) h, w)| \leq \text{constant } \|h\|_{W_2^1(\Omega)}\|w\|_{W_2^1(\Omega)},$$

$$\|F'(u, \varepsilon) h\|_{W_2^{-1}(\Omega)} \leq \text{constant } \|h\|_{W_2^1(\Omega)},$$

$$\sup_{h \in W_2^1(\Omega), \, h \in D(F)} \frac{\|F'(u,\varepsilon)h\|_{W_2^{-1}(\Omega)}}{\|h\|_{W_2^1(\Omega)}} = \|F'(u,\varepsilon)\|_{W_2^1 \to W_2^{-1}} \leq \text{constant} < \infty$$

are valid, then the operator $F'(u,\varepsilon) \colon W_2^1(\Omega) \to W_2^{-1}(\Omega)$ is bounded for any $u \in D(F)$.

**Lemma 3.3.** *For* $\alpha = 1/n - 1 \in [0,1]$ *the operator* $F'(u,\varepsilon) \colon W_2^1(\Omega) \to W_2^{-1}(\Omega)$ *is continuous in* $u$ *and satisfies the Hölder condition*

$$\|F'(u_1,\varepsilon) - F'(u_2,\varepsilon)\|_{W_2^1(\Omega) \to W_2^{-1}(\Omega)}$$

$$\leq \text{constant} \left\| \frac{\rho \varepsilon K}{\Theta n} \right\|_{C(\Omega)} \||u_1 - u_2|^{1/n-1}\|_{L_4(\Omega)}$$

$$\leq \text{constant} \left\| \frac{\rho \varepsilon K}{\Theta n} \right\|_{C(\Omega)} \|u_1 - u_2\|_{W_2^1(\Omega)}^{1/n-1}. \tag{3.4}$$

*Proof.* Since

$$|((F'(u_1,\varepsilon) - F'(u_2,\varepsilon))h, w)|$$

$$= \left| \left( \frac{\rho}{\Theta} \frac{\varepsilon K}{n} ((C^{(0)} + u_1)^\alpha - (C^{(0)} + u_2)^\alpha) h, \frac{\partial w}{\partial t} \right) \right|,$$

then taking into account the inequality

$$|(C^{(0)} + u_1)^\alpha - (C^{(0)} + u_2)^\alpha| \leq \text{constant} \, |u_1 - u_2|^\alpha$$

and the imbedding[21] of $W_2^1(\Omega)$ into $L_p(\Omega)$ for $p \leq 1$, we obtain

$$|((F'(u_1,\varepsilon) - F'(u_2,\varepsilon))h, w)|$$

$$\leq \text{constant} \left\| \frac{\rho \varepsilon K}{\Theta n} \right\|_{C(\Omega)} \||u_1 - u_2|^{1/n-1} h\|_{L_2(\Omega)} \left\| \frac{\partial w}{\partial t} \right\|_{L_2(\Omega)}$$

$$\leq \text{constant} \left\| \frac{\rho \varepsilon K}{\Theta n} \right\|_{C(\Omega)} \||u_1 - u_2|^{1/n-1}\|_{L_4(\Omega)} \|h\|_{L_4(\Omega)} \left\| \frac{\partial w}{\partial t} \right\|_{L_2(\Omega)}$$

$$\leq \text{constant} \left\| \frac{\rho \varepsilon K}{\Theta n} \right\|_{C(\Omega)} \||u_1 - u_2|^{1/n-1}\|_{L_4(\Omega)} \|h\|_{W_2^1(\Omega)} \|w\|_{W_2^1(\Omega)}.$$

Hence,

$$\|(F'(u_1,\varepsilon) - F'(u_2,\varepsilon))h\|_{W_2^{-1}(\Omega)}$$

$$\leq \text{constant} \left\| \frac{\rho \varepsilon K}{\Theta n} \right\|_{C(\Omega)} \||u_1 - u_2|^{1/n-1}\|_{L_4(\Omega)} \|h\|_{W_2^1(\Omega)},$$

and (3.4) is valid.

Using Lemma 3.3 and the well-known assertions of non-linear analysis (see Chapter 1), we arrive at the following corollaries.

**Corollary 1.** *For $\alpha \in [0, 1]$ the following representation*

$$F(u, \varepsilon) = F(u_0, \varepsilon) + \int_0^1 F'(u_0 + \xi(u - u_0), \varepsilon) \, d\xi(u - u_0) \qquad (3.5)$$

*is valid for any $u, u_0 \in D(F)$.*

**Corollary 2.** *If $\alpha \in [0, 1]$ and $u_0 = 0$, then*

$$F(u, \varepsilon) = \int_0^1 F'(\xi u, \varepsilon) \, d\xi u = A(u, \varepsilon) \, u, \qquad (3.6)$$

*where the operator $A(u, \varepsilon) = \int_0^1 F'(\xi u, \varepsilon) \, d\xi$ is defined by the equality*

$$(A(u, \varepsilon) h, w) = \left( D \frac{\partial h}{\partial x}, \frac{\partial w}{\partial x} \right) + \left( v \frac{\partial h}{\partial x}, w \right)$$

$$- \left( \left( 1 + \frac{\rho}{\Theta} \left( \varepsilon \left( K \frac{(C^{(0)} + u)^{1/n} - C^{(0)1/n}}{u} - K^{(0)} \right) \right. \right. \right.$$

$$\left. \left. \left. + K^{(0)} \right) \right) h, \frac{\partial w}{\partial t} \right) \qquad (3.7)$$

*for any $h \in D(F) \subset \overset{\circ}{W}{}^1_{2, t_0}(\Omega)$, $w \in \overset{\circ}{W}{}^1_{2, T}(\Omega)$.*

Using Corollary 2, equation (3.1) may be written in the form

$$A(u, \varepsilon) \, u = f(\varepsilon), \qquad (3.8)$$

where the operator $A(u, \varepsilon)$ maps $W_2^1(\Omega)$ into $W_2^{-1}(\Omega)$ with the domain $D(F)$ and is defined by (3.7). On the strength of the above-stated assumptions on the solution existence of the original problem, equation (3.8) has also a solution.

## 4. PERTURBATION ALGORITHM. ADJOINT EQUATION

### 4.1. Adjoint equation and computing the functional

Consider equation (2.11). It is required to find the solution of this equation and to compute the functional

$$J(C) = (C, p), \qquad (4.1)$$

where $p(x,t)$ is a prescribed function of $L_2(\Omega)$. If we assume that the solution $C^{(0)}$ of equation (2.13) is known at $D = D_0$, then the problem reduces to finding the solution of equation (3.7) and computing the correction:

$$\delta J = (u, p), \tag{4.2}$$

where $u = C - C^{(0)}$. Then we get

$$J(C) = J(C^{(0)}) + \delta J = (C^{(0)}, p) + (u, p). \tag{4.3}$$

The correction $\delta J$ may be computed approximately using the solution of the adjoint equation. Let us introduce this equation. To do this, for fixed $u \in D(F)$ consider the operator $A(u, \varepsilon)$ as an operator mapping $W_2^1(\Omega)$ into $W_2^{-1}(\Omega)$ and introduce the adjoint operator

$$A^*(u, \varepsilon): \ (W_2^{-1})^* \equiv W_2^1 \to (W_2^1)^* \equiv W_2^{-1} \tag{4.4}$$

defined by the equalities

$$(A(u, \varepsilon) h, w) = (h, A^*(u, \varepsilon) w),$$

$$(h, A^*(u, \varepsilon) w) = \left(D \frac{\partial h}{\partial x}, \frac{\partial w}{\partial x}\right) + \left(v \frac{\partial h}{\partial x}, w\right)$$

$$- \left(h, \left(1 + \frac{\rho}{\Theta}\left(\varepsilon\left(K \frac{(C^{(0)} + u)^{1/n} - C^{(0)1/n}}{u}\right.\right.\right.\right.$$

$$\left.\left.\left.\left. - K_{(0)}\right) + K^{(0)}\right)\right) \frac{\partial w}{\partial t}\right), \tag{4.5}$$

where $h \in D(F) \subset \overset{o}{W^1_{2,t_0}}(\Omega)$, $w \in \overset{o}{W^1_{2,T}}(\Omega)$, $u \in D(F)$. In the event that $w$, $D$, $v$ are sufficiently regular (for example, $w \in W_2^2(\Omega) \cap \overset{o}{W^1_{2,T}}(\Omega)$, $D, v \in C^{1,0}(\Omega)$), the expression $A^*(u, \varepsilon) w$ may be defined by

$$A^*(u, \varepsilon) w = -\left(1 + \frac{\rho}{\Theta}\left(\varepsilon\left(K \frac{(C^{(0)} + u)^{1/n} - C^{(0)1/n}}{u} - K_{(0)}\right) + K^{(0)}\right)\right)$$

$$\times \frac{\partial w}{\partial t} - \frac{\partial}{\partial x} D \frac{\partial w}{\partial x} - \frac{\partial(vw)}{\partial x}. \tag{4.6}$$

Thus, the domain $D(A^*)$ of the adjoint operator $A^*$ contains the set $W_2^2(\Omega) \cap \overset{o}{W^1_{2,T}}(\Omega)$.

Having defined the adjoint operator, consider the equation

$$A^*(u, \varepsilon) u^* = p, \tag{4.7}$$

where $u^* \in \overset{o}{W}{}^1_{2,T}(\Omega)$. A weak formulation of (4.7) may be written as

$$(h, A^*(u, \varepsilon) u^*) = (h, p) \quad \forall h \in \overset{o}{W}{}^1_{2,t_0}(\Omega), \tag{4.8}$$

where the left-hand side is defined by (4.5). Note, however, that (4.8) admits functions $h$ to belong to $W^{1,0}_2(\Omega)$. Therefore, instead of (4.8), we can consider the more general equality

$$(h, A^*(u, \varepsilon) u^*) = (h, p) \quad \forall h \in \overset{o}{W}{}^1_2(\Omega), \tag{4.9}$$

which may be used for the definition of a weak solution to equation (4.7).

**Definition.** *Let $u \in D(F)$ and $\varepsilon$ be given. A function $u^* \in \overset{o}{W}{}^1_{2,T}(\Omega)$ is said to be a weak solution of equation (4.7) if (4.9) is satisfied.*

Using the technique applied to linear parabolic equations[111], it can be shown that for $p \in L_2(\Omega)$ equation (4.7) has a unique weak solution $u^* \in \overset{o}{W}{}^1_{2,t}(\Omega)$ and

$$\|u^*\|_{W^1_2(\Omega)} \le c \|p\|_{L_2(\Omega)}, \tag{4.10}$$

where the constant $c$ is independent of $u$.

Using (4.9) and the equality

$$(A(u, \varepsilon) u, w) = (f(\varepsilon), w) \quad \forall\, w \in \overset{o}{W}{}^1_{2,T}(\Omega) \tag{4.11}$$

satisfied by the solution of equation (4.8), we come to the adjointness relationship:

$$(f(\varepsilon), u^*) = (u, p). \tag{4.12}$$

Hence, we get for the original functional

$$J(C) = (C^{(0)}, p) + (f(\varepsilon), u^*). \tag{4.13}$$

Thus, to compute the value of $J(C)$ we need to solve equation (4.7) and find $\delta J = (f(\varepsilon), u^*)$. Note, however, that $u^* = u^*(u)$, where $u$ is an unknown function. Therefore, in what follows a demand will arise for some approximations of the solution $u^*$. Some of these approximations are considered in the next subsections.

### 4.2. The simplest formula for computing the functional

Let $u^*_0$ be a solution of the equation

$$A^*_0(0, \varepsilon) u^*_0 = p, \tag{4.14}$$

where $A_0^*$ is the operator $A^*$ at $D \equiv D_0$, and the operator $A^*(0, \varepsilon)$ is defined by the equality

$$(h, A^*(0, \varepsilon) w) = -\left(\left(1 + \frac{\rho}{\Theta}\left(\varepsilon\left(K\frac{(C^{(0)})^{1/n-1}}{n} - K^{(0)}\right) + K^{(0)}\right)\right)\frac{\partial w}{\partial t}, h\right)$$

$$+ \left(D\frac{\partial w}{\partial x}, \frac{\partial h}{\partial x}\right) + \left(w, v\frac{\partial h}{\partial x}\right), \quad w \in D(A^*), \quad h \in \overset{\circ}{W_2^{1,0}}(\Omega).$$

In view of the remark on solvability of equation (4.7) it is valid to say that $u_0^* \in W_{2,T}^1(\Omega)$ if $p \in L_2(\Omega)$, and

$$\|u_0^*\|_{W_2^1(\Omega)} \le c\|p\|_{L_2(\Omega)}. \tag{4.15}$$

Let us estimate the difference $u^* - u_0^*$. From (4.7) and (4.14) we get

$$A_0^*(0, \varepsilon)(u^* - u_0^*) = (A_0^*(0, \varepsilon) - A^*(0, \varepsilon))u^* + (A^*(0, \varepsilon) - A^*(u, \varepsilon))u^*.$$

Hence, for $h \in \overset{\circ}{W_2^{1,0}}(\Omega)$ the equality

$$(A_0^*(0, \varepsilon)(u^* - u_0^*), h) = \left((D_0 - D)\frac{\partial u^*}{\partial x}, \frac{\partial h}{\partial x}\right)$$

$$+ \left(\frac{\rho\varepsilon}{\Theta}K\left(\frac{(C^{(0)} + u)^{1/n} - (C^{(0)})^{1/n}}{u}\right.\right.$$

$$\left.\left. - \frac{(C^{(0)})^{1/n-1}}{n}\right)\frac{\partial u^*}{\partial t}, h\right) \tag{4.16}$$

holds. If $D \in C^{1,0}(\bar{\Omega})$, from (4.10) and the fact that $u^* \in W_2^1(\Omega)$ we find

$$\left\|\frac{\partial^2 u^*}{\partial x^2}\right\|_{L_2(\Omega)} \le \text{constant } \|p\|_{L_2(\Omega)}.$$

Thus, for $D_0, D \in C^{1,0}(\bar{\Omega})$ (4.16) defines an equation of the form (4.14) with the solution $u^* - u_0^*$ and the right-hand side

$$\tilde{p} = -\frac{\partial}{\partial x}(D_0 - D)\frac{\partial u^*}{\partial x}$$

$$+ \frac{\rho\varepsilon K}{\Theta}\left(\frac{(C^{(0)} + u)^{1/n} - (C^{(0)})^{1/n}}{u} - \frac{(C^{(0)})^{1/n}}{n}\right)\frac{\partial u^*}{\partial t}, \quad \tilde{p} \in L_2(\Omega).$$

Hence, this equation has the solution $(u^* - u_0^*) \in \overset{\circ}{W_{2,T}^1}(\Omega)$, and

$$\|u^* - u_0^*\|_{W_2^1(\Omega)} \le C\|\tilde{p}\|_{L_2(\Omega)}$$

$$\leq C\|\delta D\|_{C^{1,0}(\Omega)}\|p\|_{L_2(\Omega)}$$

$$+ \left\|\frac{\rho\varepsilon K}{\Theta}\left(\frac{(C^{(0)}+u)^{1/n}-(C^{(0)})^{1/n}}{u}\right.\right.$$

$$\left.\left.-\frac{(C^{(0)})^{1/n}}{n}\right)\right\|_{C(\Omega)}\|p\|_{L_2(\Omega)}.$$

Notice that

$$\left|\frac{(C^{(0)}+u)^{1/n}-(C^{(0)})^{1/n}}{u}-\frac{(C^{(0)})^{1/n}}{n}\right|$$

$$=\left|\frac{1}{u}\int_0^u \frac{\partial}{\partial\xi}(C^{(0)}+\xi)^{1/n}\,d\xi - \frac{C^{(0)1/n}}{n}\right|$$

$$=\left|\frac{1}{u}\int_0^u \frac{1}{n}((C^{(0)}+\xi)^{1/n-1}-(C^{(0)})^{1/n})\,d\xi\right|$$

$$\leq \frac{1}{|u|}\int_0^{|u|} \frac{1}{n}|(C^{(0)}+(\,\mathrm{sign}\,u)\xi)^{1/n-1}-(C^{(0)})^{1/n-1}|\,d\xi$$

$$\leq \frac{1}{|u|}\int_0^{|u|} \frac{1}{n}\xi^{1/n-1}\,d\xi = |u|^{1/n-1}.$$

We have invoked here the inequality[225]

$$|x^\alpha - y^\alpha| \leq |x-y|^\alpha, \quad \alpha\in[0,1], \quad \forall\, x,y\geq 0.$$

Hence,

$$\|u^* - u_0^*\|_{W_2^1(\Omega)} \leq c\left(\|\delta D\|_{C^{1,0}(\Omega)} + \left\|\frac{\rho\varepsilon K|u|^{1/n-1}}{\Theta}\right\|_{C(\Omega)}\right)\|p\|_{L_2(\Omega)}. \quad (4.17)$$

Let us formulate the obtained results as a lemma.

**Lemma 4.1.** *Let $D_0, D \in C^{1,0}(\bar\Omega)$, $p \in L_2(\Omega)$. Then the difference norm $\|u^* - u_0^*\|_{W_2^1(\Omega)}$ is small if the values $\|\delta D\|_{C^{1,0}(\Omega)}$, $\left\|\frac{\rho\varepsilon K|C-C^{(0)}|^{1/n-1}}{\Theta}\right\|_{C(\Omega)}$ are small. The following estimate holds:*

$$\|u^* - u_0^*\|_{W_2^1(\Omega)} \leq \mathrm{constant}\left(\|\delta D\|_{C^{1,0}(\Omega)}\right.$$

$$\left.+ \left\|\frac{\rho\varepsilon K|C-C^{(0)}|^{1/n-1}}{\Theta}\right\|_{C(\Omega)}\right)\|p\|_{L_2(\Omega)}. \quad (4.18)$$

**Corollary.** *If the hypotheses of Lemma 4.1 are satisfied, one can put approximately*

$$u^* \cong u_0^*, \quad \delta J \cong \delta J_0 = (f(\varepsilon), u_0^*),$$

$$J(C) \cong J(C^{(0)}) + \delta J_0 = (C^{(0)}, p) + (f(\varepsilon), u_0^*). \tag{4.19}$$

## 4.3. Perturbation algorithm

From the above, it appears that under the hypotheses of Lemma 4.1 the functions $u_0^*$ and $u^*$ are close if $\varepsilon$ and $\delta D$ are small. Because of this, we will try to find $u^*$ with the use of the perturbation algorithm. To do this, substitute $\mu \delta D$ for $\delta D$ and $\mu \varepsilon$ for $\varepsilon$, with the assumption that $\mu \in [0, 1]$. Note that we have formally inserted the small parameter $\mu$ in the quantities of $A^*(u, \varepsilon)$ which are also small. Therefore, expansions in the small parameter $\mu$ may be considered to be justified. As a result, we get the operator $A^*(u, \varepsilon, \mu)$, where $A^*(u, \varepsilon, 1) \equiv A^*(u, \varepsilon)$. Consider the equation

$$A^*(u, \varepsilon, \mu) U^*(\mu) = p. \tag{4.20}$$

Suppose that

$$U^*(\mu) = U_0^* + \mu U_1^* + \mu^2 U_2^* + \dots \tag{4.21}$$

Substituting (4.21) into (4.20) and using the reasoning standard in the regular perturbation theory, we obtain the equations for finding the functions $U_i^* \in \overset{\circ}{W}_{2,T}^1(\Omega)$. In particular, equations for $U_0^*$ and $U_1^*$ have the form (in classical formulation):

$$-\left(1 + \frac{\rho K^{(0)}}{\Theta}\right) \frac{\partial U_0^*}{\partial t} - \frac{\partial}{\partial x} D_0 \frac{\partial U_0^*}{\partial x} - \frac{\partial(v U_0^*)}{\partial x} = p,$$

$$-\left(1 + \frac{\rho K^{(0)}}{\Theta}\right) \frac{\partial U_1^*}{\partial t} - \frac{\partial}{\partial x} D_0 \frac{\partial U_1^*}{\partial x} - \frac{\partial(v U_1^*)}{\partial x}$$
$$= \frac{\rho \varepsilon}{\Theta} \left( K \frac{(C^{(0)} + u)^{1/n} - C^{(0)1/n}}{u} - K^{(0)} \right) \frac{\partial U_0^*}{\partial t} + \frac{\partial}{\partial x} \delta D \frac{\partial U_0^*}{\partial x}.$$

In its turn, $U_1^*$ is close to the solution of the equation

$$-\left(1 + \frac{\rho K^{(0)}}{\Theta}\right) \frac{\partial \tilde{U}_1^*}{\partial t} - \frac{\partial}{\partial x} D_0 \frac{\partial \tilde{U}_1^*}{\partial x} - \frac{\partial(v \tilde{U}_1^*)}{\partial x}$$

$$= \frac{\rho \varepsilon}{\Theta} \left( K \frac{C^{(0)1/n-1}}{n} - K^{(0)} \right) \frac{\partial U_0^*}{\partial t} + \frac{\partial}{\partial x} \delta D \frac{\partial U_0^*}{\partial x}$$

if $u$ is small. Thus, under the hypotheses of Lemma 4.1 one can put (setting $\mu = 1$)

$$u^* \cong U_0^* + \tilde{U}_1^* \equiv U^*, \tag{4.22}$$

where $U^*, U_0^* \in \overset{\circ}{W}_{2,T}^1(\Omega)$ are the solution to the equations

$$-\left(1 + \frac{\rho K^{(0)}}{\Theta}\right)\frac{\partial U^*}{\partial t} - \frac{\partial}{\partial x}D_0\frac{\partial U^*}{\partial x} - \frac{\partial(vU^*)}{\partial x}$$

$$= p + \frac{\rho\varepsilon}{\Theta}\left(K\frac{C^{(0)1/n-1}}{n} - K^{(0)}\right)\frac{\partial U_0^*}{\partial t} + \frac{\partial}{\partial x}\delta D\frac{\partial U_0^*}{\partial x}, \qquad (4.23)$$

$$-\left(1 + \frac{\rho K^{(0)}}{\Theta}\right)\frac{\partial U_0^*}{\partial t} - \frac{\partial}{\partial x}D_0\frac{\partial U_0^*}{\partial x} - \frac{\partial(vU_0^*)}{\partial x} = p. \qquad (4.24)$$

From (4.22) we get the following relationship

$$J(C) \cong \tilde{J}(C^{(0)}, U^*) \equiv (C^{(0)}, p) + (f(\varepsilon), U^*) \qquad (4.25)$$

which represents a perturbation algorithm formula for computing $J(C)$ with an accuracy of the second order. It is precisely this formula that will be used below for finding an effective dispersion coefficient.

## 5. APPROXIMATE SOLUTION OF THE PROBLEM ON FINDING AN EFFECTIVE DISPERSION COEFFICIENT

Suppose that the value of the functional $J(C)$ of the solution to problem (2.5)–(2.8) is known from experimental data:

$$J(C) = (C, p^h) \cong C_H(t_i).$$

Here, the function $p_h(x, t)$ is

$$p^h(x, t) = \frac{1}{h^2}\begin{cases} 1, & \text{if } H - h \le x \le H, \quad t_i - h/2 \le t \le t_i + h/2, \\ 0, & \text{otherwise.} \end{cases} \qquad (5.1)$$

It is precisely this function that will be considered as a right-hand side of equations (4.7) and (4.14). Since

$$C_H(t_i) \cong (C^{(0)}, p^h) + (f(\varepsilon), U^*),$$

then taking into account the form of $f(\varepsilon)$ given in Subsection 4.1, we get

$$C_H(t_i) - (C^{(0)}, p^h) \cong (f, U^*) + \left(\varphi(C^{(0)}, \varepsilon), \frac{\partial U^*}{\partial t}\right)$$

$$- \left((D_0 + \delta D)\frac{\partial C^{(0)}}{\partial x}, \frac{\partial U^*}{\partial x}\right) - \left(v\frac{\partial C^{(0)}}{\partial x}, U^*\right)$$

$$+ \int_0^L \varphi(C_1(x), \varepsilon)\, U^*(x, t_0)\, dx. \qquad (5.2)$$

One can use this formula to compute $\delta D$ approximately. We assume that

$$f \equiv 0, \quad \delta D = \text{constant.} \tag{5.3}$$

Then (5.2) takes the form

$$\delta D \left( \frac{\partial C^{(0)}}{\partial x}, \frac{\partial U^*}{\partial x} \right) \cong (C^{(0)}, p^h) - C_H(t_i) + \left( \varphi(C^{(0)}, \varepsilon), \frac{\partial U^*}{\partial t} \right)$$

$$- \left( D_0 \frac{\partial C^{(0)}}{\partial x}, \frac{\partial U^*}{\partial x} \right) - \left( v \frac{\partial C^{(0)}}{\partial x}, U^* \right)$$

$$+ \int_0^L \varphi(C_1(x), \varepsilon) U^*(x, t_0) \, dx, \tag{5.4}$$

where

$$\varphi(C^{(0)}, \varepsilon) = C^{(0)} + \frac{\rho}{\Theta} (\varepsilon(K C^{(0) \, 1/n} - K_{(0)} C^{(0)}) + K^{(0)} C^{(0)}). \tag{5.5}$$

Since $C^{(0)}$ satisfies the equality

$$\left( \left( 1 + \frac{\rho K^{(0)}}{\Theta} \right) C^{(0)}, \frac{\partial w}{\partial t} \right) - \left( D_0 \frac{\partial C^{(0)}}{\partial x}, \frac{\partial w}{\partial x} \right) - \left( v \frac{\partial C^{(0)}}{\partial x}, w \right)$$

$$- \left( 1 + \frac{\rho K^{(0)}}{\Theta} \right) \int_0^L C^{(0)}(x, t_0) w(x, t_0) \, dx \quad \forall \, w \in \overset{\circ}{W}^1_{2,T}(\Omega),$$

(5.4) reads

$$\delta D \left( \frac{\partial C^{(0)}}{\partial x}, \frac{\partial U^*}{\partial x} \right) \cong (C^{(0)}, p^h) - C_H(t_i)$$

$$+ \left( \frac{\rho}{\Theta} \varepsilon(K C^{(0)1/n} - K^{(0)} C^{(0)}), \frac{\partial U^*}{\partial t} \right)$$

$$+ \frac{\rho}{\Theta} \varepsilon \int_0^L (K C_1^{1/n} - K^{(0)} C_1) U^*(x, t_0) \, dx. \tag{5.6}$$

Substituting $U^*$ with the approximating function $U_0^*$ from (4.24), we obtain the formula suitable for practical implementation:

$$\delta D \cong \left\{ (C^{(0)}, p^h) - C_H(t_i) + \left( \frac{\rho}{\Theta} \varepsilon(K C^{(0)1/n} - K^{(0)} C^{(0)}), \frac{\partial U_0^*}{\partial t} \right) \right.$$

$$\left. + \frac{\rho}{\Theta} \varepsilon \int_0^L (K C_1^{1/n} - K^{(0)} C_1) U_0^*(x, t_0) \, dx \right\} \left( \frac{\partial C^{(0)}}{\partial x}, \frac{\partial U_0^*}{\partial x} \right)^{-1}. \tag{5.7}$$

Using the relation

$$\left(\frac{\partial C^{(0)}}{\partial x}, \frac{\partial U_0^*}{\partial x}\right) = \left[\left(1 + \frac{K^{(0)}\rho}{\Theta}\right)\left(\frac{\partial C^{(0)}}{\partial t}, U_0^*\right) + v\left(\frac{\partial C^{(0)}}{\partial x}, U_0^*\right)\right] D_0^{-1},$$

and integrating by parts, represent (5.7) in the form more convenient for computations:

$$\delta D \cong \left\{ D_0 \left[\frac{\rho}{\Theta}\left(\left(\frac{KC^{(0)1/n-1}}{n} - K^{(0)}\right)\frac{\partial C^{(0)}}{\partial t}, U_0^*\right) + C_H(t_i) - (C^{(0)}, p)\right]\right.$$

$$\left. + \frac{\rho(1+\varepsilon)}{\Theta}\int_0^L (KC_1^{1/n}(x) - K^{(0)}C_1(x)) U_0^*(x, t_0)\, dx\right\}$$

$$\times \left\{\left(1 + \frac{\rho K^{(0)}}{\Theta}\right)\left(\frac{\partial C^{(0)}}{\partial t}, U_0^*\right) + v\left(\frac{\partial C^{(0)}}{\partial x}, U_0^*\right)\right\}^{-1}. \tag{5.8}$$

Taking into account the solution of equation (4.23), from (5.6) we obtain a quadratic equation for $\delta D$:

$$A(\delta D)^2 + B\delta D + C = 0,$$

where

$$A = \left(\frac{\partial C^{(0)}}{\partial x}, \frac{\partial \bar{U}_2^*}{\partial x}\right),$$

$$B = \left(\frac{\partial C^{(0)}}{\partial x}, \frac{\partial \bar{U}_0^*}{\partial x}\right) + \left(\frac{\partial C^{(0)}}{\partial x}, \frac{\partial \bar{U}_1^*}{\partial x}\right)$$

$$- \left(\frac{\rho\varepsilon}{\Theta}(KC^{(0)1/n} - K^{(0)}C^{(0)}), \frac{\partial \bar{U}_2^*}{\partial t}\right)$$

$$- \frac{\rho\varepsilon}{\Theta}\int_0^L (KC_1^{1/n}(x) - K^{(0)}C_1(x))\bar{U}_2^*(x, t_0)\, dx, \tag{5.9}$$

$$C = C_H(t_i) - (C^{(0)}, p^h) - \left(\frac{\rho\varepsilon}{\Theta}(KC^{(0)1/n} - K^{(0)}C^{(0)}), \frac{\partial(U_0^* + \bar{U}_1^*)}{\partial t}\right)$$

$$- \frac{\rho\varepsilon}{\Theta}\int_0^L (KC_1^{1/n}(x) - K^{(0)}C_1(x))(U_0^*(x, t_0) + \bar{U}_1^*(x, t_0))\, dx,$$

and functions $\bar{U}_1^*$ and $\bar{U}_2^*$ are the solutions to the equations

$$- \left(1 + \frac{\rho K^{(0)}}{\Theta}\right)\frac{\partial \bar{U}_1^*}{\partial t} - \frac{\partial}{\partial x} D_0 \frac{\partial \bar{U}_1^*}{\partial x} - \frac{\partial(v\bar{U}_1^*)}{\partial x}$$

$$= \left( \frac{\rho \varepsilon}{\Theta} \frac{K C^{(0)1/n-1}}{n} - K^{(0)} \right) \frac{\partial U_0^*}{\partial t}, \tag{5.10}$$

$$- \left( 1 + \frac{\rho K^{(0)}}{\Theta} \right) \frac{\partial \bar{U}_2^*}{\partial t} - \frac{\partial}{\partial x} D_0 \frac{\partial \bar{U}_2^*}{\partial x} - \frac{\partial (v \bar{U}_2^*)}{\partial x} = \frac{\partial^2 U_0^*}{\partial x^2},$$

respectively.

Computing $\delta D$ by formula (5.8) or from equation (5.9) and setting $D \cong D_0 + \delta D$, one can solve problem (2.5)–(2.8) (for example, by the finite-difference method). It is hoped that the obtained solution $C(x, t)$ describes the actual physical process better than the solution of linear system (2.1)–(2.2) with the coefficient $D_0$.

## 6. AN ALGORITHM FOR SOLVING THE PROBLEM

Let us formulate an algorithm for solving the problem on finding an effective dispersion coefficient for the model governing the transport of chemicals (herbicides) in a soil column with the non-linear isotherm. This algorithm involves five steps.

1. Find the solution $C^{(0)}(x, t)$ of linear problem ($\varepsilon = 0$) for a given dispersion coefficient $D_0$ in (2.5), (2.6) and $f \equiv 0$:

$$\left( 1 + \frac{K^{(0)} \rho}{\Theta} \right) \frac{\partial C^{(0)}}{\partial t} - \frac{\partial}{\partial x} D_0 \frac{\partial C^{(0)}}{\partial x} + v \frac{\partial C^{(0)}}{\partial x} = 0,$$
$$C^{(0)}(0, t) = C_0(t), \quad C^{(0)}(L, t) = 0, \quad C^{(0)}(x, t_0) = C_1(x), \tag{6.1}$$

where $L \gg H$, and $H$ is the column length.

2. Find

(a) the function $U_0^*$ as a solution to the problem

$$- \left( 1 + \frac{K^{(0)} \rho}{\Theta} \right) \frac{\partial U_0^*}{\partial t} - \frac{\partial}{\partial x} D_0 \frac{\partial U_0^*}{\partial x} - \frac{\partial (v U_0^*)}{\partial x} = p^h(x, t),$$
$$U_0^*(0, t) = U_0^*(L, t) = 0, \quad U_0^*(x, T) = 0, \tag{6.2}$$

where $p^h(x, t)$ is defined by (5.1); or

(b) the functions $\bar{U}_1^*$ and $\bar{U}_2^*$ which are the solution of equations (5.10).

3. Using the parameters $K$ and $n$, compute the correction $\delta D$ either

(a) by formula (5.8) or

(b) from equation (5.9).

4. Find an effective dispersion coefficient for the non-linear problem by

$$\tilde{D} = D_0 + \delta D.$$

5. Find the solution of non-linear problem (2.5), (2.6) for $\varepsilon = 1$:

$$\left(1 + \frac{K\rho}{\Theta n} C^{(1-n)/n}\right) \frac{\partial C}{\partial t} - \frac{\partial}{\partial x} \tilde{D} \frac{\partial C}{\partial x} + v \frac{\partial C}{\partial x} = 0,$$

$$C(0,t) = C_0(t), \qquad t_0 \leq t \leq T,$$

$$C(L,t) = 0, \qquad t_0 \leq t \leq T, \tag{6.3}$$

$$C(x,t_0) = C_1(x), \quad 0 < x \leq L.$$

For the purposes of illustration of the above algorithm we shall present the results of numerical experiments. As factual data we have used the results of the experiments on studying the migration of chlorosulphuron in soil columns.[14] We have taken as our example two types of soil with adsorption isotherms most distinct from the linear one given by

$$S = K^{(0)} C_p, \tag{6.4}$$

where $K^{(0)}$ is the distribution coefficient, $S$ the quantity of solute adsorbed per unit mass of soil, and $C_p$ the equilibrium concentration of solute in the soil water solution.

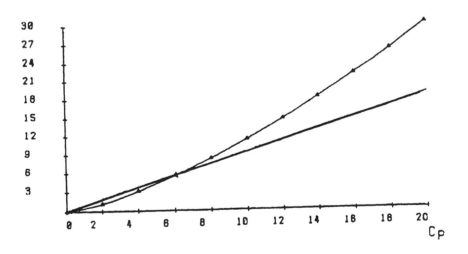

**Figure 1.** The adsorption isotherms. Chlorosulphuron, chernozem.
–◄–◄– the Freundlich isotherm with the parameters $K = 0.44$, $1/n = 1.41$;
——— the linear isotherm (6.4) with $K^{(0)} = 0.93$.

The isotherms (linear and non-linear one) used in numerical experiments are shown at Figures 1 and 2. Figures 3 and 4 show the solutions of model

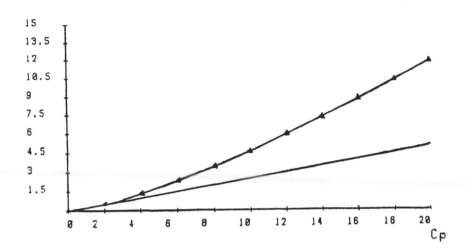

**Figure 2.** The adsorption isotherms. Chlorosulphuron, podzol.
−◄−◄−the Freundlich adsorption isotherm (2.3) with the parameters $K = 0.2$, $1/n = 1.36$;
———— the linear isotherm (6.4) with $K^{(0)} = 0.25$.

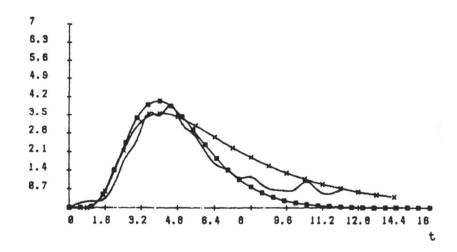

**Figure 3.** Solutions of linear and non-linear models. Chlorosulphuron, chernozem.
———— the experimental curve;
−×−×− the solution of linear model (6.1) with $K^{(0)} = 0.93$, $v = 1.63$, $\rho = 1.05$, $\Theta = 0.5$,
$D_0 = 1.37$;
−■−■− the solution of non-linear model (6.3) with $K = 0.44$, $1/n = 1.41$, $v = 1.63$,
$t_0 = 2.3$, $\bar{D} = D_0 + \delta D = 0.82$, $\rho = 1.05$, $\Theta = 0.5$.

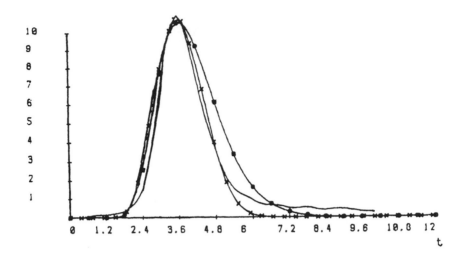

**Figure 4.** Solutions of linear and non-linear models. Chlorosulphuron, podzol.
——— the experimental curve;
$-\blacksquare-\blacksquare-$ the solution of linear model (6.1) with $K^{(0)} = 0.35$, $v = 1.48$, $\rho = 1.22$, $\Theta = 0.51$, $D_0 = 0.14$;
$-\times-\times-$ the solution of non-linear model (6.3) with $K = 0.2$, $1/n = 1.36$, $v = 1.48$, $t_0 = 2.09$, $\bar{D} = D_0 + \delta D = 0.14$, $\rho = 1.05$, $\Theta = 0.51$.

(6.3) taking into account non-linear adsorption isotherm, and of model (6.1) based on linear isotherm.

The numerical experiments proved the nonlinear model to give better results then the linear one in the case when the adsorption isotherms differ considerably from the linear isotherm.

# Chapter 9

# Applications of adjoint equations in science and technology

Adjoint equations are increasingly widespread throughout mathematics and its applications to problems of diffusion, environment protection models, theory of climate and its changes, mathematical problems of processing the information provided by satellites, mathematical models in immunology, and others. A rational approach to the solution of inverse problems and to mathematical experiment planning was evolving parallel to development of adjoint equations techniques.

In this chapter we consider applications of adjoint equations to some problems of practical interest.

## 1. ADJOINT EQUATIONS IN DATA ASSIMILATION PROBLEMS

The rational use of satellite-borne observational data for the needs of agriculture, meteorology, environment protection, etc. becomes ever more important nowadays. Mathematically this problem may be considered as a data assimilation problem, which is one of the optimal control problems. These problems attract attention of many specialists who deal with applying adjoint equations and optimal control methods to the solution of practical problems (see Marchuk[144,146,149,155], Penenko-Obraztsov[235], Marchuk-Penenko[195], Kontarev[102], Lewis-Derber[125], LeDimet[117], Navon[221], Lions[129-134], Courtier-Talagrand[48], Lorenc[139], Agoshkov[5], Agoshkov-Marchuk[13], Ipatova[89], Marchuk-Zalesny[204], Marchuk-Shutyaev[203], Zalesny-Galkin[317], and others). Simultaneously, there have been noted considerable difficulties connected with numerical solving of measurement data assimilation problems (necessity of using high power computers, instability of numerical realization and so on). In this connection it is interesting to study theoretical aspects connected with such problems in order to reveal difficulties mentioned above. In this section using adjoint equation approaches we study a class of data assimilation problems for which the equation of state has the form

$$\frac{d\varphi}{dt} + A(t)\varphi + \varepsilon f(\varphi) = f_0(t), \quad t \in (0, T),$$

where $\varepsilon \in [-\varepsilon_0, \varepsilon_0]$ is a small parameter in front of the nonlinear operator $f$. We establish the solvability conditions for problems and make some
assumptions concerning the causes of some difficulties associated with their
numerical solution. In particular, we will show that solving such problems is,
in a sense, equivalent to the study and numerical solution of some ill-posed
problems. This fact accounts for the difficulties associated with their solution
and at the same time it necessitates the analyticity of the solutions with respect to $\varepsilon$ and demonstrates that the use of the perturbation algorithm is
a possible technique of solving the problems. We also consider some iterative
methods.

## 1.1. Notation and basic assumptions

Consider the following real separable Hilbert spaces: $H, X \subset H$; $H^*, X^*$
are adjoints of $H, X$; $L_2(0, T; H)$, $Y \equiv L_2(0, T; X)$, $L_2(0, T; X^*)$ are spaces
of abstract functions $f(t)$ ($t \in [0, T]$, $T < \infty$) with values in $H, X, X^*$,
respectively;

$$W \equiv W(0, T) = \{f \in L_2(0, T; X) : df/dt \in L_2(0, T; X^*),$$

$$\|f\|_W = (\|df/dt\|^2_{L_2(0,T;X^*)} + \|f\|^2_{L_2(0,T;X)})^{1/2}\}.$$

By $C^{(0)}([0, T]; H)$ we denote the Banach space of functions $f(t)$, $t \in [0, T]$,
endowed with the norm:

$$\|f\|_{C^{(0)}([0,T];H)} = \max_{t \in [0,T]} \|f\|_H.$$

In the sequel, we'll suppose that

$$H \equiv H^*, \quad X^* \equiv X^{-1},$$

$$L_2(0, T; H) \equiv L_2^*(0, T; H) \equiv L_2(0, T; H^*), \tag{1.1}$$

$$(\cdot, \cdot)_{L_2(0,T;H)} \equiv (\cdot, \cdot),$$

$$W \subseteq L_2(0, T; X^*).$$

In view of (1.1) we also have

$$W \subset Y \subset L_2(0, T; H) \subset Y^* \subset W^*.$$

**Lemma 1.1**[135]. *If* $f(t) \in W$ *then* $f(t) \in C^{(0)}([0, T]; H)$ *and*

$$\max_{t \in [0,T]} \|f(t)\|_H \leq C\|f\|_W, \quad c = const, \tag{1.2}$$

*that is,* $W \subset C^{(0)}([0, T]; H)$.

Let $a(t; \varphi, \psi)$ be a bilinear form defined for $t \in [0, T]$, $\forall \varphi, \psi \in X$ and
satisfying the following constraints:

$$|a(t; \varphi, \psi)| \leq C\|\varphi\|_X \|\psi\|_X, \quad C = const, \tag{1.3}$$

$$\tilde{C}\|\varphi\|_X^2 \le a(t; \varphi, \varphi), \tilde{C} = \text{const} > 0, \tag{1.4}$$

$$\forall t \in [0, T], \quad \forall \varphi, \psi \in X.$$

By $A(t) \in \mathcal{L}(Y; Y^*)$ we denote the operator generated by this form:

$$(A(t)\varphi, \psi)_H = a(t; \varphi, \psi) \quad \forall \varphi, \psi \in X.$$

Let $f$ be an analytical operator on $W$ into $Y^*$, i.e. $f(\varphi) \in Y^* = L_2(0, T, X^*)$ $\forall \varphi \in W$.

Consider the evolution problem of the form

$$\begin{cases} \frac{d\varphi}{dt} + A(t)\varphi + \varepsilon f(\varphi) = f_0(t), & t \in (0, T), \\ \varphi(0) = V, \end{cases} \tag{1.5}$$

where $\varphi(0) \equiv \varphi|_{t=0}$, $\varepsilon \in [-\varepsilon_0, \varepsilon_0]$ is a small parameter, $\varepsilon_0 \in \mathbf{R}^1$, $f_0 \in Y^*$, $V \in H$. For $\varepsilon = 0$ we have the problem

$$\begin{cases} \frac{d\varphi^0}{dt} + A(t)\varphi^0 = f_0, & t \in (0, T), \\ \varphi^0(0) = V. \end{cases} \tag{1.6}$$

For this problem the following assertion holds.

**Theorem 1.1**[135]**.** *If* $f_0 \in Y^*$, $V \in H$ *then the problem (1.6) has a unique solution* $\varphi^0 \in W$ *and*

$$\|\varphi^0\|_W \le C(\|f_0\|_{Y^*} + \|V\|_H), \tag{1.7}$$

*where* $C = \text{const} > 0$ *and* $\varphi^0$ *can be represented as*

$$\varphi^0 = G_1 f_0 + G_0 V, \tag{1.8}$$

*where* $G_1 = G_1(t) \in \mathcal{L}(Y^*; W)$, $G_0 = G_0(t) \in \mathcal{L}(H; W)$.

A similar assertion holds for the adjoint problem:

$$\begin{cases} -\frac{dq^0}{dt} + A^*(t)q^0 = g_0, & t \in (0, T), \\ q^0(T) = Q_0 \quad (q(T) \equiv q|_{t=T}). \end{cases} \tag{1.9}$$

**Theorem 1.2**[135]**.** *If* $g_0 \in Y^* \subset W^*$, $Q_0 \in H$, *then the problem (1.9) has a unique solution* $q^0 \in W$ *and*

$$\|q^0\|_W \le C(\|g_0\|_{Y^*} + \|Q_0\|_H), \tag{1.10}$$

*where  $C = const > 0$ , and also*

$$q^0 = G_1^{(T)} g_0 + G_0^{(T)} Q_0, \qquad (1.11)$$

*where  $G_1^{(T)} = G_1^{(T)}(t) \in \mathcal{L}(Y^*; W)$ ,  $G_0^{(T)} = G_0^{(T)}(t) \in \mathcal{L}(H; W)$ .*

Denote by  $f'(\varphi) \in \mathcal{L}(W; Y^*)$  the Frechet derivative of the operator  $f$  and by  $f'^*(\varphi) \in \mathcal{L}(Y; W^*)$  the adjoint of  $f'(\varphi)$ . In the sequel, we shall assume that the following restrictions are imposed on the non-linear operator  $f$ .

**Supposition.**  *Let  $\varphi^0$  be a solution of the problem (1.6) for  $f_0 \in Y^*$ ,  $V \in H$ . Then  $f'(\varphi)$ ,  $f'^*(\varphi) \in \mathcal{L}(W; Y^*)$  and also*

$$\|f(\varphi)\|_{Y^*} \le C_1 < \infty,$$

$$\|f(\varphi)h\|_{Y^*} \le C_2 \|h\|_W \quad \forall h \in W, \qquad (1.12)$$

$$\|f(\varphi)q\|_{Y^*} \le C_3 \|q\|_W \quad \forall q \in W,$$

*for*

$$\forall \varphi \in S_W(\varphi^0, R_0) = \{\varphi \in W; \quad \|\varphi^0 - \varphi\|_W \le R_0\},$$

$$C_j = C_j(R_0) = const, \quad j = 1, 2, 3, \quad R_0 \in \mathbf{R}^1.$$

*Example* 1.1.  Let  $H = L_2(\Omega)$ ,  $X = W_2^1(\Omega)$ ,  $\Omega \subset \mathbf{R}^2$  and also  $f(\varphi) = \varphi^m$ ,  $m \in \mathbf{N}$ . Using known imbedding theorems it is not difficult to prove that for  $m < 3$  all the conditions of the Supposition are fulfilled.

### 1.2. Statement of the problem. The control equation

Consider the functional of the form:

$$\begin{aligned}S(\varphi) &= \tfrac{\alpha}{2}(\Lambda_Y \varphi, \varphi) + \tfrac{\beta}{2}\|\varphi(0)\|_H^2 + \tfrac{1}{2}\sum_{i=1}^{N} \int_0^T \alpha_i(\varphi_i - C_i)^2 dt \\ &\quad + \sum_{i=1}^{\tilde{N}} \tilde{\alpha}_i(\tilde{\varphi}_i - \tilde{C}_i) + (\varphi, g_0),\end{aligned} \qquad (1.13)$$

where  $\alpha$ ,  $\beta$ ,  $\alpha_i$ ,  $\tilde{\alpha}_i$ ,  $\tilde{C}_i = const$ ,  $C_i = C_i(t)$ ,  $\forall i$ ,  $\varphi_i = (\varphi, p_i)_H(t)$ ,  $\tilde{\varphi}_i = (\varphi, \tilde{p}_i)$ ,  $p_i \in X^*$ ,  $\tilde{p}_i \in Y^*$ ,  $g_0 \in Y^*$ . The operator  $\Lambda_Y$  is an isomorphism from  $Y$  on  $Y^*$ . The sets  $\{C_i(t)\}$ ,  $\{\tilde{C}_i\}$  are the measurement data. At the same time, they may be the 'sets of prescribed sanitary standards', in which case the problem may be considered to be an environment protection problem. The values  $\{C_i(t)\}$ ,  $\{\tilde{C}_i\}$  can be measured or computed, for instance, as follows:  $C_i(t) = (\varphi_{obs}, p_i)_H(t)$ ,  $\tilde{C}_i = (\varphi_{obs}, \tilde{p}_i)$   $\forall i$ , where  $\varphi_{obs}$  is the observed function. Note that a considerable number of the functionals used nowadays are specific cases of (1.13). One of the frequently used functionals is the functional of the form  $\sum_i \int_0^T \alpha_i(\varphi_i - C_i)^2 dt$ . However, we'll show in the sequel that this functional is a poor choice, since here we arrive at an ill-posed problem. The parameters $\alpha$, $\beta$, $\alpha_i$, $\tilde{\alpha}_i$ are weight coefficients or regularization parameters.

Consider the equation (1.5) with an unknown function $V \in H$ in the initial condition (i.e. with an unknown control $V$):

$$\begin{cases} \frac{d\varphi}{dt} + A(t)\varphi + \varepsilon f(\varphi) = f_0(t), & t \in (0,T), \\ \varphi(0) = V. \end{cases} \qquad (1.14)$$

The data assimilation problem can be formulated as follows: find functions $(\varphi, V)$
$\in W \times H$ such that for any $\varepsilon \in [-\varepsilon_0, \varepsilon_0]$ they satisfy equations (1.14) and on the set of solutions of equations (1.14) the functional $S(\varphi)$ attains its minimal value

$$S(\varphi) = \inf. \qquad (1.15)$$

By using calculus of variations it is not difficult to show that if such functions exist they are solutions of the following problem ($\forall \varepsilon \in [-\varepsilon_0, \varepsilon_0]$)

$$\begin{cases} \frac{d\varphi}{dt} + A(t)\varphi + \varepsilon f(\varphi) = f_0(t), & t \in (0,T), \\ \varphi(0) = V, \end{cases} \qquad (1.16)$$

$$\begin{cases} -\frac{dq}{dt} + A^*(t)q + \varepsilon f'^*(\varphi)q = K\varphi + g, & t \in (0,T), \\ q(T) = 0, \end{cases} \qquad (1.17)$$

$$q(0) + \beta V = 0, \qquad (1.18)$$

where

$$K\varphi \equiv \alpha\Lambda_Y\varphi + K_0\varphi, \quad K_0\varphi \equiv \sum_{i=1}^{N} \alpha_i(\varphi, p_i)_H p_i,$$

$$g = g_0 + \sum_{i=1}^{\tilde{N}} \tilde{\alpha}_i\tilde{p}_i - \sum_{i=1}^{N} \alpha_i p_i C_i(t).$$

On the other hand, if the problem (1.16)–(1.18) has a solution then this solution is also a solution of (1.14)–(1.15).

Our first aim is to obtain an equation for $V = -q(0)/\beta$ to investigate some properties of the operator in this equation, and to study solvability of (1.16)–(1.17). If the problems (1.16), (1.17) have solutions, then they can be represented as

$$\varphi = G_1 f_0 - \varepsilon G_1 f(\varphi) + G_0 V, \qquad (1.19)$$

$$q = G_1^{(T)} g + G_1^{(T)} K\varphi - \varepsilon G_1^{(T)} f'^*(\varphi)q. \qquad (1.20)$$

Then for $q(0) + \beta V$ the following expression holds true:

$$q(0) + \beta V = M_1 V + F_0 - \varepsilon F_1(\varphi, q), \qquad (1.21)$$

where

$$M_1 \equiv T_0 G_1^{(T)} K G_0 + \beta I = \alpha T_0 G_1^{(T)} \Lambda_Y G_0 + T_0 G_1^{(T)} K_0 G_0 + \beta I, \qquad (1.22)$$

$$I\varphi \equiv \varphi, \quad T_0 \varphi \equiv \varphi(0), \quad T_1 \varphi \equiv \varphi(T),$$

$$F_0 \equiv T_0 G_1^{(T)} (g + K G_1 f_0), \qquad (1.23)$$

$$F_1(\varphi, q) \equiv T_0 G_1^{(T)} (f'^*(\varphi) q + K G_1 f(\varphi)). \qquad (1.24)$$

If, along with (1.16), (1.17), the equation (1.18) holds, then, taking into account (1.21), it can be written in the form:

$$M_1 V + F_0 - \varepsilon F_1 = 0, \qquad (1.25)$$

which is "the equation for the control $V$". Now we can try to study the existence of a solution of (1.25) without referring to the method of obtaining this equation. In particular, for $\varepsilon = 0$ we have an abstract linear equation. If this equation is correctly solvable, then for smooth operator $F_1(\varphi(V), q(V))$ we can prove the solvability of (1.25) for any $\forall \varepsilon \in [-\varepsilon_0, \varepsilon_0]$ and $\varepsilon_0 \ll 1$. As mentioned above, the methods used in this paper are based on considering equation (1.25) as an equation with small parameter $\varepsilon$ and the smooth nonlinear operator $F_1$.

### 1.3. Solvability of some data assimilation problems

Consider the equation (1.25) for $\varepsilon = 0$

$$M_1 V^0 + F_0 = 0. \qquad (1.26)$$

This equation is equivalent to the following system:

$$\begin{cases} \dfrac{d\varphi^0}{dt} + A(t)\varphi^0 = f_0, \quad t \in (0, T), \\[2mm] \varphi^0(0) = V^0, \end{cases} \qquad (1.27)$$

$$\begin{cases} -\dfrac{dq^0}{dt} + A^*(t) q^0 = K\varphi^0 + g, \quad t \in (0, T), \\[2mm] q^0(T) = 0, \end{cases} \qquad (1.28)$$

$$q^0(0) + \beta V^0 = 0. \qquad (1.29)$$

Let us formulate some statements concerning the properties of the operator $M_1$.

**Lemma 1.2.**   *The operator $M_1$ on a domain $D(M_1) \equiv H$ into $H$ is symmetric and positive definite for $\beta_1 > 0$:*

$$(M_1 \rho, \rho)_H \geq \beta_1 \|\rho\|_H^2 \quad \forall \rho \in H, \qquad (1.30)$$

*where $\beta_1 = \beta + c_0 \alpha, c_0 = const > 0$.*

*Proof.* For $\rho, \tilde{\rho} \in H$ and for $\psi \equiv G_1^{(T)} KG_0\rho$, $\varphi \equiv G_0\rho$, $\tilde{\varphi} \equiv G_0\tilde{\rho}$ we have

$$(M_1\rho, \tilde{\rho})_H = (T_0 G_1^{(T)} KG_0\rho, \tilde{\rho})_H + \beta(\rho, \tilde{\rho})_H =$$

$$= (-\frac{d\psi}{dt} + A^*(\psi)\psi, \tilde{\varphi}) + \beta(\rho, \tilde{\rho})_H = (KG_0\rho, G_0\tilde{\rho}) + \beta(\rho, \tilde{\rho})_H =$$

$$= (G_0\rho, KG_0\tilde{\rho}) + \beta(\rho, \tilde{\rho})_H = (\rho, M_1\tilde{\rho})_H .$$

Since $(G_0\rho, \Lambda_Y G_0\rho) \geq c\|G_0\rho\|_W^2 \geq c_0\|\rho\|_H^2$, $c, c_0 = \mathrm{const} > 0$, then $(M_1\rho, \rho)_H \geq \beta\|\rho\|_H^2$.

**Corollary.** *For $\beta_1 > 0$ the equation (1.26) is everywhere and correctly solvable in $H$ and*

$$\|V\|_H \leq \|F_0\|_H / \beta_1. \tag{1.31}$$

By virtue of Lemma 1.2 we conclude that the following statement holds.

**Lemma 1.3.** *If $\beta_1 > 0$ then problem (1.27)–(1.29) has a unique solution $(\varphi^0, q^0, V^0) \in W \times W \times H$.*

Now consider the equation (1.25) and corresponding problem (1.16)–(1.18). Let us describe the following iterative process:

$$M_1 V^{(n+1)} = \varepsilon F_1(\varphi^{(n)}, q^{(n)}) - F_0, \tag{1.32}$$

$$\begin{cases} \dfrac{d\varphi^{(n+1)}}{dt} + A(t)\varphi^{(n+1)} = f_0 - \varepsilon f(\varphi^{(n)}), \quad t \in (0, T), \\ \varphi^{(n+1)}|_{t=0} = V^{(n+1)}, \end{cases} \tag{1.33}$$

$$\begin{cases} -\dfrac{dq^{(n+1)}}{dt} + A^*(t)q^{(n+1)} = K\varphi^{(n+1)} - \varepsilon f'^*(\varphi^{(n)})q^{(n)} + g, \\ q^{(n+1)}(T) = 0, \quad n = 0, 1, \dots \end{cases} \tag{1.34}$$

If $\beta_1 > 0$, then it is easy to prove that there exists a sufficiently small $\varepsilon_0 = \varepsilon_0(\beta_1) > 0$ such that for $\varepsilon \in [-\varepsilon_0, \varepsilon_0]$ process (1.32)–(1.34) is convergent[4,11]:

$$\|\varphi^{(n+1)} - \varphi\|_W \to 0, \quad \|q^{(n+1)} - q\|_W \to 0,$$

$$\|V^{(n+1)} - V\|_H \to 0, \quad n \to \infty$$

and the limit functions $\varphi, V, q$ are analytic in $\varepsilon$. Now using available small parameter methods, we come to the following theorem.

**Theorem 1.3.** *Let $\beta_1 > 0$. Then there exists a sufficiently small value $\varepsilon_0 = \varepsilon_0(\beta_1, R_0)$ such that for $\forall \varepsilon \in [-\varepsilon_0, \varepsilon_0]$ the problem (1.16)–(1.18) has* *a unique solution analytic with respect to* $\varepsilon$:

$$\varphi = \varphi^0 + \sum_{i=1}^{\infty} \varepsilon^i \varphi_i, \quad q = q^0 + \sum_{i=1}^{\infty} \varepsilon^i q_i, \quad V = V^0 + \sum_{i=1}^{\infty} \varepsilon^i V_i, \qquad (1.35)$$

*where* $\{\varphi_i\}$, $\{q_i\}$, $\{V_i\}$ *can be computed by the small parameter method, i.e. by successive solution of the systems of the form:*

$$
\begin{cases}
\dfrac{d\varphi^0}{dt} + A(t)\varphi^0 = f_0, \quad t \in (0, T); \quad \varphi^0(0) = V^0, \\[2mm]
-\dfrac{dq^0}{dt} + A^*(t)q^0 = K\varphi^0 + g, \quad t \in (0, T); q^0(T) = 0, \\[2mm]
q^0(0) + \beta V^0 = 0,
\end{cases}
\qquad (1.36)
$$

$$
\begin{cases}
\dfrac{d\varphi_1}{dt} + A(t)\varphi_1 = -f(\varphi^0), \quad t \in (0, T); \quad \varphi_1(0) = V_1, \\[2mm]
-\dfrac{dq_1}{dt} + A^*(t)q_1 = K\varphi_1 - f'^*(\varphi_0)q^0, \quad t \in (0, T); q_1(T) = 0, \\[2mm]
q_1(0) + \beta V_1 = 0,
\end{cases}
\qquad (1.37)
$$

$$\ldots$$

*where* $\varphi \in S_W(\varphi^0, R_0)$, $R_0 = const > 0$.

Thus, the data assimilation problem under consideration for finding the initial distribution function $V$ is equivalent to solving equation (1.25) with the operator

$$M_1 = M_0 + \beta I + \alpha T_0 G_1^{(T)} \Lambda_Y G_0, \qquad (1.38)$$

where $M_0 = T_0 G_1^{(T)} K_0 G_0$. It is easy to see that the operator $M_0$ is linear, symmetric, continuous, and obviously degenerate if the span of the functions $\{p_i\}$ does not coincide with the whole space $Y^*$. And if we formulate an algorithm to solve the problem on finding $V$ without taking into account these facts we are doomed to failure while trying to implement the algorithm numerically. Now the role of the operators $\beta I$, $\alpha T_0 G_1^{(T)}$, and $\Lambda_Y G_0$ in equation (1.25) becomes obvious: they are regularizers with regularization parameters $\alpha$ and $\beta$. It should be noted that in some papers devoted to the solution of similar problems the foregoing facts were ignored and the functionals were considered for $\alpha = 0$, $\beta = 0$. Thus, as a rule, the data assimilation problems lead to equations with continuous degenerate operators (unless appropriate regularizers are used). The above-mentioned peculiarity of the problems involved requires suitable methods for their numerical solution. At the same time, a new problem arises: the problem of choosing optimal values of regularization parameters $\alpha$, $\beta$, ... and also a more general problem of choosing regularizing operators $\Lambda_Y$, the norm $\|\varphi(0)\|$ in the functional $S(\varphi)$, and so on.

Thus, we have to find suitable numerical algorithms for solving problems of the form (1.16)–(1.18). Here iterative algorithms may prove to be more efficient in comparison with perturbation algorithm (1.36), (1.37). In the next subsection we'll consider one of such iterative algorithms.

*Example* 1.2. Let $\varepsilon = 0$, $\alpha = 0$, $\alpha_i = 0$ $\forall i$. Then to solve the problem (1.16)–(1.18) we should solve successively the following subproblems:

$$\begin{cases} -\dfrac{dq}{dt} + A^*(t)q = g, & t \in (0,T); q(T) = 0, \\ V = -q(0)/\beta; \\ \dfrac{d\varphi}{dt} + A(t)\varphi = f_0, & t \in (0,T); \quad \varphi(0) = V. \end{cases} \tag{1.39}$$

Let

$$H = L_2(\Omega), \quad \Omega \subset R^n, \quad X = \overset{\circ}{W_2^1}(\Omega), \quad A(t) \equiv -\Delta = -\sum_{i=1}^{n} \frac{\partial^2}{\partial x_i^2}.$$

Introduce eigenfunctions $\{\Phi_i\}$ and eigenvalues $\{\lambda_i\}$ of the operator $A = -\Delta$, i.e. $-\Delta\Phi_i = \lambda_i\Phi_i$, $i = 1, 2, \ldots$ Assume that $\|\Phi_i\|_{L_2(\Omega)} = 1$. Then the solutions of subproblems (1.39) have the form:

$$q(x,t) = \sum_{i=1}^{\infty} \left( \int_t^T e^{-\lambda_i(t'-t)} (\Phi_i, g)_{L_2(\Omega)}(t') dt' \right) \Phi_i(x),$$

$$V(x) = -\frac{1}{\beta} \sum_{i=1}^{\infty} \left( \int_0^T e^{-\lambda_i t'} (\Phi_i, g)_{L_2(\Omega)}(t') dt' \right) \Phi_i(x),$$

$$\varphi(x,t) = \sum_{i=1}^{\infty} \varphi_i(t)\Phi_i(x),$$

where

$$\varphi_i(t) = \int_0^T e^{-\lambda_i(t-t')}(\Phi_i, f_0)_{L_2(\Omega)}(t') dt' - \frac{1}{\beta}\left( \int_0^T e^{-\lambda_i t'}(\Phi_i, g)_{L_2(\Omega)}(t') dt' \right) e^{-\lambda_i t}.$$

In particular, if $g = \sum_{i=1}^{\tilde{N}} \tilde{\alpha}_i \tilde{p}_i(x)$, where

$$\tilde{p}_i(x) = \begin{cases} \dfrac{1}{mes(\Omega_i)}, & x \in \Omega_i \subset \Omega, \\ 0, & x \notin \Omega_i, \end{cases}$$

$\tilde{\alpha}_i = const$, $mes(\Omega_i \cap \Omega_j) = \emptyset$ for $i \neq j$ , then

$$\varphi(x,t) = \sum_{i=1}^{\infty} \left[ \int_0^t e^{-\lambda_i(t-t')}(\Phi_i, f_0)_{L_2(\Omega)}(t') dt' - \right.$$

$$- \frac{1}{\beta} \sum_{j=1}^{\tilde{N}} \tilde{\alpha}_j (\Phi_i, \tilde{p}_j)_{L_2(\Omega)} \left( \frac{1 - e^{-\lambda_i T}}{\lambda_i} \right) e^{-\lambda_i t}] \Phi_i(x).$$

### 1.4. Linear data assimilation problem

From the above consideration we see that on each step of the iterative process one needs to solve a linear data assimilation problem given by the equation

$$M_1 V + F_0 = 0 \qquad (1.40)$$

or by the following system of main and adjoint equations:

$$\begin{cases} \dfrac{d\varphi}{dt} + A(t)\varphi = f, \quad t \in (0, T), \\[2mm] \varphi(0) = V, \end{cases} \qquad (1.41)$$

$$\begin{cases} -\dfrac{dq}{dt} + A^*(t)q = K\varphi + g, \quad t \in (0, T), \\[2mm] q(T) = 0, \end{cases} \qquad (1.42)$$

$$q(0) + \beta V = 0. \qquad (1.43)$$

Thus we have to find suitable numerical algorithms for solving problems of the form (1.41)–(1.43).

Let us apply the method of steepest descent to solve the equation (1.40):

$$\begin{aligned} & V^{(0)} \equiv 0, \quad r^{(0)} = F_0, \\ & w^{(n)} = M_1 r^{(n)}, \\ & \tau_{n+1} = \|r^{(n)}\|_H^2 / (w^{(n)}, r^{(n)})_H, \\ & V^{(n+1)} = V^{(n)} - \tau_{n+1} r^{(n)}, \\ & r^{(n+1)} = r^{(n)} - \tau_{n+1} w^{(n)}, \\ & n = 0, 1, 2, \ldots \end{aligned} \qquad (1.44)$$

Since

$$\beta \le \frac{(M_1 \rho, \rho)_H}{\|\rho\|_H^2} \le \beta + C, \qquad (1.45)$$

then using well-known results of the iterative processes theory we conclude that

$$[V^{(n)} - V] \le \left( \frac{C}{C + 2\beta} \right)^n [V] \to 0, \quad n \to \infty, \qquad (1.46)$$

where

$$[V] = (M_1 V, V)_H^{1/2}. \qquad (1.47)$$

In terms of (1.41)–(1.43) the process (1.65) is given by the following steps.

STEP 0. (Calculation of $F_0$). Solve the problem:

$$i) \quad \frac{d\varphi}{dt} + A(t)\varphi = f, \quad t \in (0, T) \quad \varphi(0) = 0;$$

$$ii) \quad -\frac{dq}{dt} + A^*(t)q = K\varphi + g, \quad t \in (0, T), \quad q(T) = 0; \qquad (1.48)$$

$$iii) \quad F_0 = q(0).$$

STEP 1. For given $r^{(n)}$ $(r^{(0)} \equiv F_0, V^{(0)} \equiv 0)$ solve the problem given by

$$i) \quad \frac{d\varphi}{dt} + A(t)\varphi = 0, \quad t \in (0, T), \quad \varphi(0) = r^{(n)};$$

$$ii) \quad -\frac{dq}{dt} + A^*(t)q = K\varphi, \quad t \in (0, T), \quad q(T) = 0; \qquad (1.49)$$

$$iii) \quad w^{(n)} = q(0) + \beta r^{(n)}.$$

STEP 2. Calculate $V^{(n+1)}$, $r^{(n+1)}$ :

$$\tau_{n+1} = \|r^{(n)}\|_H^2 / (w^{(n)}, r^{(n)})_H,$$

$$V^{(n+1)} = V^{(n)} - \tau_{n+1} r^{(n)},$$

$$r^{(n+1)} = r^{(n)} - \tau_{n+1} w^{(n)} \qquad (1.50)$$

and return to STEP 1 with new function $r^{(n+1)}$. If we construct the function $V^{(n+1)}$ of needed accuracy we can take $V \cong V^{(n+1)}$.

*Remark* 1.3. If $\tau_n = \tau$ $\forall n$ in (1.50) then it is easy to see that Steps 0–2 represent the well-known simplest iteration process for solving (1.40).

## 1.5. Data assimilation problem for the linear quasi-geostrophic ocean model

Let us apply the general scheme discussed above to the linear quasi-geostrophic ocean model[16]. Let $\Omega \subset R^2$ be a bounded domain with piecewise smooth boundary $\partial\Omega$, $Q \equiv \Omega \times (0, T)$, $T < \infty$, $\Gamma \equiv \partial\Omega \times (0, T)$. Consider the ocean circulation problem described by the following equation:

$$L\varphi \equiv \frac{\partial}{\partial t}\left(\Delta\varphi - \frac{1}{R^2}\varphi\right)$$

$$+ \left(\vec{U}, \nabla\right)\Delta\varphi + \beta_0\frac{\partial\varphi}{\partial x} + \nu\Delta\varphi - \mu\Delta^2\varphi = -F \quad in \quad Q$$

$$\varphi = \Delta\varphi = 0 \quad on \quad \Gamma, \tag{1.51}$$

$$\varphi = V \quad at \quad t = 0, \quad (x, y) \in \Omega,$$

where

$$\Delta = \partial^2/\partial x^2 + \partial^2/\partial y^2, \quad \vec{U} = (U_1, U_2), \quad U_1 = -\frac{\partial\tilde{\varphi}}{\partial y}, \quad U_2 = \frac{\partial\tilde{\varphi}}{\partial x},$$

$$R, \beta_0, \mu = const > 0, \quad \nu = const \geq 0,$$

and $\tilde{\varphi}$ is a given approximation of $\varphi$.

We assume that

$$\|\Delta U_1\|^2_{L_2(\Omega)} + \|\Delta U_2\|^2_{L_2(\Omega)} \leq const < \infty, \quad \forall t \in [0, T].$$

Introduce the functional $S(\varphi)$ :

$$S(\varphi) = \frac{\beta}{2}\|\varphi(0)\|^2_H + \frac{1}{2}\sum_{i=1}^{N}\alpha_i(\varphi_i - \varphi_{i,obs})^2, \tag{1.52}$$

where $\beta, \alpha_i = const > 0, \{\varphi_{i,obs}\}$ are the observational data,

$$\varphi_i = (\varphi, g_i)_{L_2(Q)} = \varphi(x_i, y_i, t_i),$$

$$g_i = \delta(x - x_i)\delta(y - y_i)\delta(t - t_i).$$

Consider the following problem:

$$L\varphi \equiv \frac{\partial}{\partial t}\left(\Delta\varphi - \frac{1}{R^2}\varphi\right)$$

$$+(\vec{U}, \nabla)\Delta\varphi + \beta_0\frac{\partial\varphi}{\partial x} + \nu\Delta\varphi - \mu\Delta^2\varphi = -F \quad in \quad Q,$$

$$\varphi = \Delta\varphi = 0 \quad on \quad \Gamma, \tag{1.53}$$

$$\varphi = V \quad at \quad t = 0, \quad (x, y) \in \Omega,$$

$$S(\varphi) = inf,$$

where we suppose that $\vec{U} = 0$ on $\partial\Omega$ $\forall t$. We can use the following spaces to investigate (1.53):

$$H = \left\{w : w \in \overset{\circ}{W_2^1}(\Omega), \|w\|_H = \left(\|\nabla w\|^2 + \frac{1}{R^2}\|w\|^2\right)^{1/2}\right\},$$

$$X = \{w : w \in \overset{\circ}{W}{}_2^1 (\Omega) \cap W_2^2(\Omega), \quad \|w\|_X = \|\Delta w\|\}.$$

Using the reasoning stated above, one can conclude the following. If the solution of (1.53) is smooth, then it satisfies the equations:

$$L\varphi \equiv \frac{\partial}{\partial t}\left(\Delta\varphi - \frac{1}{R^2}\varphi\right)$$

$$+(\vec{U}, \nabla)\Delta\varphi + \beta_0\frac{\partial\varphi}{\partial x} + \nu\Delta\varphi - \mu\Delta^2\varphi = -F \quad in \quad Q,$$

$$\varphi = \Delta\varphi = 0 \quad on \quad \Gamma,$$

$$\varphi = V \quad at \quad t = 0, \quad (x, y) \in \Omega$$

$$L^*q \equiv \frac{\partial}{\partial t}\left(\Delta q - \frac{1}{R^2}q\right) \tag{1.54}$$

$$+\Delta div(\vec{U}q) + \beta_0\frac{\partial q}{\partial x} - \nu\Delta q + \mu\Delta^2 q = \sum_{i=1}^{N}\alpha_i(\varphi_i - \varphi_{i,obs})g_i \quad in \quad Q,$$

$$q = \Delta q = 0 \quad on \quad \Gamma,$$

$$q = 0 \quad at \quad t = T, \quad (x, y) \in \Omega$$

$$\beta V + q|_{t=0} = 0 \quad in \quad \Omega.$$

Consider the problem (1.54). The algorithm (1.44) in terms of equations (1.54) is the following one:

$$L\varphi^{(n)} = -F \quad in \quad Q, \quad \varphi^{(n)} = \Delta\varphi^{(n)} = 0 \quad on \quad \Gamma,$$

$$\varphi^{(n)} = V^{(n)} \quad at \quad t = 0,$$

$$L^*q^{(n)} = \sum_{i=1}^{N}\alpha_i(\varphi_i^{(n)} - \varphi_{i,obs})g_i \quad in \quad Q,$$

$$q^{(n)} = \Delta q^{(n)} = 0 \quad on \quad \Gamma, \quad q^{(n)} = 0 \quad at \quad t = T, \tag{1.55}$$

$$V^{(n+1)} = V^{(n)} - \tau_{n+1}\left(q^{(n)}\Big|_{t=0} + \beta V^{(n)}\right).$$

If we put $\tau_n = \tau \ \forall n$, we obtain the simplest iteration process to solve (1.54).

## 2. APPLICATION OF ADJOINT EQUATIONS FOR SOLVING THE PROBLEM OF LIQUID BOUNDARY CONDITIONS IN HYDRODYNAMICS

Consider the problem of 'boundary functions' on liquid (open) parts of boundaries using optimal control approaches applying to the following quasi-geostrophic model of the ocean circulation[7]:

$$A\varphi \equiv \frac{1}{2\alpha}\Delta\varphi + (\vec{U}, \nabla)\varphi + \frac{H\beta}{l}\frac{\partial\varphi}{\partial x} = F \text{ in } \Omega, \ \varphi = 0 \text{ on } \Gamma_0, \varphi = V \text{ on } \Gamma_1,$$

$$(2.1)$$

where $(x, y) \in \Omega \subset R^2, \partial\Omega = \Gamma_0 \cup \Gamma_1, \vec{U} = (U_1, U_2), U_1 = -\partial H/\partial y, U_2 = \partial H/\partial x; l = l_0 + \beta y, l_0, \beta, \alpha = \text{const} > 0, F(x, y) \in L_2(\Omega), H(x, y) \in C^{(2)}(\Omega)$. Assume that the boundary function $V$ is unknown but we can use the observation data $\{v_i\}, \{u_i\}$, which are approximate values of '$(p_i, g/l \cdot \partial\varphi/\partial x)$', '$-(p_i, g/l \cdot \partial\varphi/\partial y)$', respectively, $i = 1, \ldots, N$. Here $\{p_i\}$ are some 'weight' functions in $L_2(\Omega), g = \text{const} > 0, (\cdot, \cdot) \equiv (\cdot, \cdot)_{L_2(\Omega)}, \|\cdot\| \equiv \|\cdot\|_{L_2(\Omega)}$. Introduce the functional $J(\varphi)$ of the form

$$J(\varphi) = \frac{\alpha_0}{2}\|\varphi\|_W^2 + \frac{1}{2}\sum_{i=1}^{N}\alpha_i\left(\left(v_i - \left(p_i, \frac{g}{l}\frac{\partial\varphi}{\partial x}\right)\right)^2 + \left(u_i + \left(p_i, \frac{g}{l}\frac{\partial\varphi}{\partial y}\right)\right)^2\right),$$

$$(2.2)$$

where $\alpha_0, \alpha_i = \text{const} > 0$. The space $W$ is defined as follows: $W \equiv W_{2,0}^{1/2}(\Gamma_1) = \{\psi: \psi = w \text{ on } \Gamma_1 \ \forall w \in W_2^1(\Omega); w = 0 \text{ on } \Gamma_0\}, (\psi_1, \psi_2)_W \equiv (\nabla E\psi_1, \nabla E\psi_2), \|\psi\|_W = \|\nabla E\psi\|_{L_2(\Omega)}$. Here $E$ is the operator of harmonic extension of $\psi \in W$ from $\Gamma_1$ onto $\Omega$, i.e.: $\Delta(E\psi) = 0$ in $\Omega, E\psi = 0$ on $\Gamma_0, E\psi = \psi$ on $\Gamma_1$. Let us consider the following problem: given $F \in (W_2^1(\Omega))^*, \{u_i\}, \{v_i\}$, find $\varphi, V \in W_2^1(\Omega) \times W$ such that

$$A\varphi = F \text{ in } \Omega, \ \varphi = 0 \text{ on } \Gamma_0, \ \varphi = V \text{ on } \Gamma_1, \ J(\varphi) = \inf. \qquad (2.3)$$

It is easy to write down the variational equations corresponding to the problem (2.3) (see Agoshkov[7]):

$$A\varphi = F \quad \text{in } \Omega, \qquad \varphi = 0 \quad \text{on } \Gamma_0, \qquad \varphi = V \quad \text{on } \Gamma_1,$$

$$A^*q \equiv \frac{1}{2\alpha}\Delta q - \text{div}(\vec{U}q) - \frac{\partial}{\partial x}\left(\frac{H\beta}{l}q\right) = \text{div}\vec{G} \text{ in } \Omega, q = 0 \text{ on } \partial\Omega, \quad (2.4)$$

$$\alpha_0\frac{\partial}{\partial n}(EV) - \frac{1}{2\alpha}\frac{\partial q}{\partial n} + (\vec{G}\cdot\vec{n}) = 0 \quad \text{on } \Gamma_1,$$

where

$$\vec{G} = (G_1, G_2), \vec{n} = (n_x, n_y), |\vec{n}| = 1,$$

$$G_1 = -\frac{g}{l}\sum_{i=1}^{N}\alpha_i\left(v_i - \left(p_i, \frac{g}{l}\frac{\partial\varphi}{\partial x}\right)\right)p_i, \ G_2 = \frac{g}{l}\sum_{i=1}^{N}\alpha_i\left(u_i + \left(p_i, \frac{g}{l}\frac{\partial\varphi}{\partial y}\right)\right)p_i.$$

For (2.4) we can derive the equation for 'the control $V$':

$$BV = \tilde{g}, \quad \tilde{g} \in W^*  \qquad (2.5)$$

with the operator $B : W \rightarrow W^*$. The properties of this operator are given by the following lemma.

**Lemma 2.1.** *The operator $B$ is (i) symmetric, i.e.* $(BV, \tilde{V})_{L_2(\Gamma_1)} = (V, B\tilde{V})_{L_2(\Gamma_1)} \ \forall \ V, \tilde{V} \in W$, *(ii) positive as* $\alpha_0 > 0$ : $(BV, V)_{L_2(\Gamma_1)} \geq \beta \|V\|_W^2$, *(iii) bounded, when problem (2.1) has a unique solution for given $F, V \in (W_2^1(\Omega))^* \times W$.*

*Proof.* Let $< V, \tilde{V} >$ be the form given by

$$< V, \tilde{V} > = \sum_{i=1}^{N} \alpha_i \left( \left( \int_0^H p_i \mathrm{d}z, \frac{g}{l} \frac{\partial \varphi_V}{\partial x} \right) \cdot \left( \int_0^H p_i \mathrm{d}z, \frac{g}{l} \frac{\partial \varphi_{\tilde{V}}}{\partial x} \right) \right.$$

$$\left. + \left( \int_0^H p_i \mathrm{d}z, \frac{g}{l} \frac{\partial \varphi_V}{\partial y} \right) \cdot \left( \int_0^H p_i \mathrm{d}z, \frac{g}{l} \frac{\partial \varphi_{\tilde{V}}}{\partial y} \right) \right),$$

where $\varphi_V, \varphi_{\tilde{V}}$ are the solution of (2.1) with boundary functions $V, \tilde{V}$ on $\Gamma_1$, respectively, and with $F \equiv 0$. Then we have $(BV, \tilde{V})_{L_2(\Gamma_1)} = \alpha_0(\nabla E V, \nabla E \tilde{V}) + < V, \tilde{V} > = (V, B\tilde{V})_{L_2(\Gamma_1)}$, i.e. the operator $B$ is symmetric. Moreover, we have $(BV, V)_{L_2(\Gamma_1)} \geq \alpha_0(\nabla E V, \nabla E V) = \alpha_0 \|V\|_W^2 > 0 \ \forall V \neq 0, \alpha_0 > 0$. If the boundary value problem (2.1) has a unique solution $\varphi$ then for the solution $(q, \varphi, V)$ of the problem (2.4) the following relations hold true:

$$\|q\|_{W_2^1(\Omega)} \leq C \|\varphi\|_{W_2^1(\Omega)} \leq \tilde{C}\|V\|_W, \ C, \tilde{C} = \text{const}, \ |(BV, \tilde{V})_{L_2(\Gamma_1)}| \leq$$

$$\mathbf{C}\|V\|_W \cdot \|\tilde{V}\|_W, \|B\|_{W \rightarrow W^*} \leq \mathbf{C} = \text{const} < \infty,$$

i.e. $B$ is bounded.

**Corollary.** *If the data of (2.1) are such that $\tilde{g} \in \mathcal{R}(B)$ then the equation (2.5) (and the system (2.4)) has a unique solution. In particular, if $\alpha_0 > 0$ then $\mathcal{R}(B) = W^*$.*

Using the properties of $B$ stated above we can formulate various iterative procedures to solve (2.4) (and (2.5)). One of them is the method of steepest descent:

$$w_n = \tilde{B}^{-1} r_n, \ g_n = B w_n$$

$$\tau_{n+1} = (r_n, w_n)_{L_2(\Gamma_1)} / (g_n, w_n)_{L_2(\Gamma_1)}, \ V_{n+1} = V_n - \tau_{n+1} w_n,  \qquad (2.6)$$

$$V_0 \equiv 0, \ r_{n+1} = r_n - \tau_{n+1} g_n, \ r_0 \equiv -\tilde{g}, \ n = 0, 1, \ldots$$

$$(r_n = BV_n - \tilde{g}, \tilde{B}\psi \equiv \partial E\psi/\partial n \text{ on } \Gamma_1).$$

Steps of this iterative process in terms of 'initial differential problems' are the following:

**Step 0**: (Calculation of $\tilde{g} = -r_0$):

$$A\varphi = F \text{ in } \Omega, \ \varphi = 0 \text{ on } \partial\Omega; A^*q = \text{div}\vec{G} \text{ in } \Omega, q = 0 \text{ on } \partial\Omega;$$

$$\tilde{g} = \frac{1}{2\alpha}\frac{\partial q}{\partial n} - (\vec{G} \cdot \vec{n}) \text{ on } \Gamma_1. \tag{2.7}$$

**Step 1**: For given $r_n$ we solve the problem

$$\Delta\psi = 0 \text{ in } \Omega, \ \psi = 0 \text{ on } \Gamma_0, \ \frac{\partial\psi}{\partial n} = r_n \text{ on } \Gamma_1 \tag{2.8}$$

and set $w_n = \psi$ on $\Gamma_1$.

**Step 2**: We solve the following problems:

$$A\varphi = 0 \text{ in } \Omega, \ \varphi = 0 \text{ on } \Gamma_0, \varphi = w_n \text{ on } \Gamma_1; A^*q = \text{div}\vec{G} \text{ in } \Omega, q = 0 \text{ on } \partial\Omega; \tag{2.9}$$

$$\Delta\psi = 0 \text{ in } \Omega, \ \psi = 0 \text{ on } \Gamma_0, \psi = w_n \text{ on } \Gamma_1,$$

and calculate $g_n$: $g_n = \alpha_0\dfrac{\partial\psi}{\partial n} - \dfrac{1}{2\alpha}\dfrac{\partial q}{\partial n} + (\vec{G} \cdot \vec{n})$ on $\Gamma_1$.

**Step 3**: We calculate $\tau_{n+1}, V_{n+1}, r_{n+1}$ by (2.6).
Then we repeat all calculations beginning from Step 1 with the new function $r_{n+1}$.

**Theorem 2.1.** *The algorithm (2.6) converges as $n \to \infty$ and the estimate holds:* $[V - V_n] \leq (C/(C + 2\alpha_0))^n [V]$, *where* $[V] = (BV, V)^{1/2}_{L_2(\Gamma_1)}$.

*Proof.* The statements of this theorem follow from the Corollary and results of the iterative process theory.

*Example* 2.1. Let the data of (2.3) be: $H = 1.0, \beta = 0.1, l_0 = 4.8, \alpha_0 = 0.1, \alpha = 5.0, \alpha_1 = 3.5 \cdot h^2 \cdot H \ (h = 1/I, I \in \mathbf{N}), N = I^2, g = 9.81, \Omega = (0,1) \times (0,1), p_i \equiv p_{ij}(x, y) = \omega_i(x)\omega_j(y)$, where $\omega_i(t)$ is a piecewise linear function such that $\omega_j(j \cdot h) = 1/h, \omega_j(i \cdot h) = 0$ for $i \neq j, \ i, j = 0, 1, \dots, I$. Assume $F, \ u_i \equiv u_{ij}, \ v_i \equiv v_{ij}$ to be given by:

$$F(x, y) = \left(\frac{5\pi^2}{8\alpha}\sin(\frac{\pi x}{2}) + \frac{H\beta\pi}{2l}\cos(\frac{\pi x}{2})\right)\sin(\pi y),$$

$$u_{ij} = -g\pi\cos(\pi j \cdot h)\sin(\pi i \cdot h/2)/(l_0 + \beta j \, h),$$

$$v_{ij} = g\pi\sin(\pi j \cdot h)\cos(\pi i \cdot h/2)/(2(l_0 + \beta j \, h)).$$

As $I \to \infty, \alpha_0 \to 0$, then the exact solution of (2.3) is $\varphi = \sin(\pi y)\sin(\pi x/2)$, $V = 0$ at $x = 0, V = \sin(\pi y)$ at $x = 1$. The problems (2.7)–(2.9) have been approximated by finite difference methods (mesh steps in $x, y$ are equal to $h$). The numerical values of $\tau_n$, $\varepsilon_n \equiv \|\tilde{B}(V_{n+1} - V_n)\|_{L_2(\Gamma_1)}$ are given below for $I = 25$:

$$\tau_1 = 0.03310, \; \varepsilon_1 = 0.06221; \; \tau_2 = 0.07234, \; \varepsilon_2 = 0.03215; \; \tau_3 = 0.04918,$$

$$\varepsilon_3 = 0.01644; \; \tau_4 = 0.15058, \varepsilon_4 = 0.02004; \; \tau_5 = 0.05974, \; \varepsilon_5 = 0.00548;$$

$$\tau_6 = 0.24126, \; \varepsilon_6 = 0.00204; \; \tau_7 = 0.26621, \; \varepsilon_7 = 0.00062.$$

*Example 2.2.* Assume that $\alpha_0 = 0.01$ in the data of Example 2.1. Then

$$\tau_1 = 0.03423, \; \varepsilon_1 = 0.06292; \; \tau_2 = 0.07286, \; \varepsilon_2 = 0.03256; \; \tau_3 = 0.05421,$$

$$\varepsilon_3 = 0.01800; \; \tau_4 = 0.13552, \; \varepsilon_4 = 0.02043.$$

*Example 2.3.* Assume that $\alpha_0 = 0.5$. Then

$$\tau_1 = 0.02884, \; \varepsilon_1 = 0.05834; \; \tau_2 = 0.07016, \; \varepsilon_2 = 0.03118; \; \tau_3 = 0.03620,$$

$$\varepsilon_3 = 0.01201; \; \tau_4 = 0.18764, \; \varepsilon_4 = 0.01621.$$

We can see that one needs 3-4 iterations by (2.7)-(2.9) to obtain the approximate solution $V_n$ with an accuracy of 1-3%.

## 3. SHAPE OPTIMIZATION USING ADJOINT EQUATION APPROACHES

Let a (linear or non-linear) problem be written in the form

$$\tilde{A}(\tilde{x}, \tilde{\varphi}(\tilde{x})) = 0, \; \tilde{x} = (\tilde{x}_1, \ldots, \tilde{x}_n) \in \tilde{\Omega}. \tag{3.1}$$

Here $\partial\tilde{\Omega}$ is the unknown boundary of a domain $\tilde{\Omega} \subset \mathbf{R}^n$, which is to be found, along with the solution $\tilde{\varphi}(\tilde{x})$. Consider one of the approaches for solving this problem[9,91,92,120,230,236].

Let $\tilde{\Omega}$ be an image of a fixed 'standard' domain $\Omega \subset \mathbf{R}^m$ with points $x = (x_1, \ldots, x_n)$, the mapping being given by the explicit formula

$$\tilde{x} = \Phi(v, x) \equiv (\Phi_1(v, x), \ldots, \Phi_n(v, x)), \tag{3.2}$$

depending on the vector $v = (v_1, \ldots, v_m)^T \in \mathbf{R}^m$ of unknown parameters. We assume that the mapping of $\tilde{\Omega}$ onto $\Omega$ is one-to-one and is regular enough so that one can perform an ordinary change of variables

$$\tilde{x} = \Phi(v, x), \; \varphi(x) \equiv \tilde{\varphi}(\Phi(v, x)),$$

$$\tilde{\Omega} \to \Omega, \; A(\tilde{x}, \tilde{\varphi}(\tilde{x})) \to A(x, v, \varphi(x)). \tag{3.3}$$

On such a change of variables, all the functions under consideration (coefficients, the solution $\varphi(x)$, etc.) will be defined on a fixed domain and the operator $A$ will depend on the vector $v$ of unknown parameters which determine the sought-for domain $\widetilde{\Omega}$.

Let, on transformation (3.2), problem (3.1) read: find $\varphi \in W, v \in \mathbf{R}^m$ such that

$$A(x, v, \varphi) = 0, \tag{3.4}$$

where the operator $A$ maps $W$ into $Y^*$, $W$ and $Y$ are some functional spaces densely enclosed into the basic Hilbert space $H = H^*$; $W^*$, $Y^*$ are adjoints of $W$, $Y$, respectively.

A cost function (or functional) can be introduced using the variables $\tilde{x}$ (followed by transformation (3.2)) or the variables $x$. We consider the second way. Thus, let a cost function have the form

$$S(v, \varphi) = \frac{\alpha}{2}(B(v - v_{ob}), v - v_{ob})_2 + \frac{\beta}{2}\int_\Omega |C\varphi - \phi_{ob}|^2 dx + \frac{\gamma}{2}\int_{\partial\Omega} |D\varphi - \psi_{ob}|^2 d\Gamma,$$

$$\tag{3.5}$$

where $\alpha, \beta, \gamma = \text{const} \geq 0$, $v_{ob} \in \mathbf{R}^m$ is a prescribed vector, $\phi_{ob}, \psi_{ob}$ are known functions defined on $\Omega$ and $\partial\Omega$, respectively, and $C$ and $D$ are linear operators. Here $B$ is a $m \times m$–matrix, which is symmetric and positive definite, $(Bv, v)_2 \equiv \sum_{i,j=1}^{m} B_{ij} v_i v_j$. The functional $S(v, \varphi)$ is assumed to be bounded on $\mathbf{R}^m \times W$.

The problem on finding the optimal shape of the boundary $\partial\widetilde{\Omega}$ can be formulated as follows: find functions $\varphi \in W, v \in \mathbf{R}^m$ such that

$$A(x, v, \varphi(x)) = 0,$$
$$S(v, \varphi) = \inf_v. \tag{3.6}$$

Let, in the sequel, the operator $A(x, v, \varphi(x)) \equiv A(v, \varphi)$ be bounded and have continuous partial Frechet derivatives in $v$ and $\varphi$ (for $(v, \varphi) \in \mathbf{R}^m \times W$):

$$A_v(v, \varphi) : \mathbf{R}^m \to Y^*, \quad A_\varphi(v, \varphi) : W \to Y^*,$$

with the operator $A_\varphi(v, \varphi)$ being continuously invertible.

Assume that $(v, \varphi)$ is a solution of problem (3.6). Then the variations $\delta S, \delta v, \delta \varphi$ satisfy the relations:

$$0 = \delta S(v, \varphi; \delta v, \delta \varphi) =$$

$$\alpha(B(v - v_{ob}), \delta v)_2 + (K\varphi - g, \delta\varphi)_H , \tag{3.7}$$

$$A_v(v, \varphi)\delta v + A_\varphi(v, \varphi)\delta\varphi = 0,$$

where the operator $K : W \to W^*$ and the function $g \in W^*$ are defined by

$$(K\varphi, \psi)_H \equiv \beta \int_\Omega C\varphi C\psi dx + \gamma \int_{\partial\Omega} D\varphi D\psi d\Gamma,$$

$$(g, \psi)_H \equiv \beta \int_\Omega \phi_{ob} C\psi dx + \gamma \int_{\partial\Omega} \psi_{ob} D\psi d\Gamma, \ \forall \varphi, \psi \in W.$$

Let $q$ be a solution of the following adjoint equation

$$A_\varphi^*(v, \varphi)q = K\varphi - g, \tag{3.8}$$

where $A_\varphi^*(v, \varphi) : Y \to W^*$ is the operator adjoint to $A_\varphi(v, \varphi)$. Then, taking into account the fact that $\delta v$ is arbitrary, from (3.7), (3.8) we get the equation

$$\alpha B(v - v_{ob}) - A_v^*(v, \varphi)q = 0,$$

where $A_v^*(v, \varphi) : Y \to \mathbf{R}^m$ is the operator adjoint to $A_v(v, \varphi)$:

$$(q, A_v(v, \varphi)\psi)_H \equiv (A_v^*(v, \varphi)q, \psi)_2.$$

Therefore, in terms of main and adjoint equations, problem (3.6) reads as follows: find $\varphi \in W, q \in Y, v \in \mathbf{R}^m$ such that

$$A(v, \varphi) = 0,$$

$$A_v^*(v, \varphi)q = K\varphi - g, \tag{3.9}$$

$$\alpha B(v - v_{ob}) - A_v^*(v, \varphi)q = 0.$$

Thus, the problem on finding the optimal shape of the boundary $\partial\widetilde{\Omega}$ reduces to studying and solving the set of nonlinear equations (3.9). If a solution of system (3.9) is found (possibly under additional restrictions), then the sought-for boundary $\partial\widetilde{\Omega}$ is defined by formula (3.2): $\partial\widetilde{\Omega} = \{\tilde{x} = \Phi(v, x); \ x \in \partial\Omega\}$.

Let us prove the existence of a solution of (3.9). Assume that the operator $A(v, \varphi)$ is analytic in the neighbourhood of a point $(v_0, \varphi_0)$:

$$A(v, \varphi) = A(v_0, \varphi_0) + A_v V + A_\varphi \Phi + \sum_{i+j \geq 2} A_{ij} V^i \Phi^j,$$

where

$$A_v = A_v(v_0, \varphi_0), \ A_\varphi = A_\varphi(v_0, \varphi_0),$$

$$A_{ij} = \frac{1}{i!j!} \frac{\partial^{i+j} A(v_0, \varphi_0)}{\partial v^i \partial \varphi^j}; \ V \equiv v - v_0, \ \Phi \equiv \varphi - \varphi_0.$$

Let us rewrite (3.9) in the form:

$$A_v V + A_\varphi \Phi = F,$$

$$A_\varphi^* q - K\Phi = G, \qquad (3.10)$$

$$\alpha BV - A_v^* q = \tilde{g},$$

where $F = -A(v_0, \varphi_0) - \sum_{i+j \geq 2} A_{ij} V^i \Phi^j$, $G = K\varphi_0 - g + (A_\varphi^* - A_\varphi^*(v, \varphi))q$,

$\tilde{g} = (A_v^*(v, \varphi) - A_v^*)q$.

From (3.10) we get the equation for $V$:

$$AV = \tilde{g} + A_v^* A_\varphi^{*-1}(KA_\varphi^{-1}F + G), \qquad (3.11)$$

where

$$AV = \alpha BV + A_v^* A_\varphi^{*-1} K A_\varphi^{-1} A_v V.$$

Some properties of the operator $A$ are given by the following lemma.

**Lemma 3.1[9].** *The operator* $A : \mathbf{R}^m \to \mathbf{R}^m$ *is symmetric and positive definite for* $\alpha > 0$:

$$(AV, V)_2 = \alpha(BV, V)_2 + \beta \int_\Omega |CA_\varphi^{-1} A_v V|^2 dx$$

$$+ \gamma \int_{\partial\Omega} |DA_\varphi^{-1} A_v V|^2 d\Gamma \equiv \|V\|_A^2. \qquad (3.12)$$

Assume that $v_0, \varphi_0$ satisfy the equation $A(v_0, \varphi_0) = 0$ and $v_{ob}, \varphi_{ob}, \psi_{ob}$ have the form

$$v_{ob} = v_0,$$

$$\Phi_{ob} = C\varphi_0 + \varepsilon \Phi_{ob}^1, \qquad (3.13)$$

$$\psi_{ob} = D\varphi_0 + \varepsilon \psi_{ob}^1,$$

where $\varepsilon \in [-\varepsilon_0, \varepsilon_0]$ is a small parameter, and $\Phi_{ob}^1$ and $\psi_{ob}^1$ are prescribed functions. It is easily seen that the solution of system (3.9) for $\varepsilon = 0$ is $\varphi = \varphi_0$, $q = 0$, $v = v_0$, and the solution of (3.10) is the vector $(\varphi, q, v) = (0, 0, 0)$. Moreover, the first Frechet derivative of the non-linear operator of problem (3.10) at the point $(0, 0, 0)$ is the linear operator in the left-hand side of system (3.10). Using the assertions of Lemma 3.1, the above restrictions on the operator $A$, and the well-known theorems of non-linear analysis (in particular, the theorem on implicit operators), one can establish the existence of a solution of the problems under consideration.

**Theorem 3.1[9].** *Let a pair $v_0, \varphi_0$ satisfy the following restrictions: (1) $A(v_0, \varphi_0) = 0$, (2) the operator $A(v, \varphi)$ is analytic in a neighbourhood of $v_0, \varphi_0$ and the operator $A_\varphi(v_0, \varphi_0)$ is continuously invertible, (3) relations (3.13) are valid. Then for $\alpha > 0$ there exists a sufficiently small $\varepsilon_0 > 0$ such that for any $\varepsilon \in [-\varepsilon_0, \varepsilon_0]$ problem (3.9) has a unique solution analytic in $\varepsilon$:*

$$\varphi = \varphi_0 + \varepsilon\Phi_1 + \varepsilon^2\Phi_2 + \cdots, \, q = \varepsilon q_1 + \varepsilon^2 q_2 + \cdots,$$

$$v = v_0 + \varepsilon V_1 + \varepsilon^2 V_2 + \cdots,$$

(3.14)

*where the functions $\Phi_1, q, V_1$ satisfy a system of the form*

$$A_v V_1 + A_\varphi \Phi_1 = 0,$$

$$A_\varphi^* q_1 - K\Phi_1 = -g_1,$$

(3.15)

$$\alpha B V_1 - A_v^* q_1 = 0$$

*with*

$$(g_1, \psi)_H \equiv \beta \int\limits_\Omega \Phi_{ob}^1 C\psi dx + \gamma \int\limits_{\partial\Omega} \psi_{ob}^1 D\psi d\Gamma, \, \forall \psi \in W.$$

The other functions $\Phi_i, q_i, V_i, i \geq 2$, in (3.14) may be determined by the perturbation method.

## 4. GLOBAL TRANSPORT OF POLLUTANTS

Model computations on pollutant transport are usually performed for long periods; therefore, to make the results more reliable, we used information on wind velocity fields and other meteoelements based on observation data. Power of the sources and pollutant distribution function are assigned as the functions of spatial coordinates and time. The model considers the Earth as a sphere and uses the spherical system of coordinates $(\lambda, \psi, z)$ where $\lambda$ is a longitude, $\psi$ is a supplement to latitude and $z$ is a height counted from the underlying surface. Write the main pollutant transport equation as follows:

$$\frac{\partial\varphi}{\partial t} + \frac{u}{a\sin\psi}\frac{\partial\varphi}{\partial\lambda} + \frac{v}{a}\frac{\partial\varphi}{\partial\psi} + (w - w_g)\frac{\partial\varphi}{\partial z}$$

$$- \frac{\partial}{\partial z}\nu\frac{\partial\varphi}{\partial z} - \frac{1}{a^2\sin^2\psi}\frac{\partial}{\partial\lambda}\mu\frac{\partial\varphi}{\partial\lambda} - \frac{1}{a^2\sin\psi}\frac{\partial}{\partial\psi}\mu\sin\psi\frac{\partial\varphi}{\partial\psi} = f. \quad (4.1)$$

Here $\varphi = \varphi(\lambda, \psi, z, t)$ is a pollutant concentration, $\mathbf{u} = (u, v, w - w_g)^T$ – wind velocity vector with components in the directions $\lambda$, $\psi$, $z$ respectively, $w_g$ – gravitation subsidence velocity, $\mu$, $\nu$ – turbulence diffusion coefficients in horizontal and vertical directions, $f = f(\lambda, \psi, z, t)$ is a function of allocation and power of sources, and $a$ is the average radius of the Earth.

We solve this problem in the domain $\Omega \times (0, T)$, where $\Omega = S \times (b + h, H)$, $S = \{(\lambda, \psi) : 0 < \lambda < 2\pi, \ 0 < \psi < \pi\}$, $b = b(\lambda, \psi)$ is a function describing the relief of underlying surface, $H$ is an upper boundary of the domain, and $h$ is a height of the atmopshere surface layer.

Let us formulate a boundary condition of problem (4.1) at the level of a height of the atmosphere surface layer. As is known from the Obukhov–Monin similarity theory, turbulent flow of a passive impurity in the atmosphere surface layer may be considered as constant in height. Then the following relationship is valid at $z \leq b + h$:

$$\frac{\partial \varphi}{\partial z} = \frac{\varphi_*}{z} \eta(\xi), \tag{4.2}$$

$$\varphi - \varphi_0 = \varphi_* \int_{\xi_0}^{\xi} \frac{\eta(\xi)}{\xi} d\xi \equiv \varphi_* f_\varphi(\xi, \xi_0), \tag{4.3}$$

$$\nu(\xi) = \frac{u_* \kappa z}{\eta(\xi)}, \quad u_* = \frac{\kappa |\mathbf{u}|}{f_u(\xi, \xi_0)} \equiv c_u |\mathbf{u}|, \tag{4.4}$$

where $\varphi_*$ is a scale for a change of pollutant concentration, $\eta$, $f_u$, $f_\varphi$ are universal functions, $\xi = z/L$ is a dimensionless length characterizing the atmosphere stability, the zero index means that corresponding values are taken at $z = b + z_0$, where $z_0$ is a roughness parameter, $L$ is a turbulent layer scale, $\kappa$ is the Karman's constant, $u_*$ is a scale of wind velocity, and $c_u = \kappa / f_u(\xi, \xi_0)$. Obtain by (4.2) and (4.3)

$$\varphi_* = \frac{\varphi - \varphi_0}{f_\varphi(\xi, \xi_0)}, \tag{4.5}$$

$$\frac{\partial \varphi}{\partial z} = \frac{\eta(\xi)}{z} \frac{\varphi - \varphi_0}{f_\varphi(\xi, \xi_0)}. \tag{4.6}$$

Multiply equation (4.6) by $\nu$ to obtain

$$\nu \frac{\partial \varphi}{\partial z} = \alpha(\varphi - \varphi_0),$$

where $\alpha = c_u c_\varphi |\mathbf{u}|$, $c_\varphi = \kappa / f_\varphi(\xi, \xi_0)$. Examine this relationship at $z = b + h$.

$$\nu \frac{\partial \varphi}{\partial z} = \alpha(\varphi - \varphi_0) \quad \text{at} \quad z = b + h. \tag{4.7}$$

Assume this condition as a sought boundary condition for the main problem. Notice that the parameter $\alpha$ characterizes interaction of a pollutant with underlying surface.

Substitute $\Omega = S \times (b + h, H)$ for the definition domain $\Omega$ since the atmosphere surface layer may be excluded through its parameterization. The function $\varphi_0$ is thus the only unknown function in boundary condition (4.7) and we use the following expedient to determine it. Write the impurity balance equation in the neighborhood of underlying surface

$$-(\nu\frac{\partial\varphi}{\partial z})_0 + (\beta_i - w_g)\varphi_0 = \sum_{k=1}^{K} Q_{0k}\delta(x - x_k)\delta(y - y_k), \qquad (4.8)$$

where $K$ is the number of all surface sources with coordinates $(x_k, y_k)$, $Q_{0k}$ is a power of each source, i. e. amount of pollutant falling out at a unit area of underlying surface in a unit of time, $\beta_i$ $(i = 1, 2)$ – coefficient characterizing interaction of a pollutant with underlying surface ($\beta_1 = 0.01$ m/s corresponds to land, $\beta_2 = 1$ m/s – to sea surface).

Equation (4.8) is written in Cartesian coordinates $x$, $y$ for convenience; with regard to (4.7) it takes the form

$$\varphi_0 = \frac{\sum_{k=1}^{K} Q_{0k}\delta(x - x_k)\delta(y - y_k) + c_u c_\varphi |\mathbf{u}_{b+h}|\varphi_{b+h}}{\beta_i - w_g + c_u c_\varphi |\mathbf{u}_{b+h}|}, \qquad (4.9)$$

where $\mathbf{u}_{b+h} = \mathbf{u}|_{z=b+h}$, $\varphi_{b+h} = \varphi|_{z=b+h}$.

Let us use the following boundary condition for problem (4.1) at the top of the atmosphere at $z = H$:

$$\frac{\partial\varphi}{\partial z} = 0 \quad \text{at} \quad z = H. \qquad (4.10)$$

Choose the initial condition

$$\varphi = \bar{\varphi} \quad \text{at} \quad t = 0, \qquad (4.11)$$

where $\bar{\varphi} = \bar{\varphi}(\lambda, \psi, z)$ is a background concentration of a pollutant. We assume $\bar{\varphi} = 0$ hereafter for simplicity. Periodicity conditions are set for all functions on lateral boundaries in horizontal coordinates:

$$\varphi(0, \psi, z, t) = \varphi(2\pi, \psi, z, t),$$

$$\varphi(\lambda, -\psi, z, t) = \varphi(\lambda + \pi, \psi, z, t), \qquad (4.12)$$

$$\varphi(\lambda, \pi + \psi, z, t) = \varphi(\lambda + \pi, \pi - \psi, z, t).$$

Then assume that the function $\varphi(\lambda, \psi, z, t)$ as a solution of the pollutant transport problem in the form (4.1)–(4.7), (4.10)–(4.12), is continuous in $\Omega \times [0, T]$ and differentiable in $t$. Besides, let at each $t$ the function $\varphi(\lambda, \psi, z, t)$ belong to the set of functions $D(A)$ from real Hilbert space $L_2(\Omega)$ continuous and differentiable in $\Omega$, so that they satisfy the condition

$$\frac{\partial}{\partial z}\nu\frac{\partial\varphi}{\partial z} + \frac{1}{a^2\sin^2\psi}\frac{\partial}{\partial\lambda}\mu\frac{\partial\varphi}{\partial\lambda} + \frac{1}{a^2\sin\psi}\frac{\partial}{\partial\psi}\mu\sin\psi\frac{\partial\varphi}{\partial\psi} \in L_2(\Omega).$$

One more important question. Since vertical resolution in global models is not sufficient to restore the fields in lower layers of the atmosphere with proper accuracy, we compute the meteorological characteristics of the boundary layer using the following parameterization of the atmosphere planetary boundary layer. The following external parameters are determined with the help of the model at the nodes of the calculation grid in the horizontal plane at the first level through known values of fields of velocity and temperature:

$$Ro = \frac{|\mathbf{u}_{g0}|}{lz_0}, \quad S_T = \frac{\beta\delta\hat{\theta}}{l|\mathbf{u}_{g0}|},$$

$$(4.13)$$

$$\eta_x = \frac{\kappa^2}{l}\frac{\partial u_g}{\partial z} = -\frac{\beta\kappa^2}{l^2}\frac{\partial\hat{\theta}}{\partial y}, \quad \eta_y = \frac{\kappa^2}{l}\frac{\partial v_g}{\partial z} = \frac{\beta\kappa^2}{l^2}\frac{\partial\hat{\theta}}{\partial x},$$

where $Ro$ is the Kibbel–Rossbi number, $S_T$ is a stratification parameter, $u_g$, $v_g$ are the components of geostrophic wind velocity $\mathbf{u}_g$, $|\mathbf{u}_{g0}|$ is a modulus of geostrophic wind velocity close to underlying surface, $l$ is the Coriolis' parameter, $\eta_x$, $\eta_y$ are baroclinicity parameters, $\hat{\theta}$ is the potential temperature, $\beta = g/\hat{\theta}$ is a buoyancy parameter, $g$ is the gravity acceleration, $\delta\hat{\theta}$ is a difference between the values of the potential temperature at the boundary of the planetary boundary layer and at the underlying surface.

By the values of $Ro$, $S_T$, $\eta_x$, $\eta_y$ we find the values: $C_g = u_*/|\mathbf{u}_g|$, coefficient of "geostrophic resistance"; $\alpha$, an angle between turbulent friction stress at underlying surface and $\mathbf{u}_{g0}$; and $\mu = h_0/L_0$, dimensionless "internal" stratification parameter.

The following notation is accepted here: $h_0 = \kappa u_*/l$ is the "internal" height scale of a boundary layer; $L_0 = -c_p\rho u_*^3/(\kappa\beta q_0)$ is a Monin–Obukhov length scale; $c_p$ is a heat capacity of the air; $\rho$ is the air density; $q_0$ is a heat flux at the Earth surface. By the values $C_g$, $\alpha$, $\mu$ the value $q_0$ is calculated and then by $C_g$ and $\mathbf{u}_g$ the value $u_*$ and the coefficients of turbulent exchange at the heights $z \geq h$ are computed through the formula

$$\nu(z) = \frac{\kappa^2 u_*^2}{l^2} \times \begin{cases} h/(1+10\mu h) & \text{for} \quad \mu \geq 0 \\ h & \text{for} \quad -2.33 \leq \mu \leq 0 \\ (-0.07/\mu)^{-1/3}h^{4/3} & \text{for} \quad \mu \leq -2.33, \end{cases} \quad (4.14)$$

where $h$ is a height of a surface layer.

The components of a velocity vector may be written as follows:

$$u = |\mathbf{u}_{g0}|\cos\alpha + \frac{u_*}{\kappa}\alpha_1,$$

$$(4.15)$$

$$v = -|\mathbf{u}_{g0}|\sin\alpha + \frac{u_*}{\kappa}\alpha_2,$$

where $\alpha_1$ and $\alpha_2$ are dimensionless velocity "defects".

Pollutants with small velocities of gravitational subsidence are of special interest for global pollutant transport, since they are suspended for a long period and are transported by the air masses at large distances. In this case the theory of turbulence, used in general circulation models for the description of turbulent exchange with temperature and humidity, is applicable for the pollutants. In particular, the following model may be used to determine the horizontal turbulence coefficient as the first approximation:

$$\mu = k_1^2 \Delta S |D_N|, \quad D_N = (D_T^2 + D_S^2)^{1/2},$$

$$D_T = -\frac{1}{a \sin \psi} \frac{\partial u}{\partial \lambda} - \frac{1}{2} \frac{\partial v}{\partial \psi}, \quad D_S = -\frac{1}{a \sin \psi} \frac{\partial v}{\partial \lambda} + \frac{1}{2} \frac{\partial u}{\partial \psi}, \quad (4.16)$$

where $\Delta S$ is an area of an elementary cell of a grid domain, and $k_1$ is a dimensionless parameter.

Examine now the functionals from the solution $\varphi$ in the form

$$J = \int_0^T dt \int_\Omega p\varphi \, d\Omega , \quad (4.17)$$

where $p = p(\lambda, \psi, z, t)$ is an assigned function from $L_2(\Omega \times (0, T))$. If, for example,

$$p = \begin{cases} p_0, & (\lambda, \psi, z) \in \omega \\ 0, & (\lambda, \psi, z) \in \Omega \backslash \omega \end{cases} \quad p_0 > 0, \quad \omega \subset \Omega, \quad (4.18)$$

then the functional $J$ represents total concentration of a pollutant in a selected subdomain $\omega$ of the domain $\Omega$ weighted with a weight $p_0$. The region $\omega$ corresponds to the zone where the pollution is estimated. We may obtain different integral characteristics of pollution fields depending on the assignment of $\omega$. So, the problem is reduced to the estimation of the functionals of the (4.17) type defined on a set of state functions satisfying the original problem (4.1)–(4.7), (4.10)–(4.12); it is necessary to know the fields of the function $\varphi$, real information on pollutant sources in the domain $\Omega \times (0, T)$ and the initial state of atmosphere pollution in the domain $\Omega$.

An approach based on adjoint equations allows us to estimate the extent of potential danger of atmosphere pollution in the region $\omega$ from all the sources located in the domain $\Omega \times (0, T)$ with the prescribed scenarios of meteorological regime. Construct a problem adjoint to the original problem:

$$-\frac{\partial \varphi^*}{\partial t} - \frac{u}{a \sin \psi} \frac{\partial \varphi^*}{\partial \lambda} - \frac{v}{a} \frac{\partial \varphi^*}{\partial \psi} - (w - w_g) \frac{\partial \varphi^*}{\partial z} - \left( \frac{\partial}{\partial z} \nu \frac{\partial \varphi^*}{\partial z} \right)$$

$$+ \frac{1}{a^2 \sin^2 \psi} \frac{\partial}{\partial \lambda} \mu \frac{\partial \varphi^*}{\partial \lambda} + \frac{1}{a^2 \sin \psi} \frac{\partial}{\partial \psi} \mu \sin \psi \frac{\partial \varphi^*}{\partial \psi} \Bigg) = p, \quad (4.19)$$

$$\varphi^* = 0 \quad \text{at} \quad t = T,$$

$$\nu \frac{\partial \varphi^*}{\partial z} = \alpha \varphi^* \quad \text{at} \quad z = b + h,$$

$$\nu \frac{\partial \varphi^*}{\partial z} = 0 \quad \text{at} \quad z = H$$

$$\varphi^*(0, \psi, z, t) = \varphi^*(2\pi, \psi, z, t),$$

$$\varphi^*(\lambda, -\psi, z, t) = \varphi^*(\lambda + \pi, \psi, z, t),$$

$$\varphi^*(\lambda, \pi + \psi, z, t) = \varphi^*(\lambda + \pi, \pi - \psi, z, t),$$

$$(4.20)$$

where $p$ is a function determining functional (4.17).

Assume that the solution $\varphi^*(\lambda, \psi, z, t)$ of problem (4.19)–(4.20) is continuous in $\Omega \times [0, T]$ and is a function differentiable in $t$. Besides, let the function $\varphi^*(\lambda, \psi, z, t)$ belong at each $t$ to the set of functions $D(A^*) = D(A)$ from $L_2(\Omega)$. Assume again that other functions and parameters of the problem are sufficiently smooth which provides for the existence of the unique solution of problem (4.19)–(4.20). As follows from general theory, problem (4.19)–(4.20) is correct while being solved from $t = T$ up to the time to $t = 0$. That is why $\varphi^* = 0$ at $t = T$ was chosen as an "initial" condition in (4.20).

Perform some transformations. Make an inner product of equation (4.1) and $\varphi^*$ in $L_2(\Omega)$ and that of equation (4.19) and $\varphi$, subtract the results from each other and integrate in $t$ on the interval $[0, T]$. Integrating by parts with regard to boundary conditions, we arrive at another representation of the same functional $J$

$$J = \int_0^T dt \int_\Omega f \varphi^* \, d\Omega + \int_0^T dt \int_0^{2\pi} d\lambda \int_0^\pi \alpha \varphi_0 \varphi^* \bigg|_{z=b+H} d\psi. \qquad (4.21)$$

In many cases the second term in formula (4.21) may be neglected. Since we consider this very case, formula (4.21) may be written as

$$J = \int_0^T dt \int_\Omega f \varphi^* \, d\Omega. \qquad (4.22)$$

So, total amount of pollutant in the subdomain $\omega$ may be obtained using the solution $\varphi^*$ of adjoint problem (4.19)–(4.20). This function is a weight function affecting contribution of every pollution source $f$ to the atmosphere pollution over selected subdomain $\omega$. It characterizes thus an extent of danger of polluting the atmosphere in this subdomain by a source which may be located anywhere in the domain $\Omega \times (0, T)$. In other words, the input of the source to the functional is equal to the product of power of a discharge and the value of $\varphi^*$ at each $t$ in the domain where the source is located.

More or less dangerous zones may be separated and mapped in the domain $\Omega \times (0, T)$ using the values of the function $\varphi^*$, as related to the atmosphere pollution in the subdomain $\omega$. To do so, the function $\varphi^*$ is normed by its maximum value $\varphi^*_{\max}$.

## 5. PROBLEMS OF CLIMATE CHANGE SENSITIVITY IN VARIOUS REGIONS OF THE WORLD

In modelling mathematically the climate change for separate regions of the Earth's continents, the sensitivity theory of chosen functionals is important as related to the continents, World Ocean, initial data, external sources and internal parameters of the problem. The problem of climate sensitivity allows on the basis of real data to evaluate retrospectively a quality of the models and find new mechanisms responsible for formation of climate.

We will discuss in this section general approaches to the estimation of sensitivity on the basis of simple theoretical models. These results may be generalized then for the most complex statements of problems.

Let us consider the thermal interaction of atmosphere with the World Ocean and continents. Consider three-dimensional model domain $\Omega$ in a spherical system of coordinates $(\lambda, \psi, z)$, where $\lambda$ is the longitude, $\psi$ is the latitude and $z$ is the height measured from the Earth surface, which is assumed to be spherical, as well as that of the ocean. Let $h_1$ be the height of atmosphere layer, $h_2$ – thickness of active ocean layer, $h_3$ – thickness of soil layer, $S$ – the Earth's surface, $S = S_1 \cup S_2 \cup S_3$, $S_1$ – part of the Earth covered with ice and/or snow, $S_2$ – oceanic surface, and $S_3$ – continental surface free of ice and snow.

The model's domain consists of three domains: the spherical layer $\Omega_1$ of the atmosphere (troposphere), the domain $\Omega_2$ of the upper layer of the World Ocean and the domain $\Omega_3$ of the upper layer of the soil.

Here
$$\Omega_1 = \{(\lambda, \psi, z) \quad : \quad (\lambda, \psi) \in S, \quad 0 < z < h_1\},$$
$$\Omega_2 = \{(\lambda, \psi, z) \quad : \quad (\lambda, \psi) \in S_2, \quad -h_2 < z < 0\},$$
$$\Omega_3 = \{(\lambda, \psi, z) \quad : \quad (\lambda, \psi) \in S_3, \quad -h_3 < z < 0\},$$
$S = S_1 \cup S_2 \cup S_3$ is the Earth surface, $S_1$ is a part of S covered by snow and ice, $S_2$ is an ocean surface, and $S_3$ is continental surface free of snow and ice.

Examine the following problem for the temperature field $T(\lambda, \psi, z, t)$ at the interval $\Omega_t = (0, T)$:

$$\alpha \frac{\partial T}{\partial t} + \text{div}(\alpha \mathbf{u} T) - \frac{\partial}{\partial z}(\bar{\nu} \frac{\partial T}{\partial z}) - \bar{\mu} \Delta T = \varepsilon \tag{5.1}$$

with initial condition
$$T = T_0(\mathbf{r}) \quad \text{at} \quad t = 0, \tag{5.2}$$

where $\mathbf{r} = (\lambda, \psi, z)^T$, $\bar{\nu} = \alpha\nu$, $\bar{\mu} = \alpha\mu$, and $\nu$, $\mu$ are coefficients of vertical and horizontal turbulent exchange respectively assumed to be dependent on the height $z$ since they are different for atmosphere and ocean. The function $\nu$ depends also on horizontal coordinates and time. The meaning of other variables in equation (5.1): $\varepsilon$ is the source of radiation energy, the function which differs from zero just in the atmosphere, that is, in the domain $\Omega_1$; $\alpha = c_p \rho(z)$ where $c_p$ is the specific heat of a medium, $\rho(z)$ is standard density of atmosphere (at $z > 0$) and of ocean (at $z < 0$); $\mathbf{u}(\mathbf{r}, t)$ is a velocity vector of wind in the atmosphere $\Omega_1$ and of ocean currents in $\Omega_2$.

Assume that the values $\mathbf{u}$, $\nu$ and $\mu$ are assigned in the definition domain of the problem with $\mathbf{u}(\mathbf{r}, t) = 0$, $\mu = 0$ in $\Omega_3$. Assume also that the vector function $\mathbf{u}(\mathbf{r}, t)$ satisfies in atmosphere and ocean the simplest continuity equation

$$\text{div}(\rho \mathbf{u}) = 0 \tag{5.3}$$

and its normal component equals zero at the lateral surface of the World Ocean surface as well as at the surfaces $z = 0$, $z = h_1$, $z = -h_2$. Naturally, for the atmosphere, if needed, the continuity equation can be considered with regard to changing with time density of air masses.

Equation (5.1) is considered in the domain $\Omega = \Omega_1 \cup \Omega_2 \cup \Omega_3$ at $t \in \Omega_t = (0, \bar{t})$. Boundary conditions must be formulated at the boundary of the domain $\Omega$ at $z = 0$, $z = h_1$, $z = -h_2$, $z = -h_3$ and at the atmosphere-continent contact surfaces, which will provide for the existence of a unique solution to the initial boundary value problem.

Conditions of temperature equality and jump of heat fluxes are assigned at the surfaces dividing atmosphere and ocean, atmosphere and continent $(z = 0)$:

$$[T] = 0, \quad \left[ \nu \frac{\partial T}{\partial z} \right] = F(\lambda, \psi, z) \quad \text{at} \quad z = 0, \quad (\lambda, \psi) \in S_2 \cup S_3, \tag{5.4}$$

where $[f] = f|_{z=-0} - f|_{z=+0}$ is a jump of the function $f$ in $z$ at the point $z = 0$, and $F(\lambda, \psi, z)$ is assumed to be a function known from observations.

The following condition is assigned at the boundary between the atmosphere and the part of the Earth's surface covered with ice and/or snow:

$$\bar{\nu} \frac{\partial T}{\partial z} + F = 0 \quad \text{at} \quad z = 0, \ (\lambda, \psi) \in S_1. \tag{5.5}$$

The condition of no heat flux is assigned at the top boundary of atmosphere (at $z = h_1$) and at the lower boundary of active oceanic and soil layers:

$$\bar{\nu} \frac{\partial T}{\partial z} = 0 \quad \text{at} \quad z = h_1, \ z = -h_2, \ z = -h_3. \tag{5.6}$$

Finally, for simplicity, assume the boundaries between the continents and the World Ocean as cylindrical and

$$\bar{\mu} \frac{\partial T}{\partial n} = 0 \quad \text{at} \quad \partial \Omega_2, \tag{5.7}$$

where $\partial \Omega_2$ is the lateral surface of $\Omega_2$ and $n$ is normal to $\partial \Omega_2$.

So, problem (5.1)–(5.7) is posed completely and we will suggest that initial data of the problem determine its unique solution which belongs to the Hilbert space $L_2(\Omega \times \Omega_t)$. Assume that this solution is a sufficiently smooth function so that all the further transformations are justified.

The following functional is the most interesting for us:

$$J = \int\limits_0^{\bar{t}} dt \int\limits_S F^*(\lambda, \psi, t) T(\lambda, \psi, 0, t) dS, \qquad (5.8)$$

where $F^*(\lambda, \psi, t)$ is some weight function connected with a temperature field at the surface $z = 0$. For example, if we want to determine an average temperature in some selected region of a continent $\omega$ at $z = 0$ in the interval $\bar{t} - \tau \leq t \leq \bar{t}$, then we choose for $F^*$ the function

$$F^*(\lambda, \psi, t) = \begin{cases} 1/(\tau \text{mes}\,\omega), & \text{if } (\lambda, \psi) \in \omega,\ \bar{t} - \tau \leq t \leq \bar{t} \\ 0, & \text{otherwise}, \end{cases} \qquad (5.9)$$

where $\text{mes}\,\omega$ means, as usual, an area of the region $\omega$. Then functional (3.8) will be rewritten in the form:

$$J = \frac{1}{\tau} \int\limits_{\bar{t}-\tau}^{\bar{t}} dt \left( \frac{1}{\tau \text{mes}\,\omega} \int\limits_\omega T(\lambda, \psi, 0, t) dS \right). \qquad (5.10)$$

Expression (5.10) represents an average temperature over the interval $\bar{t} - \tau \leq t \leq \bar{t}$ for selected region $\omega$. These types of functionals are the most interesting in the theory of climate change.

Since the functional of the problem is defined in the form (5.8) or (5.10), we may formulate an adjoint problem to (3.1)–(3.7) using the methods we discussed in preceding chapters. It will have the form:

$$- \alpha \frac{\partial T^*}{\partial t} - \text{div}(\alpha \mathbf{u} T^*) - \frac{\partial}{\partial z} (\bar{\nu} \frac{\partial T^*}{\partial z}) - \bar{\mu} \Delta T^* = 0 \qquad (5.11)$$

with "initial" condition

$$T^*(\mathbf{r}, t) = 0 \quad \text{at} \quad t = \bar{t}. \qquad (5.12)$$

Equation (5.11) is considered in the domain $\Omega = \Omega_1 \cup \Omega_2 \cup \Omega_3$ at $t \in \Omega_t$. As for the surface $z = 0$ as well as for another "special" surface, where the necessary condition of differentiability of the solution $T^*$ and its derivatives is not met, one must determine additional boundary conditions:

$$[T^*] = 0, \quad \left[ \bar{\nu} \frac{\partial T^*}{\partial z} \right] = F^* \quad \text{at} \quad z = 0, \quad (\lambda, \psi) \in S_2 \cup S_3, \qquad (5.13)$$

where $F^*$ is defined by functional (5.8) or, in particular, (5.9).

The following condition is assigned at the boundary between the atmosphere and the part of the Earth's surface covered with ice and/or snow:

$$\bar{\nu} \frac{\partial T^*}{\partial z} + F^* = 0 \quad \text{at} \quad z = 0,\ (\lambda, \psi) \in S_1. \qquad (5.14)$$

The conditions at the top boundary of atmosphere (at $z = h_1$) and at the lower boundary of active oceanic and soil layers are as follows:

$$\bar{\nu}\frac{\partial T^*}{\partial z} = 0 \quad \text{at} \quad z = h_1, \ z = -h_2, \ z = -h_3. \tag{5.15}$$

We assume at the boundaries between the continents and the World Ocean

$$\bar{\mu}\frac{\partial T^*}{\partial n} = 0 \quad \text{at} \quad \partial\Omega_2, \tag{5.16}$$

where $n$ is an external normal to the boundary cylindrical surface "ocean–land".

Next, examine another form of the functional (5.8), expressed through the solution $T^*$ of the adjoint problem. Multiply equation (5.1) by $T^*$ and equation (5.11) by $T$, integrate in time at $\Omega_t$ and over $\Omega$ with regard to initial data (5.2), (5.12) and boundary conditions (5.4)–(5.7), (5.13)–(5.16), and subtract the results from each other. Integrate by parts to obtain

$$J = \int_\Omega \alpha T^*(\mathbf{r}, 0) T_0(\mathbf{r}) \, d\Omega + \int_0^{\bar{t}} dt \int_S F(\lambda, \psi, t) T^*(\lambda, \psi, 0, t) \, dS$$

$$+ \int_0^{\bar{t}} dt \int_{\Omega_1} \varepsilon T^*(\lambda, \psi, z, t) \, d\Omega_1. \tag{5.17}$$

Notice, that the left-hand part of this relationship coincides exactly with the functional $J$ defined by formula (5.8).

Relationship (5.17) describes a sensitivity of the functional $J$ to the location of various regions and input parameters of the problem, since the weight functions $T^*$ represent an influence of initial data $T_0(\mathbf{r})$, heat fluxes $F(\lambda, \psi, t)$ at the boundary $z = 0$ and radiation sources in the atmosphere $\varepsilon$.

The methods of adjoint equations can be also applied to other practical problems in a wide class of various problems in science and technology: from modelling the nuclear reactors, problems of global changes, to the theory of measurements and planning the experiments[134,144,149,155,183,242].

# Bibliography

[1] Agoshkov, V. I., Convergence rate estimates of perturbation algorithms, *Preprint no.30*, OVM, Moscow, 1982.

[2] Agoshkov, V. I., Projective-difference method in perturbation algorithms, *Preprint no.38*, OVM, Moscow, 1982.

[3] Agoshkov, V. I., Adjoint equations in the $N$-th order perturbation algorithms, *Sopryazhennye Uravneniya i Teoriya Vozmushenii v Zadachakh Matematicheskoi Fiziki*, OVM, Moscow, 1985, 62.

[4] Agoshkov, V. I., The $N$-th order perturbation algorithms for functionals of solutions to nonlinear problems and convergence rate estimates, *Sopryazhennye Uravneniya i Teoriya Vozmushenii v Zadachakh Matematicheskoi Fiziki*, OVM, Moscow, 1989, 3.

[5] Agoshkov, V. I., Solvability of a class of insensitive optimal control problems and application of perturbation algorithms, *Sopryazhennye Uravneniya, Algoritmy Vozmushenii i Optimalnoe Upravlenie*, IVM RAN, 1993, 2.

[6] Agoshkov, V. I., Control theory approaches in data assimilation processes, inverse problems and hydrodynamics, *Computer Mathematics and its Applications*, 1, 21, 1994.

[7] Agoshkov, V. I., Application of mathematical methods for solving the problem of liquid boundary conditions in hydrodynamics, *Report INM-95-39*, INM, Moscow, 1995.

[8] Agoshkov, V. I., Functional approaches to solving some inverse problems for the second-order abstract equations, *J. Inv. Ill-Posed Problems*, 3, 259, 1995.

[9] Agoshkov, V. I., On problems of optimal shape design, *Report* INM-95-59, INM, Moscow, 1995.

[10] Agoshkov, V. I. and Buleev, S. N., Finite element approximations in sensitivity analysis of pollution process in the Gagliary Bay, in *Proc. of the 9-th Int. Conf. of Finite Elements in Fluids. New Trends and Applications*, Padova, Venezia, 1995, 1273.

[11] Agoshkov, V. I. and Ipatova, V. M., On solvability of an insensitive control problem, *Sopryazhennye Uravneniya, Algoritmy Vozmushenii i Optimalnoe Upravlenie*, IVM RAN, 1993, 15.

[12] Agoshkov, V. I. and Maggio, F., A new method for the solution of an inverse problem for the wave equation, *Report APPMATH-93-4*, CRS4, Cagliari, 1993.

[13] Agoshkov, V. I. and Marchuk, G. I., On solvalibility and numerical solution of data assimilation problems, *Russ. J. Numer. Anal. Math. Modelling*, 8, 1, 1993.

[14] Agoshkov, V. I. and Mishneva, A. P., On finding the dispersion coefficient in non-linear parabolic equation, *Preprint no.200*, OVM, Moscow, 1988.

[15] Agoshkov, V. I., Popykin, A. I., and Shikhov, S. B., On small perturbation theory for the transport equation, *Sopryazhennye Uravneniya i Teoriya Vozmushenii v Zadachakh Matematicheskoi Fisiki*, OVM, Moscow, 1985, 76.

[16] Agoshkov, V. I. and Trufanov, O. D., The solvability and numerical solution of data assimilation problem for the linear quasi-geostrophic model, *Report INM-95-51*, INM, Moscow, 1995.

[17] Aloyan, A. E., Iordanov, D. L., and Penenko, V. V., Numerical model of pollution transport in atmospheric boundary layer, *Meteorologiya i gidrologiya*, **8**, 32, 1981.

[18] Asachenkov, A. L., An algorithm for solving the inverse problems with the use of the adjoint equation theory, *Preprint no.119*, OVM, Moscow, 1986.

[19] Ashyralyev, A. and Sobolevskii, P. E., *Well-Posedness of Parabolic Difference Equations*, Birkhäuser, Basel, 1994.

[20] Avdoshin, C. M., Belov, V. V., and Maslov, V. P., *Mathematical Aspects of Computational Media Synthesis*, MIEM, Moscow, 1984 (in Russian).

[21] Baiochi, K. and Capelo, A., *Variational and Quasi-variational Inequalities*, John Wiley, New York, 1984.

[22] Bakhvalov, N. S. and Panasenko, G. P., *Averaging the Processes in Periodical Media. Mathematical Problems of Composite Mechanics*, Nauka, Moscow, 1984 (in Russian).

[23] Bakushinsky, A. B. and Goncharsky, A. V., *Ill-Posed Problems: Theory and Applications*, Kluwer, Dordrecht, 1994.

[24] Balakrishnan, A. V., *Introduction to Optimization Theory in a Hilbert Space*, Springer, Berlin, 1971, chap. 1.

[25] Balakrishnan, A. V., Parameter estimation in stochastic differential systems: Theory and Application, in *Developments in Statistics*, Krishnaiah, P. K., Ed., Academic Press, New York, 1978, 1.

[26] Balakrishnan, A. V., *Stochastic Differential Systems. Filtering and Control*, Springer, Berlin, 1973, chap. 2.

[27] Balakrishnan, A. V., A note on the Marchuk–Zuev identification problem, in *Vistas in Applied Mathematics. Numerical Analysis. Atmospheric Sciences. Immunology*, Balakrishnan, A. V., Dorodnitsyn, A. A., and Lions, J. L., Eds., Optimization Software, New York, 1986, 291.

[28] Bates, J. R., Semazzi, F. N., and Higgins, R. W., Integration of the shallow water equations on the sphere using a vector semi-Lagrangean scheme with a multigrid solver, *Month. Weather Rev.*, **118**, 1615, 1990.

[29] Bellman, R., *Dynamic Programming*, Princeton Univ. Press, New Jersey, 1957, chap. 1.

[30] Bellman, R., *Perturbation Techniques in Mathematics, Physics and Engineering*, Holt, New York, 1964, chap. 2.

[31] Bellman, R. and Kalaba, R. E., *Quasilinearization and Nonlinear Boundary-Value Problems*, American Elsevier Publishing Company, New York, 1965, chap. 2.

[32] Bensoussan, A., Lions, J. L., and Papanicolau, G., *Asymptotic Methods in Periodic Structures*, North Holland, Amsterdam, 1978, chap. 1.

[33] Berkovitz, L. P., *Optimal Control Theory*, Springer, New York, 1974.

[34] Bloch, C., Sur la théorie des perturbations des états liés, *Nuclear Physics*, **6**, 329, 1958.

[35] Boast, C. W., Modelling the movement of chemicals in soils by water, *Soil Science*, **115**, 224, 1973.

[36] Bocharov, G. A., Study of asymptotical stability of the equilibrium state in the mathematical model of antivirus T-cell immune response on the basis of perturbation theory, *Sopryazhennye Uravneniya i Teoriya Vozmushenii v Zadachakh Matematicheskoi Fiziki*, OVM, Moscow, 1985, 116.

[37] Bocharov, G. A., Adjoint Equations and Sensitivity Analysis of Mathematical Models, *Paper 2858-B94*, VINITI, Moscow, 1994.

[38] Bogolubov, N. N. and Mitropolskii, Yu. A., *Asymptotical Methods in Nonlinear Wave Theory*, Fizmatgiz, Moscow, 1958 (in Russian).

[39] Boltyanskii, V. G., *Mathematical Methods of Optimal Control*, Nauka, Moscow, 1969 (in Russian).

[40] Boltyanskii, V. G., *Optimal Control of Discrete Systems*, Nauka, Moscow, 1973 (in Russian).

[41] Bukhgeim, A. L., The Volterra operator equations in scales of the Banach spaces, *Doklady Acad. Nauk SSSR*, **242**, 272, 1978.

[42] Butkovskii, A. G., *Control Methods for Systems with Distributed Parameters*, Nauka, Moscow, 1975 (in Russian).

[43] Cacuci, D. G., Weber, C. F., Oblow, E. M., and Marable, J. H., Sensitivity theory for general systems of nonlinear equations, *Nucl. Sci. Eng.*, **75**, 88, 1980.

[44] Carleman, I., Applications de la théories des équations intégrales singulières aux équations differentielles de la dynamique, *Arkiv. Mat. Astronom. Fys.*, **22**, 1, 1932.

[45] Chernousko, F. L. and Banichuk, V. P., *Variational Problems of Mechanics and Control*, Nauka, Moscow, 1973 (in Russian).

[46] Ciarlet, P. G., *Introduction to Numerical Linear Algebra and Optimization*, Cambridge Univ. Press, Cambridge, 1989.

[47] Cioranescu, D. and Donato, P., Exact internal controllability in perforated domains, *J. Math. Pures et Appl.*, **68**, 185, 1989.

[48] Courtier, P. and Talagrand, O., Variational assimilation of meteorological observations with the adjoint vorticity equation, *Quart. J. Roy. Meteorol. Soc.*, **113**, 1329, 1987.

[49] Davidson, J. M. and Chang, R. K., Transport of picloram in relation to soil physical conditions and pore-water velocity, *Soil Sci. Soc. Amer. Proc.*. **36**, 257, 1972.

[50] Davis, G. B., A Laplace transform technique for the analytical solution of a diffusion-convection equation over a finite domain, *Appl. Math. Modelling*, **9**, 69, 1985.

[51] Denisov, A. M. and Tuikina, S. P., On approximate solution of an inverse problem of adsorption dynamics, *Vestnik MGU*, **3**, 27, 1983.

[52] Derber, J. C., The variational 4-D assimilation of analysis using filtering models as constraints, Ph.D. Thesis, Univ. of Wisconsin, Madison, 1985.

[53] Derber, J. C., Variational four dimensional analysis using the quasigeostrophic constraint, *Month. Weather Rev.*, **115**, 998, 1987.

[54] Dulin, V. A., *Nuclear Reactor Criticity Perturbations and Group Constant Corrections*, Atomizdat, Moscow, 1979 (in Russian).

[55] Dunford, N. and Schwartz, J. T., *Linear Operators*, John Wiley, New York, 1958.

[56] Dymnikov, V. P., *Numerical Methods in Geophysical Hydrodynamics*, OVM, Moscow, 1984 (in Russian).

[57] Dymnikov, V. P. and Aloyan, A. E., Monotone schemes for solving the transport equation in problems of weather prediction, ecology and climate theory, *Izvestiya Acad. Nauk SSSR, Fizika Atmosfery i Okeana*, **26**, 1237, 1990.

[58] Dymnikov, V. P. and Filatov, A. N., *Foundations of Mathematical Theory of Climate*, VINITI, Moscow, 1994 (in Russian).

[59] Ehrlich, R. and Hurwitz, H., Multigroup methods for neutron diffusion problems, *Nucleonics*, **12**, 23, 1954.

[60] Eneev, T. M., On application of the gradient method in optimal control problems, *Kosmicheskie Issledovaniya*, **4**, 5, 1966.

[61] Ermakov, S. M., On optimal plans of regressive experiments, *Trudy MIAN*, **3**, 252, 1970.

[62] Ermakov, S. M., *Monte Carlo Method and Adjacent Questions*, Nauka, Moscow, 1975 (in Russian).

[63] Ermakov, S. M. and Makhmudov, A. A., On plans of regressive experiments minimizing the systematic error, *Zavod. Lab.*, **7**, 854, 1977.

[64] Faddeev, L. D., On the Friedrichs model in the theory of perturbation of continuous spectrum, *Trudy MIAN*, **73**, 292, 1964.

[65] Fedorenko, R. P., *Approximate Solution of Optimal Control Problems*, Nauka, Moscow, 1978 (in Russian).

[66] Feldbaum, A. A., *Fundamentals of Optimal Automatic System Theory*, Nauka, Moscow, 1966 (in Russian).

[67] Filatov, A. N., *Asymptomatical Methods in Differential and Integro-differential Equation Theory*, FAN, Tashkent, 1974 (in Russian).

[68] Filatov, A. N. and Sharova, L. V., *Integral Inequalities and Nonlinear Wave Theory*, Nauka, Moscow, 1976 (in Russian).

[69] Filatov, A. N. and Shershkov, V. V., *Asymptotical Methods in Atmospheric Models*, Gidrometeoizdat, Leningrad, 1988 (in Russian).

[70] Friedrichs, K. O., *Perturbation of Spectra in Hilbert Space*, American Math. Society, Providence, 1965, 51.

[71] Friedrichs, K. O., Uber die Spectralzerlegung eines Integraloperators, *Math. Ann.*, **115**, 249, 1938.

[72] Friedrichs, K. O., On the perturbation of continuous spectra, *Comm. Pure Appl. Math.*, **1**, 361, 1948.

[73] Friedrichs, K. O. and Rejto, P. A., On a perturbation through which the discrete spectrum becomes continuous, *Comm. Pure Appl. Math.*, **15**, 219, 1962.

[74] Gabasov, R. and Kirillova, F. M., To question on the extension of the Pontryagin maximum principle to discrete systems, *Avtomatika i Telemekhanika*, **27**, 11, 1966.

[75] Gantmakher, F. R., *Matrix Theory*, Nauka, Moscow, 1967 (in Russian).

[76] Gates, W. L., AMIP: The atmospheric model intercomparison project, *Bull. Amer. Meteorol. Soc.*, **73**, 12, 1992.

[77] Gelfand, I. M. and Fomin, S. V., *Calculus of Variations*, Fizmatgiz, Moscow, 1963 (in Russian).

[78] Genuchten, M. Jh. and Wierenga, P. J., Mass transfer studies in sorbing porous media. Analytical solutions, *Soil Sci. Soc. Amer. J.*, **40**, 473, 1976.

[79] Glasstone, S. and Edlund, M. C., *The Elements of Nuclear Reactor Theory*, Van Nostrand Co., New York, 1952, chap. 13.

[80] Glowinski, R., *Numerical Methods for Nonlinear Variational Problems*, Springer, New York, 1984.

[81] Glowinski, R. and Li, C. H., On the numerical implementation of the Hilbert uniqueness method for the exact boundary controllability of the wave equation, *C. R. Acad. Sci.*, **311**, 136, 1990.

[82] Glowinski, R., Li, C. M., and Lions, J. L., A numerical approach to the exact boundary controllability of the wave equations, *Jap. J. Appl. Math.*, **7**, 1, 1990.

[83] Glowinski, R. and Lions, I. L., Exact and approximate controllability for distributed parameter systems, *Acta Numerica*, **1**, 269, 1994.

[84] Green, R. E., Prediction of the pesticide mobility in soils (the dispersion and adsorbtion effects), *Prognozirovanie Povedeniya Pestitsidov v Okruzhayushei Srede*, Gidrometeoizdat, Leningrad, 1984, 81.

[85] Hall, M. C. G., Application of adjoint sensitivity theory to an atmospheric general circulation model, *J. Atm. Sci.*, **43**, 2644, 1986.

[86] Hudson, V. C. L. and Pym, J. S., *Application of Functional Analysis and Operator Theory*, Academic Press, New York, 1980.

[87] Iordanov, D. L., Penenko, V.V., and Aloyan, A. E., Parametrization of stratified planetary boundary layer for numerical simulation of atmospheric processes, *Izvestiya Acad. Nauk, Fizika Atmosfery i Okeana*, **14**, 815, 1978.

[88] Ipatova, V. M., On conservation laws, *Differentsialnye Uravneniya*, **5**, 903, 1992.

[89] Ipatova, V. M., A data assimilation problem for the model of general ocean circulation in quasi-geostrophic approximation, *Paper 233-1392*, VINITI, Moscow, 1992.

[90] Ivanov, V. K., On ill-posed problems, *Matem. Sbornik*, **61**, 211, 1963.

[91] Jameson, A., Aerodynamic design via control theory, *Journal of Scientific Computing*, **3**, 233, 1988.

[92]   Jameson, A., Optimum Aerodynamic Design Using CFD and Control Theory, *Paper 95-1729*, AIAA, Princeton, 1995.

[93]   Kadomtsev, B. B., On the influence function in transport theory, *Doklady Acad. Nauk SSSR*, **113**, 541, 1957.

[94]   Kalinichev, A. I. and Zolotarev, P. P., Longitudinal diffusion and nonlinear isotherms in chromatography, in *Kinetika i Dynamika Fizicheskoi Adsorbtsii*, Dubinin, M. M. and Radushkevitch, L. B., Eds., Nauka, Moscow, 1973, 189.

[95]   Kalitkin, N. N., *Numerical Methods*, Nauka, Moscow, 1978 (in Russian).

[96]   Kato, T., On the convergence of the perturbation method, *J. Fac. Sci.*, **6**, 198, 1951.

[97]   Kato, T., Perturbation of continuous spectra by trace operators, *Proc. Jap. Acad.*, **33**, 260, 1957.

[98]   Kato, T., *Perturbation Theory for Linear Operators*, Springer, Berlin, 1963, chap. 7.

[99]   Khisamutdinov, A. I., Importance access in transport theory, *Zhurnal Vychislitelnoi Matematiki i Matematicheskoi Fiziki*, **10**, 990, 1970.

[100]  Kolmogorov, A. N. and Fomin, S. V., *Elements of Theory of Functions and Functional Analysis*, Nauka, Moscow, 1981 (in Russian).

[101]  Kondratiev, K. Ya., *Meteorological Satellites*, Gidrometeoizdat, Leningrad, 1963 (in Russian).

[102]  Kontarev, G. R., The adjoint equation technique applied to meteorological problems, *Technical report No. 21*, European Centre for Medium Range Weather Forecasts, Reading, 1980.

[103]  Krasovskii, N.N., *Control Theory of Motion*, Nauka, Moscow, 1968 (in Russian).

[104]  Krasovskii, N.N., *Control of Dynamical System*, Nauka, Moscow, 1985 (in Russian).

[105]  Krein, S. G., *Linear Equations in Banach Spaces*, Nauka, Moscow, 1971 (in Russian).

[106]  Krein, S.G., On correctness classes for some problems, *Doklady Acad. Nauk SSSR*, **114**, 1162, 1957.

[107]  Kuroda, S. T., Perturbation of continuous spectra by unbounded operators, *J. Math. Soc. Jap.*, **11**, 247, 1959.

[108]  Ladyzhenskaya, O. A., *Mathematical Questions of Viscous Incompressible Fluid Dynamics*, Nauka, Moscow, 1970 (in Russian).

[109]  Ladyzhenskaya, O. A., *Boundary Value Problems of Mathematical Physics*, Nauka, Moscow, 1973 (in Russian).

[110]  Ladyzhenskaya, O. A. and Faddeev, L. D., On perturbation theory of continuous spectrum, *Doklady Acad. Nauk SSSR*, **120**, 1187, 1958.

[111]  Ladyzhenskaya, O. A., Solonnikov, V. A., and Uraltseva, N. N., *Linear and Quasilinear Equations of Parabolic Type*, American Math. Society, Providence, 1968, chap. 2.

[112]  Ladyzhenskaya, O. A. and Uraltseva, N. N., *Linear and Quasilinear Elliptic Equations*, Nauka, Moscow, 1973 (in Russian).

[113]  Landau, L. D. and Lifshitz, E. M., *Quantum Mechanics*, Fizmatgiz, Moscow, 1963 (in Russian).

[114] Lavrentiev, M.M., On statement of some ill-posed problems of mathematical physics, *Nekotorye Voprosy Vychislitelnoi i Prikladnoi Matematiki*, Nauka, Novosibirsk, 1966, 258.

[115] Lavrentiev, M. M. and Vasiliev, V. G., On statement of some ill-posed problems of mathematical physics, *Sibirskii Matem. Zhurnal*, **7**, 559, 1966.

[116] Lax, P. D. and Phillips, R. S., Scattering theory, *Bull. Amer. Math. Soc.*, **70**, 130, 1964.

[117] Le Dimet, F. X., A general formalism of variational analysis, *Report OK-73091-22-1*, CIMMS, Norman, 1982.

[118] Le Dimet, F. X. and Talagrand, O., Variational algorithms for analysis and assimilation of meteorological observations: theoretical aspects, *Tellus A*, **38**, 97, 1986.

[119] Lee, E. B. and Markus, L., *Foundations of Optimal Control Theory*, John Wiley, New York, 1967.

[120] Le Tallec, P. and Halard, M., Second order methods for the optimal design of non-linear structures, in *Computational Methods in Applied Sciences*, Hirsch, Ch., Ed., Elsevier Science Publishers, New York, 1992, 247.

[121] Letov, A. M., *Flight Dynamics and Control*, Nauka, Moscow, 1969 (in Russian).

[122] Le Veque, R. J., Time-split methods for partial differential equations, *Report STAN-CS-82-904*, Stanford University, Stanford, 1982.

[123] Levitus, S. and Oort, A. N., Global analysis of oceanographic data, *Bull. Amer. Meteorol. Soc.*, **58**, 1270, 1977.

[124] Lewins, J., *Importance. The Adjoint Function*, Pergamon Press, New York, 1965, chap. 3.

[125] Lewis, J. and Derber, J., The use of adjoint equations to solve a variational adjustment problem with advective constraints, *Tellus A*, **37**, 309, 1985.

[126] Liestra, M. and Dekkers, W. A., Some models for the adsorption kinetics of pesticides in soil, *J. Environ. Sci. Health*, **12**, 85, 1977.

[127] Lighthill, M. J., A technique for rendering approximate solutions to physical problems uniformly valid, *Phil. Mag.*, **40**, 1179, 1949.

[128] Lindstrom, F. I. and Boersma, L., Theory of chemical transport with simultaneous sorption in a water saturated porous medium, *Soil Science*, **110**, 1, 1970.

[129] Lions, J. L., Sur les sentinelles des systems distributes. Le cas des conditions initials incompletes, *C. R. Acad. Sci.*, **307**, 819, 1988.

[130] Lions, J. L., Exact controllability, stabilization and perturbation for distributed systems, *SIAM Rev.*, **30**, 1, 1988.

[131] Lions, J. L., Insensitive controls, *Computational Mathematics and Applications, Proceedings of 8-th France–USSR–Italy Joint Symposium*, Pavia, October 2–6, 1989, Publicazioni No.730, Pavia, 1989, 285.

[132] Lions, J. L., *Sur le Controle Optimal de Systémes Gouvernes par des Equation aux Derivees Partielles*, Dunod, Paris, 1968.

[133] Lions, J. L., *Contrôllabilité Exacte Perturbations et Stabilisation de Systèmes Distribués*, Masson, Paris, 1988, 32.

[134] Lions, J. L., *El Planeta Tierra*, Espasa, Madrid, 1990, chap. 1.

[135] Lions, J. L. and Magenes, E., *Problémes aux Limites non Homogenes et Applications*, Dunod, Paris, 1968, chap. 3.

[136] Lions, J. L., Temam, R., and Wang, S., Models for the coupled atmosphere and ocean, *Computational Mechanics Advances*, **1**, 3, 1993.

[137] Lobarev, I. V., Perturbation method for investigation of simple eigenvalues of the integral transport equation, *Sopryazhennye Uravneniya i Teoriya Vozmushenii v Zadachakh Matematicheskoi Phisiki*, OVM, Moscow, 1985, 136.

[138] Lomov, S. A., *Introduction to General Singular Perturbation Theory*, Nauka, Moscow, 1981 (in Russian).

[139] Lorenc, A. C., Optimal nonlinear objective analysis, *Quart. J. Roy. Meteor. Soc.*, **114**, 205, 1988.

[140] Lurie, K. A., *Optimal Control in Problems of Mathematical Physics*, Nauka, Moscow, 1977 (in Russian).

[141] Lusternik, L. A. and Sobolev, V. I., *Elements of Functional Analysis*, Nauka, Moscow, 1965 (in Russian).

[142] Lyapunov, A. M., *General Problem on Motion Stability*, Kharkov Math. Society, Kharkov, 1892 (in Russian).

[143] Lyapunov, A. M., *Collected Works*, Gostekhizdat, Moscow, 1956 (in Russian).

[144] Marchuk, G., I., *Numerical Methods for Nuclear Reactor Computations*, Consultants Bureau, London, 1959.

[145] Marchuk, G. I., *Introduction into the Methods of Numerical Analysis*, Cremonese, Roma, 1973.

[146] Marchuk, G. I., *Numerical Methods in Weather Prediction*, Acad. Press, New York, 1974.

[147] Marchuk, G. I., *Methodes de Calcul Numerique*, Mir, Moscow, 1980.

[148] Marchuk, G. I., *Mathematical Models in Immunology*, Optimization Software, New York, 1983.

[149] Marchuk, G. I., *Mathematical Models in Environmental Problems*, North Holland, Amsterdam, 1986.

[150] Marchuk, G. I., Splitting and Alternating Direction Methods, in *Handbook of Numerical Analysis*, **1**, Ciarlet, P. G. and Lions, J. L., Eds., North Holland, Amsterdam, 1990, 197.

[151] Marchuk, G. I., Application of adjoint equations for solving the problems of mathematical physics, *Advances of Mechanics*, **1**, 3, 1981.

[152] Marchuk, G. I., On statement of inverse problems, *Doklady Acad. Nauk SSSR*, **156**, 503, 1964.

[153] Marchuk, G. I., Equation for importance of information from meteorological satellites and statement of inverse problems, *Kosmicheskie Issledovaniya*, **2**, 462, 1964.

[154] Marchuk, G. I., Main and adjoint equations of atmosphere and ocean dynamics, *Meteorologiya i Gidrologiya*, **2**, 9, 1974.

[155] Marchuk, G. I., Adjoint Equations and Analysis of Complex Systems, Nauka, Moscow, 1992 (in Russian).

[156] Marchuk, G. I., Methods of long-range weather forecast on the basis of solutions of main and adjoint equations, *Meteorologiya i Gidrologiya*, **3**, 17, 1974.

[157] Marchuk, G. I., Environment and problems of optimal siting of plants, *Doklady Acad. Nauk SSSR*, **227**, 1056, 1976.

[158] Marchuk, G. I., A survey of nuclear reactor computation methods, *Atomnaya Energiya*, **11**, 356, 1961.

[159] Marchuk, G. I., On the problem of mathematical modelling of ecological situations in water basins, *Sovremennye Problemy Matematicheskoi Fiziki i Vychislitelnoi Matematiki*, Nauka, Moscow, 1982, 254.

[160] Marchuk, G. I., Environment and optimization problems, *Preprint no.1*, Vychislit. Tsentr SO AN SSSR, Novosibirsk, 1975.

[161] Marchuk, G. I., To the problem of environment protection, *Vychislitelnye Metody v Matematicheskoi Fizike, Geofizike i Optimalnom Upravlenii*, Nauka, Novosibirsk, 1978, 20.

[162] Marchuk, G. I., Economical criteria for planning, protection and restoration of environment, *Zhurnal Vychislitelnoi Matematiki i Matematicheskoi Fiziki*, **20**, 1365, 1980.

[163] Marchuk, G. I., *Some Mathematical Problems of Environment Protection. Complex Analysis and Applications*, Nauka, Moscow, 1981 (in Russian).

[164] Marchuk, G. I., Environment and optimization problems, *Trudy MIAN*, **166**, 123, 1984.

[165] Marchuk, G. I., Climate change modelling and problem of long-range weather forecast, *Meteorologiya i Gidrologiya*, **7**, 25, 1979.

[166] Marchuk, G. I., Some application of splitting-up methods to the solution of mathematical physics problems, *Appl. Math.*, **13**, 2, 1968.

[167] Marchuk, G. I., L'etablissment d'un modele de changements de climat et le probleme de la prevision meteorologique a long term, *La Meteorologie*, **16**, 103, 1979.

[168] Marchuk, G. I., Formulation of the theory of perturbations for complicated models, *Geofisica Internacional*, **15**, 103, 1975.

[169] Marchuk, G. I., Mathematical issues of industrial effluent optimization, *J. Meteorol. Soc. Jap.*, **60**, 481, 1982.

[170] Marchuk, G. I., Perturbation theory and the statement of inverse problems, *Lect. Notes Comput. Sci.*, **4**, 159, 1973.

[171] Marchuk, G. I., Applications of adjoint equations to problems of global change and environment protection, *Computer Mathematics and its Applications*, **1**, 1, 1994.

[172] Marchuk, G. I. and Agoshkov, V. I., *Introduction aux Methodes des Elements Finis*, Mir, Moscow, 1985.

[173] Marchuk, G. I. and Agoshkov, V. I., Conjugate operators and algorithms of perturbation in non-linear problems: 1. Principles of construction of conjugate operators, *Sov. J. Numer. Anal. Math. Modelling*, **1**, 21, 1988.

[174] Marchuk, G. I. and Agoshkov V. I., Conjugate operators and algorithms of perturbation in non-linear problems: 2. Perturbation algorithms, *Sov. J. Numer. Anal. Math. Modelling*, **2**, 115, 1988.

[175] Marchuk, G. I. and Agoshkov, V. I., Adjoint equations in nonlinear problems and applications, *Funktsionalnye i Chislennye Metody Matematicheskoi Fiziki*, Naukova Dumka, Kiev, 1988, 138.

[176] Marchuk, G. I. and Agoshkov, V. I., Symmetrization of nonstationary transport equation and variational principle, *Preprint no.222*, Vychislit. Tsentr SO AN SSSR, Novosibirsk, 1980.

[177] Marchuk, G. I., Agoshkov, V. I., and Shutyaev, V. P., Adjoint equations and perturbation algorithms in applied problems, *Vychislitelnye Protsessy i Sistemy*, Nauka, Moscow, 1986, 5.

[178] Marchuk, G. I., Agoshkov, V. I., and Shutyaev, V. P., *Adjoint Equations and Perturbation Algorithms*, OVM, Moscow, 1986 (in Russian).

[179] Marchuk, G. I., Agoshkov, V. I., and Shutyaev, V. P., *Adjoint Equations and Perturbation Methods in Nonlinear Problems of Mathematical Physics*, Nauka, Moscow, 1993 (in Russian).

[180] Marchuk, G. I. and Belskaya, J. N., On application of adjoint equations to computation of radiation protection, *Voprosy Fiziki Zashity Reaktorov*, Gosatomizdat, Moscow, 1963, 99.

[181] Marchuk, G. I. and Drobyshev, Yu. P., Some questions of linear measurement theory, *Avtometriya*, **3**, 24, 1967.

[182] Marchuk, G. I., Dymnikov, V. P., Kurbatkin, G. P., and Sarkisyan, A. S., Programme "Sections" and the World ocean monitoring, *Meteorologiya i Gidrologiya*, **6**, 9, 1984.

[183] Marchuk, G. I. and Ermakov, S. M., On some problems of the planning experiment theory, *Matematicheskie Metody Planirovaniya Eksperimenta*, Nauka, Novosibirsk, 1981, 3.

[184] Marchuk, G. I. and Kagan, B. A., *Ocean Tides: Mathematical Models and Numerical Algorithms*, Pergamon Press, Oxford, 1984.

[185] Marchuk, G. I. and Kurbatkin, G. P., Physical and mathematical aspects of analysis and weather forecast, *Meteorologiya i Gidrologiya*, **11**, 25, 1977.

[186] Marchuk, G. I., Kuzin, V. I., and Obraztsov, N. N., *Numerical Modelling of Distribution in a Water Basin*, Computing Centre Publ., Novosibirsk, 1979.

[187] Marchuk, G. I., Kuzin, V. I., and Skiba, Yu. N., Projective-difference method for computing the adjoint functions for the model of heat transport in the system "atmosphere-ocean-soil", *Aktualnye Problemy Vychislitelnoi i Priklednoi Matematiki*, Nauka, Moscow, 1983, 149.

[188] Marchuk, G. I. and Lebedev, V. I., *Numerical Methods in the Theory of Neutron Transport*, Harwood Academic Publishers, New York, 1986.

[189] Marchuk, G. I. and Lykosov, V. N., Diagnostical computation of vertical mixing coefficients in upper boundary ocean layer, *Matematicheskoe Modelirovanie Protsessov v Pogranichnykh Sloyakh Atmosfery i Okeana*, OVM, Moscow, 1989, 4.

[190] Marchuk, G. I. and Mikhailov, G. A., Solution of transport theory problems by the Monte Carlo method, *Teoreticheskie i Prikladnye Problemy Rasseyaniya Sveta*, Nauka i Technika, Minsk, 1971, 43.

[191] Marchuk, G. I., Mikhailov, G.A., and Nazaraliev, M.A., *The Monte-Carlo Methods in Atmospheric Optics*, Springer, Berlin, 1980.

[192] Marchuk, G. I. and Orlov, V. V., To the theory of adjoint functions, *Neitronnaya Fizika*, Gosatomizdat, Moscow, 1961, 30.

[193] Marchuk, G. I. and Penenko, V. V., Study of sensitivity of discrete models of atmosphere and ocean dynamics, *Izvestiya Acad. Nauk SSSR, Fizika Atmosfery i Okeana*, **15**, 1123, 1979.

[194] Marchuk, G. I. and Penenko, V. V., Some applications of optimization methods to environment problems, *Vychislitelnye Metody v Prikladnoi Matematike*, Nauka, Novosibirsk, 1982, 5.

[195] Marchuk, G. I. and Penenko, V. V., Application of optimization method to the problem of mathematical simulation of atmospheric processes and environment, in *Modelling and Optimization of Complex Systems: Proc. of the IFIP-TC7 Work Conf.*, Marchuk, G. I., Ed., Springer, Berlin, 1978, 240.

[196] Marchuk, G. I., Penenko, V. V., and Protasov, A. V., Variational principle in a small-parameter model of the atmosphere dynamics, *Variatsionno-raznostnye Metody v Matematicheskoi Fizike*, Vychisl. Tsentr SO AN SSSR, Novosibirsk, 1978, 213.

[197] Marchuk, G. I. and Skiba, Yu. N., Numerical computation of the adjoint problem for the model of thermal interaction of the atmosphere with the ocean and continents, *Izvestiya Acad. Nauk SSSR, Fizika Atmosfery i Okeana*, **12**, 459, 1976.

[198] Marchuk, G. I. and Skiba, Yu. N., The computation of spatial-temporal influence functions for monthly-averaged anomalies of the surface air temperature in bounded regions, *Dynamika Atmosfery i Okeana*, OVM, 1990, 35.

[199] Marchuk, G. I. and Skiba, Yu. N., Numerical calculation of the conjugate problem for a model of the thermal interaction of the atmosphere with the oceans and continents, *Atmos. Ocean. Phys.*, **12**, 279, 1976.

[200] Marchuk, G. I. and Skiba, Yu. N., Role of adjoint equations in estimating monthly mean air surface temperature anomalies, *Atmosphera*, **5**, 119, 1992.

[201] Marchuk, G. I., Skiba, Yu. N., and Protsenko, I. G., A method for computing the evolution of random hydrological fields on the basis of adjoint equations, *Izvestiya Acad. Nauk SSSR, Fizika Atmosfery i Okeana*, **21**, 115, 1985.

[202] Marchuk, G. I. and Shaidurov, V. V., *Difference Methods and Their Extrapolation*, Springer, New York, 1983.

[203] Marchuk, G. I. and Shutyaev, V. P., Iteration methods for solving a data assimilation problem, *Russ. J. Numer. Anal. Math. Modelling*, **9**, 265, 1994.

[204] Marchuk, G. I. and Zalesny, V. B., A numerical technique for geophysical data assimilation problem using Pontryagin's principle and splitting-up method, *Russ. J. Numer. Anal. Math. Modelling*, **8**, 4, 1993.

[205] Maslov, V. P., *Perturbation Theory and Asymptotical Methods*, MGU, Moscow, 1965 (in Russian).

[206] Maslov, V. P., Perturbation theory when a point spectrum becomes continuous, *Doklady Acad. Nauk SSSR*, **109**, 267, 1956.

[207] Mignot, F. and Puel, J. P., Optimal control in some variational inequalities, *SIAM J. Control. Opt.*, **22**, 466, 1984.

[208] Mikhailov, G. A., The use of approximate solutions of the adjoint problem for approvement of the Monte Carlo algorithms, *Zhurnal Vychislitelnoi Matematiki i Matematicheskoi Fiziki*, 9, 1145, 1969.

[209] Mikhailov, G. A., *Some Questions of the Monte Carlo Method Theory*, Nauka, Novosibirsk, 1974 (in Russian).

[210] Mikhailov, V. P., *Partial Differential Equations*, Nauka, Moscow, 1983 (in Russian).

[211] Millman, M. N. and Keller, J. B., Perturbation theory of nonlinear boundary value problems, *J. of Math. Phys.*, **10**, 342, 1969.

[212] Mironenko, E. V., Mathematical description of the pesticide adsorption dynamics, *Metody i Problemy Ekotoksikologicheskogo Modelirovaniya i Prognozirovaniya*, Pushkino, 1979, 138.

[213] Mishneva, A. P., On finding the effective dispersion coefficient with the use of experimental data and the theory of adjoint equations and perturbation algorithms, *Sopryazhennye Uravneniya i Algoritmy Vozmushenii*, OVM, 1988, 101.

[214] Mishneva, A. P., The effective dispersion coefficient prediction when varying the parameters of the experiment, *Sopryazhennye Uravneniya i Algoritmy Vozmushenii*, OVM, 1988, 112.

[215] Moiseev, N. N., *Elements of the Optimal System Theory*, Nauka, Moscow, 1975 (in Russian).

[216] Moiseev, N. N., *Asymptotical Methods of Nonlinear Mechanics*, Nauka, Moscow, 1981 (in Russian).

[217] Monin, A. S. and Yaglom, A. M., *Statistical Hydromechanics*, Nauka, Moscow, 1965 (in Russian).

[218] Moser, J., Stürüngstheorie des kontinuierlichen Spektrums für gewöhnliche Differentialgleichungen zweiter Ordnung, *Math. Ann.*, **125**, 366, 1953.

[219] Musaelyan, Sh. A., Parametrization problem for the process of transport of solar radiation to the system "ocean-atmosphere" and the long-range weather forecast, *Meteorologiya i Gidrologiya*, **10**, 9, 1974.

[220] Natanson, I. P., *Theory of Functions of Real Variables*, Nauka, Moscow, 1974 (in Russian).

[221] Navon, I. M. and Legler, D. M., Conjugate-gradient methods for large-scale minimization in meteorology, *Month. Weather Rev.*, **115**, 1479, 1987.

[222] Nayfeh, A. H, *Perturbation Methods*, John Wiley, New York, 1973, chap. 1.

[223] Nazaraliev, M. A., Study of approximate importance function by the Monte Carlo method, *Metody Monte Carlo i ikh Primeneniya*, Vychislit. Tsentr SO AN SSSR, Novosibirsk, 1971, 116.

[224] Nekrutkin, V. V., The direct and adjoint Neumann-Ulam schemes for solving nonlinear integral equations, *Zhurnal Vychislitelnoi Matematiki i Matematicheskoi Fiziki*, **14**, 1409, 1974.

[225] Nikolskii, S. M., *Approximation of Functions of Many Variables and Imbedding Theorems*, Nauka, Moscow, 1977 (in Russian).

[226] Obraztsov, N. N., Mathematical modelling and optimization in environment protection problems, *Preprint no.85*, OVM, Moscow, 1985.

[227] Obraztsov, N. N., On pollution transport with a point sourse, *Preprint no.66*, OVM, Moscow, 1983.

[228] Oleinik, O. A., Boundary value problems for partial differential equations with small parameter and the Cauchy problem for nonlinear equations, *Uspekhi Matem. Nauk*, **10**, 229, 1955.

[229] Olver, P. J., *Application of Lie Groups to Differential Equations*, Springer, Berlin, 1986.

[230] Osipov, Yu. and Suetov, A. P., On a problem of J.-L.Lions, *Soviet Math. Dokl.*, **29**, 487, 1984.

[231] Ovsyannikov, L. V., *Group Analysis of Differential Equations*, Nauka, Moscow, 1978 (in Russian).

[232] Pankratiev, Yu. D., To the question on perturbations of linear functionals, *Sistemnyi Analiz i Issledovaniye Operatsii*, Vychislit. Tsentr SO AN SSSR, Novosibirsk, 1977, 5.

[233] Penenko, V. V., Computational aspects of the atmosphere dynamics modelling and estimation of the effect of various factors, *Nekotorye Problemy Vychislitelnoi i Prikladnoi Matematiki*, Nauka, Novosibirsk, 1975, 61.

[234] Penenko, V. V. and Aloyan, A. E., *Models and Methods for Environment Protection Problems*, Nauka, Novosibirsk, 1985 (in Russian).

[235] Penenko, V. and Obraztsov, N. N., A variational initialization method for the fields of the meteorological elements, *Sov. J. Meteorol. Hydrol.*, **11**, 1, 1976.

[236] Pironneau, O., *Optimal Shape Design for Elliptic Systems*, Springer, New York, 1984.

[237] Poincaré, H., *Les Méthodes Nouvelles de la Mécanique Céleste*, Gauthier–Villars, Paris, 1892, 8.

[238] Pokhozhaev, S. I., Normal solvability of nonlinear equations in uniformly convex Banach spaces, *Funktsionalnyi Analiz i ego Prilozheniya*, **3**, 80, 1969.

[239] Pontryagin, L. S., Mathematical theory of optimal processes and differential games, *Trudy MIAN*, **169**, 119, 1985.

[240] Pontryagin, L. S., *Selected Works*, Nauka, Moscow, 1988 (in Russian).

[241] Pontryagin, L. S., Boltyanskii, V. G., Gamkrelidze, R. V., and Mischenko, E. F., *The Mathematical Theory of Optimal Processes*, John Wiley, New York, 1962, chap.1.

[242] Pupko, V. Ya., Zrodnikov, A. V., and Likhachev, Yu. I., *The Adjoint Function Method in Physics and Engineering*, Energoatomizdat, Moscow, 1984 (in Russian).

[243] Rahnema, F., Internal interface perturbation in neutron transport theory, *Nucl. Scien. and Eng.*, **86**, 76, 1984.

[244] Rayleigh, L. (Stratt, J. W.), *Theory of Sound*, Mc Millan, London, 1926, 1.

[245] Prokhorov, V. M., *Nuclear Pollutant Migration in Soil*, Energoizdat, Nauka, 1981 (in Russian).

[246] Rellich, F., Störungthorie des Spektralzerlegung, *Math. Ann.*, **117**, 346, 1936.

[247] Rellich, F., Störungthorie der Spectralzerlegung, in *Proc. Internat. Congr. Mathematicians*, American Math. Society, Providence, 1952, 606.

[248] Rellich, F., *Perturbation Theory of Eigenvalue Problems*, Gordon and Breach Sci. Pub., New York, 1969, chap.2.

[249] Riesz, F. and Sz.–Nagy, B., *Functional Analysis*, Frederik Ungar, New York, 1955, chap.3.

[250] Romanov, V. G., *Inverse Problems of Mathematical Physics*, Nauka, Moscow, 1984 (in Russian).

[251] Rose, D. A. and Passioura, J. B., The analysis of experiments on hydrodynamic dispersion, *Soil Science*, **111**, 252, 1971.

[252] Rosenblum, M., Perturbation of the continuous spectrum and unitary equivalence, *Pacific J. Math.*, **7**, 997, 1957.

[253] Rozhdestvenskii, B. L. and Yanenko, N. N., *Systems of Quasilinear Equations*, Nauka, Moscow, 1968 (in Russian).

[254] Sadokov, V. P. and Shteynbok, D. B., Application of adjoint functions to the analysis and forecast of the temperature anomalies, *Sov. J. Meteorol. Hydrol.*, **8**, 6, 1976.

[255] Sanchez–Palencia, E. *Nonhomogeneous Media and Vibration Theory*, Springer, Berlin, 1980, chap.2.

[256] Sarkisyan, A. S. and Ibraev, R., A note on modelling the world ocean climate, *Ocean Modelling*, 89, 10, 1990.

[257] Schrödinger, E., Quantisierung als Eigenwertproblem, *Ann. Phys.*, **80**, 437, 1926.

[258] Schwartz, L., *Théorie des Distributions*, Hermann, Paris, 1966, chap.3.

[259] Schweich, D. and Sardin, M., Adsorption partition, ion exchange and chemical reaction in batch reactions or in columns – a review, *J. Hydrology*, **50**, 1, 1981.

[260] Sedunov, E. V., On the experiment plans minimizing a systematical error, *Matematicheskie Metody Planirovaniya Eksperimenta*, Nauka, Novosibirsk, 1981, 102.

[261] Shikhov, S. B., *Questions of Mathematical Nuclear Reactor Theory*, Atomizdat, Moscow, 1972 (in Russian).

[262] Shutyaev, V. P., Properties of adjoint operators arisen in perturbation algorithms for a quasilinear elliptic problem, *Sopryazhennye Uravneniya i Algoritmy Vozmushenii*, OVM, Moscow, 1988, 119.

[263] Shutyaev, V. P., On justification of perturbation algorithm in a nonlinear hyperbolic problem, *Matem. Zametki*, **49**, 155, 1991.

[264] Shutyaev, V. P., An algorithm for computing functionals for a class of nonlinear problems using the conjugate equation, *Sov. J. Numer. Anal. Math. Modelling*, **6**, 169, 1991.

[265] Shutyaev, V. P., Properties of solutions of the adjoint equation in a nonlinear hyperbolic problem, *Differentsialnye Uravneniya*, **28**, 706, 1992.

[266] Shutyaev, V. P., On a class of insensitive control problems, *Control and Cybernetics*, **23**, 257, 1994.

[267] Shutyaev, V. P., Some properties of the control operator in the problem of data assimilation and iterative algorithms, *Russ. J. Numer. Anal. Math. Modelling*, **10**, 357, 1995.

[268] Shutyaev, V. P. and Parmuzin, E. I., Numerical analysis of iterative algorithms for solving a data assimilation problem, *Report INM-95-47*, INM, Moscow, 1995.

[269] Shutyaev, V. P. and Seleznev, S. G., Quasilinear partial differential equations and applications in problems of mathematical physics, *Preprint no.258*, OVM, Moscow, 1990.

[270] Shutyaev, V. P. and Seleznev, S. G., Adjoint equations and perturbation algorithms for quasilinear motion equation, *Preprint no.259*, OVM, Moscow, 1990.

[271] Simon, J., Differentiation with respect to the domain in boundary value problems, *Numerical Functional Analysis and Optimization*, **2**, 649, 1980.

[272] Smagorinsky, J., General circulation experiments with the primitive equations: 1. The basic experiment, *Month. Weather Rev.*, **91**, 99, 1963.

[273] Smedstad, O. M. and O'Brien, J. J., Variational data assimilation and parameter estimation in an equatorial Pacific Ocean model, *Prog. Oceanolog.*, **26**, 179, 1991.

[274] Sobolev, S. L., *Application of Functional Analysis in Mathematical Physics*, American Math. Society, Providence, 1963, chap. 1.

[275] Sobolev, S. L., *Introduction to the Theory of Cubature Formulas*, Nauka, Moscow, 1971 (in Russian).

[276] Spiridonov, Yu. Ya., Shestakov, V., G., Matveev, Yu. M., and Spiridonova, S. G., The herbicide adsorption by main soil components, *Agrokhimiya*, **3**, 83, 1984.

[277] Stoker, L. J, *Water Waves. The Mathematical Theory with Applications*, Interscience Publishers, New York, 1957, chap. 1.

[278] Strang, G., On the construction and comparison of difference schemes, *SIAM J. Numer. Anal.*, **5**, 23, 1968.

[279] Strang, G., *Linear Algebra and its Applications*, Academic Press, New York, 1976, chap. 1.

[280] Stumbur, E. A., *Application of Perturbation Theory in Nuclear Reactor Physics*, Nauka, Moscow, 1976 (in Russian).

[281] Sz.-Nagy, B., Perturbations des tranformations lineaires fermeés, *Acta Sci. Math. Szeged.*, **14**, 125, 1951.

[282] Talagrand, O. and Courtier, P., Variational assimilation of meteorological observations with the adjoint vorticity equation. Part I: Theory, *Quart. J. Roy. Meteorol. Soc.*, **113**, 1311, 1987.

[283] Tikhonov, A. N., On solution of ill-posed problems and regularization method, *Doklady Acad. Nauk*, **151**, 501, 1963.

[284] Tikhonov, A. N., On regularization of ill-posed problems, *Doklady Acad. Nauk*, **153**, 49, 1963.

[285] Tikhonov, A. N. and Arsenin, V. Ya., *Methods for Solving the Ill-posed Problems*, Nauka, Moscow, 1974 (in Russian).

[286] Tikhonov, A. N., Leonov, A. S., and Yagola, A. G., *Nonlinear Ill-posed Problems*, Nauka, Moscow, 1995 (in Russian).

[287] Tikhonov, A. N., Vasilieva, A. B., and Sveshnikov, A. G., *Differential Equations*, Nauka, Moscow, 1980 (in Russian).

[288] Titchmarsh, E. C., Some theorems on perturbation theory, *J. Analys. Math.*, **4**, 187, 1954.

[289] Trenogin, V. A., Development and applications of the Lusternik-Vishik asymptotical method, *Uspekhi Matem. Nauk*, **25**, 123, 1970.

[290] Trenogin, V. A., *Functional Analysis*, Nauka, Moscow, 1980 (in Russian).

[291] Tuikina, S. R., Numerical solution of some inverse problems of adsorption dynamics, *Vestnik MGU*, 1, 31, 1985.

[292] Tziperman, E. and Thacker, W. C., An optimal-control/adjoint-equations approach to studying the oceanic general circulation, *Phys. Oceanogr.*, **19**, 1471, 1989.

[293] Usachev, L. N., The importance nuclear reactor equation and perturbation theory, *Reaktorostroenie i Teoriya Reaktorov*, AN SSSR, Moscow, 1955, 251.

[294] Usachev, L. N., Perturbation theory for the reproduction coefficient and other values of various nuclear reactor processes, *Atomnaya Energiya*, **15**, 472, 1963.

[295] Usachev, L. N. and Bobkov, Yu., G., *Perturbation Theory and Planning the Experiments in the Problem of Nuclear Data for Reactors*, Atomizdat, Moscow, 1980 (in Russian).

[296] Vainberg, M. M., *Functional Analysis*, Prosveshenie, Moscow, 1979 (in Russian).

[297] Vainberg, M. M. and Trenogin, V. A., *Bifurcation Theory for Solutions of Nonlinear Equations*, Nauka, Moscow, 1969 (in Russian).

[298] Van Dyke, M. D., *Perturbation Methods in Fluid Mechanics*, Academic Press, New York, 1964.

[299] Vasiliev, V. G., One dimensional inverse problem, *Nekotorye Metody i Algoritmy Interpretatsii Geofisicheskikh Nabludenii*, Nauka, Moscow, 1967, 35.

[300] Vasiliev, F. P., *Numerical Methods for Solving the Extremal Problems*, Nauka, Moscow, 1988 (in Russian).

[301] Vasilieva, A. B. and Butuzov, V. F., *Asymptotic Expansions of Solutions to Singular Perturbed Equations*, Nauka, Moscow, 1973 (in Russian).

[302] Vishik, M. I. and Lusternik, L. A., Solution of perturbation problems for matrices and selfadjoint and nonselfadjoint differential equations, *Uspekhi Matem. Nauk*, **15**, 3, 1960.

[303] Vishik, M. I. and Lusternik, L.A., Regular degeneracy and boundary layer for linear differential equations with a small parameter, *Uspekhi Matem. Nauk*, **12**, 3, 1957.

[304] Vishik, M. I. and Lusternik, L. A., Asymptotic behaviour of solutions of differential equations with large or oscillating coefficients and boundary conditions, *Uspekhi Mathem. Nauk*, **15**, 27, 1960.

[305] Vishik, M. I. and Lusternik, L. A., Some questions of perturbations of boundary value problems for partial differential equations, *Doklady Acad. Nauk SSSR*, **129**, 1203, 1959.

[306] Vishik, M. I. and Lusternik, L. A., Perturbation of eigenvalues and eigenfunctions for some nonselfadjoint operators, *Doklady Acad. Nauk SSSR*, **130**, 251, 1960.

[307] Vladimirov, V. S., *Equations of Mathematical Physics*, Nauka, Moscow, 1981 (in Russian).

[308] Vladimirov, V. S., Mathematical problems of one-velocity transport theory, *Trudy MIAN*, Moscow, 1961, 3.

[309] Vladimirov, V. S., *Generalized Functions in Mathematical Physics*, Nauka, Moscow, 1976 (in Russian).

[310] Vladimirov, V. S. and Volovich, I. V., Conservation laws for nonlinear equations, *Doklady Acad. Nauk SSSR*, **279**, 843, 1984.

[311] Vladimirov, V. S. and Volovich, I. V., Conservation laws for nonlinear equations, *Aktualnye Problemy Vychislitelnoi Matematiki i Matematicheskogo Modelirovaniya*, Nauka, Novosibirsk, 1985, 147.

[312] Voevodin, V. V. and Kuznetsov, Yu. A., *Matrices and Computations*, Nauka, Moscow, 1984 (in Russian).

[313] Vorobiyov, G. I., Susuev, V.V., and Pavlushkin, L. T., Study of the fertilizer migration in podzol, *Vestnik Selskokhozyaistvennykh Nauk*, **10**, 73, 1980.

[314] Whitham, G. B., *Linear and Nonlinear Waves*, John Wiley, New York, 1974.

[315] Wilkinson, J., *The Algebraic Eigenvalue Problem*, Clarendon Press, Oxford, 1965.

[316] Yosida, K., *Functional Analysis*, Springer, Berlin, 1971.

[317] Zalesny, V. B. and Galkin, N. A., The temperature investigation in the thermodynamical model of the Arabian Sea, *Okeanologiya*, **35**, 514, 1995.

[318] Zhu, K., Navon, I. M., and Zou, X., Variational data assimilation with a variable resolution finite-element shallow-water equation model, *Month. Weather Review*, **122**, 946, 1994.

[319] Zolotarev, P.P., Lukshin, A. V., and Ryutin, A. A., On mathematical modelling of adsorption dynamics, *Vestnik MGU*, **3**, 56, 1983.

[320] Zou, J., Hsieh, W. W., and Navon, I. M., Sequential open-boundary control by data assimilation in a limited-area model, *Month. Weather Rev.*, **123**, 2905, 1995.

[321] Zou, J. and Holloway, G., Improving steady-state fit of dynamics to data using adjoint equation with gradient proconditioning, *Month. Weather Rev.*, **123**, 199, 1995.

[322] Zou, X., Navon, I. M., and LeDimet, F. X., Incomplete observations and control of gravity waves in variational data assimilation, *Tellus A*, **44**, 273, 1992.

[323] Zuev, S. M., *Statistical Parameter Estimation in Mathematical Models of Disease*, Nauka, Moscow, 1988 (in Russian).

# Index

### A

adjoint equation, 37, 65
adjoint operator, 5, 65
adjointness relationship, 66

### B

Banach space, 1
boundary conditions, 238

### C

Cauchy–Bunyakovsky inequality, 2
chemical exchange processes, 197
class $\mathcal{D}$, 16
climate change sensitivity, 251
conservation law, 59
conserved current, 62
control equation, 228
control problem, 168
convection model, 201
convection–diffusion equation, 202
convergence rate estimate, 136
correct solvability, 44, 47

### D

data assimilation problem, 225
dispersion equation, 202
dual space, 1

### E

effective dispersion coefficient, 218
eigenvalue problem, 163
equation for control, 230
Euler equation, 88
extremum of functional, 81

### F

functional, 1, 192
Freundlich equation, 205

### G

Gâteaux derivative, 11
global transport of pollutants, 245

### H

Hilbert space, 3
Hölder condition, 211
hydrodynamics, 238

### I

infinitesimal operator, 89
insensitive functional, 103
invariant functional, 89

# V

# W

Printed and bound by CPI Group (UK) Ltd, Croydon, CR0 4YY

23/10/2024

01778237-0010